RESTRUCTURED ELECTRIC POWER SYSTEMS

A complete list of titles in the IEEE Press Series on Power Engineering appears
at the end of this book.

RESTRUCTURED ELECTRIC POWER SYSTEMS

Analysis of Electricity Markets with Equilibrium Models

EDITED BY

XIAO-PING ZHANG

Mohamed E. El-Hawary, *Series Editor*

A JOHN WILEY & SONS, INC., PUBLICATION

Published by John Wiley & Sons, Inc., Hoboken, New Jersey.
Published simultaneously in Canada.

For general information on our other products and services or for technical support, please contact our Customer Care Department within the United States at (800) 762-2974, outside the United States at (317) 572-3993 or fax (317) 572-4002.

Wiley also publishes its books in a variety of electronic formats. Some content that appears in print may not be available in electronic formats. For more information about Wiley products, visit our web site at www.wiley.com.

Library of Congress Cataloging-in-Publication Data:

Zhang, Xiao-Ping.
 Restructured electric power systems : analysis of electricity markets with equilibrium models / Xiao-Ping Zhang.
     p. cm.
  ISBN 978-0-470-26064-7 (cloth)
1. Electric utilities–Rates. 2. Electric power systems. 3. Equilibrium (Economics)
4. Marketing–Mathematical models. I. Title.
  HD9685.A2.Z43 2010
  333.793′2–dc22

                                                                    2009043674

Printed in the United States of America.

10 9 8 7 6 5 4 3 2 1

CONTENTS

2 RESTRUCTURED ELECTRIC POWER SYSTEMS AND ELECTRICITY MARKETS

Kwok W. Cheung, Gary W. Rosenwald, Xing Wang, and David I. Sun

7   USING MARKET SIMULATIONS FOR ECONOMIC ASSESSMENT OF TRANSMISSION UPGRADES: APPLICATION OF THE CALIFORNIA ISO APPROACH   241

Mohamed Labib Awad, Keith E. Casey, Anna S. Geevarghese, Jeffrey C. Miller, A. Farrokh Rahimi, Anjali Y. Sheffrin, Mingxia Zhang, Eric Toolson, Glenn Drayton, Benjamin F. Hobbs, and Frank A. Wolak

PREFACE

Electricity market deregulation is driving the power energy production from a monopolistic structure into a competitive market environment. The liberalization of the energy production has brought the issue of market equilibrium into the electricity power industry. Many studies have been performed in order to adjust the available equilibrium analysis methods to fit to the electricity market rules and sensitivities. With the development of electricity markets, one of the challenging and yet important task is to analyze the electricity market behavior and market power in order to improve the efficiency of electricity markets.

Special contributions of this book are to overview the latest developments in analyzing and assessing electricity market behavior and market power, that is, the electricity market equilibrium models, and discuss the application of such models in practical analysis of electricity markets. The topics of this book reflect the recent research and development of the electricity market equilibrium models, and foresee the future applications of such models and computational techniques in electricity market analysis:

- Fundamentals of electric power systems such as system structure and evolution, analytical techniques for system operation and control, and their consequence in electricity market environments.
- State-of-the-art electricity market design, and operations drawn from the real electricity markets.
- Problems of electricity market behavior and market power are reviewed and electricity market equilibrium models for analyzing market behavior, and market power are outlined.
- Mathematical programs with equilibrium constraints (MPEC) and equilibrium problems with equilibrium constraints (EPEC) are presented, the state of the art techniques for computing the electricity market equilibrium problems are discussed, and the challenges and recent advances in solving the electricity market equilibrium problems are discussed.
- Applications of the electricity market equilibrium models in electricity market modeling and analysis are presented.

Chapter 1 discusses the fundamentals of electric power systems. The structure and evolution of electric power systems are outlined. New developments include the integration of renewable generation sources into electric power systems, new operating and control paradigm such as microgrids, virtual power plants, plug-in hybrid electric vehicles, and the development of super power grids, which will have a significant impact on the operation of electric power systems as well as electricity markets. Then the concepts, analytical methods, and tools for operation and control of electric power systems are presented where the implications of these in electricity market environments are also briefly discussed. Finally real-time control of electric power systems via SCADA/EMS systems and the future trend of system operation and control are discussed, which is closely related to the development of future electricity markets.

In Chapter 2, the history of electric power systems deregulation is reviewed while the structure and the evolution of electricity markets are discussed. Then Chapter 2 addresses the key market design objectives and fundamental market design principles, especially the state-of-the art standard market design (SMD) framework, and also the operation of electricity markets and the criteria for its success. In addition, computational tools for electricity markets operations are presented. The treatment in this chapter reflects the current practice of electricity market structure, design, and operations, drawn from design and operation of the real electricity markets.

In Chapter 3, in connection with the electricity market development, the implication of market power is discussed. Then different electricity market equilibrium models for analyzing market behavior of participants and market power, which are related to the development of mathematical programs with equilibrium constraints (MPEC) and equilibrium problems with equilibrium constraints (EPEC) in mathematical programming, are overviewed; the challenges in the computing electricity market equilibrium are outlined; and recent advances in solving the electricity market equilibrium problems are discussed, and future research needs are also presented.

As in most fields, any attempt to develop a tractable model must abstract away from at least some of the detail. However, the choices in electricity markets are particularly difficult in part because experience with electricity markets is still accumulating and in part because there are several features of electricity markets are not features of other markets. Chapter 4 discusses the formulation of electricity market equilibrium models, distinguishing the physical, commercial, and economic models. It outlines the uses of such models, qualified in the light of the many assumptions that must be made for them to be tractable.

Most existing Nash-Cournot models of competition among electricity generators assume that firms behave purely *a la* Cournot or Bertrand with respect to transmission decisions by the independent system operator. Such models are unrealistic for markets in which interfaces connecting subnetworks are frequently saturated but the congestion pattern within individual subnetworks is less predictable. In order to deal with such situations, Chapter 5 proposes two approaches for dealing with them. The first is a hybrid Bertrand-Cournot model of these markets in which firms are assumed to behave *a la* Cournot regarding inter-subnetwork transmission quantities, but *a la* Bertrand regarding intra-subnetwork transmission prices. A second approach is a Bertrand-type model where transmission lines that are congested most of the time are designated as "common knowledge constraint" and treated as equality constraints by all market participants including the ISO and all generation firms. Under affine demand functions and quadratic costs, the market equilibrium of these models becomes mixed linear complementarity problems with bisymmetric positive semi-definite matrices.

In Chapter 6, the electricity market equilibrium analysis is performed with the aid of a nonlinear primal-dual interior point algorithm to solve the linear SFE bid-based electricity market model with a full AC network representation. This algorithm is based upon the AC transmission model, fully taking into consideration all the operating aspects such as the generation capacity limits, bus voltage limits, transmission line constraints, network losses, transformer tap-ratio control, and especially the effect of the reactive power. In the market equilibrium algorithm proposed, the impact of the electricity network control such as voltage control, transformer tap-ratio control on the market equilibrium is examined.

In Chapter 7, in response to the new requirements that restructured power markets place upon transmission planning, a method for assessing the economic benefits of transmission upgrades has been proposed by the California Independent System Operator (CAISO). Economic effects considered include reductions in the cost of building and operating power plants along with changes in market prices. The methodology accounts for how transmission upgrades mitigate market power by increasing the size of a supplier's geographic market, considering historical patterns of bidding behavior. Five principles underlie the methodology: consideration of multiple perspectives (consumers, generators, transmission operators, and society at large); full network representation; market-based pricing, accounting for strategic behavior by generators; modeling of uncertainty, including the value of transmission as insurance against extreme events; and recognition of how supply, demand-side, and transmission resources can substitute for each other. The methods used in the first full-scale application, to the proposed Palo Verde-Devers 2 (PVD2) upgrade, are summarized, along with results. Novel methods for modeling market power and for specifying probabilities of future scenarios and analyzing the effect of uncertainty are summarized and applied. Mitigation of market power accounts for a substantial portion of the benefits of that project.

The materials are derived mainly from the research and industrial development in which the authors have been heavily involved. The book will be a very useful reference for electrical power engineers, university professors, and undergraduate and postgraduate students in the subject area of electrical power systems, power system economics, and energy policy. The book can be used for postgraduate courses and industry courses as well.

Finally I am most grateful for the timely cooperation of all the contributors, in particular Ross Baldick, Benjamin Hobbs, Shmuel Oren, and David Sun for their enthusiasm in this book and their timely inputs. Without them, this book would not exist. I very much appreciate the staff from the IEEE Press and Wiley for their patience and good-natured support during the preparation of the book. I would also like to thank very much Dr. Mohamed El-Hawary, Series Editor of Power Engineering, for his kind advice and suggestions during the process of preparing this book. Last but not at least, I thank very much my wife, Zhong, my daughter, Dorothy, and my son, George for their patience, understanding and support during the development of the book.

Xiao-Ping Zhang
Birmingham, UK
June 2010

CONTRIBUTORS

Mohamed Labib Awad, Ph.D., was formerly Lead Grid Planning Engineer in the Department of Grid Planning (DGP) at the CAISO, and is now a member of the Faculty of Engineering at the University of Cairo.

Ross Baldick is Professor of Electrical and Computer Engineering at The University of Texas at Austin. He received his B.Sc. (in physics and mathematics) and B.E. (in electrical engineering) degrees from the University of Sydney, Australia, and his M.S. and Ph.D. (both in electrical engineering and computer sciences) from the University of California, Berkeley. In 1991–1992 he was a post-doctoral fellow at the Lawrence Berkeley Laboratory researching electric transmission policy. In 1992–1993 he was an Assistant Professor at Worcester Polytechnic Institute. Dr. Baldick has been a Research Fellow at the Harvard Electricity Policy Group of the John F. Kennedy School of Government, Harvard University, and a Visiting Researcher at the University of California Energy Institute. He has published more than forty refereed journal articles and has research interests in electric power. His current research involves optimization and economic theory applied to electric power system operations, the public policy and technical issues associated with electric transmission under deregulation, and the robustness of the electricity system to terrorist interdiction. Dr. Baldick is a Fellow of the IEEE.

Keith E. Casey, Ph.D., is Director of the Department of Market Monitoring at the CAISO. He received a Ph.D. in Environmental and Resource Economics from the University of California, Davis in 1997. Prior to working at the CAISO, he conducted post-doctoral research and taught environmental economics at the University of California, Davis.

Kwok W. Cheung received his B.S. from National Cheng Kung University, Taiwan, in 1986, his M.S. from University of Texas at Arlington, in 1988, and his Ph.D. from Rensselaer Polytechnic Institute, Troy, NY, in 1991, all in Electrical Engineering. He joined AREVA T&D Inc. (formerly ESCA Corp.) in 1991. He worked on dynamic security assessment. Since 1995 he has been developing wholesale electricity market applications and systems. He is currently the R&D Director of Market Management Systems Worldwide. His current interests include optimization-based market applications, power system security and intelligent systems. Dr. Cheung is a senior member of IEEE and a registered Professional Engineer of the State of Washington.

Glenn Drayton received the B.Sc (Hons) degree in operations research and the Ph.D. degree in management science from the University of Canterbury, Christchurch,

New Zealand. He is CEO of Drayton Analytics, the developers of the PLEXOS power system simulation software. His experience covers market design, market modeling, operations research and statistics, and applied microeconomic analysis, with particular focus on the electricity sector.

Anna S. Geevarghese was formerly Senior Production Cost Analyst and Market Surveillance Committee (MSC) Liaison, at the CAISO DMA and later became Product Director of Market Analytics at Global Energy Decisions, Software Division. Her M.S. degree is in Engineering Economic Systems from Stanford University.

Benjamin F. Hobbs is Professor in the Deptartment of Geography and Environmental Engineering (DoGEE) of The Johns Hopkins University. He is a member of the California ISO Market Surveillance Committee, and Scientific Advisor to the ECN Policy Studies Unit. He holds a Ph.D. in Environmental System Engineering from Cornell University and is a Fellow of the IEEE.

Jeffrey C. Miller (M.) was formerly Manager, Regional Transmission at the CAISO Department of Grid Planning.

Shmuel S. Oren is the Earl J. Isaac Chair Professor in the department of Industrial Engineering and Operations Research at the University of California at Berkeley and is the Berkeley site director of the Power System Engineering Research Center (PSERC). He has 25 years of academic and consulting experience in the electric power industry and he published numerous articles on aspects of electricity market design, resource optimization and risk management. Dr. Oren has been a consultant to many private and government organizations in the US and abroad. He is currently an adviser to the Market Oversight Division of the Public Utility Commission of Texas (PUCT), to the Energy Division of the California Public Utility Commission (CPUC) and to the market monitor of ISO-NE. He holds a Ph.D in Engineering Economic Systems from Stanford and is a Fellow of the IEEE and of INFORMS.

A. Farrokh Rahimi, Ph.D. (S.M.) was a Principal Market Engineer, CAISO Market and Product Development, and is now Vice President, Market Design and Consulting, Open Access Technology International, Inc. He earned a Ph.D. from M.I.T. in electric power engineering.

Gary W. Rosenwald received his B.S. and Ph.D. in Electrical Engineering from the University of Washington in 1992 and 1996, respectively. He worked at ABB on unit commitment and market participant software solutions before joining AREVA T&D Inc. in 2001. His current focus is on development and implementation of deregulated electricity market solutions. He is a member of the IEEE.

Anjali Y. Sheffrin, Ph.D., is Chief Economist and Director, CAISO Market and Product Development. She has 26 years of management experience in the electric utility industry. In her current position she manages the direction of market design, develops infrastructure and regulatory policy, and designs new products and services for the CAISO wholesale electricity markets. Prior to joining the CAISO, Dr. Sheffrin was Manager of Power System Planning at the Sacramento Municipal Utility District. She managed a department of 40 engineers, economists, statisticians, and financial analysts responsible for preparing strategic business plans for transmission, and generation, and demand-side projects. She received a Ph.D. in Economics from the University of California, Davis in 1981.

David I. Sun has actively participated in the design and implementation of many leading energy and transmission markets worldwide. They include OASIS, New Zealand, Australia, New England, PJM, ERCOT, MISO, SPP, and the North China Grid (NCG) in China. Prior to working on competitive electricity reforms, David's experiences include development of power system security and optimization applications. He has numerous professional publications, including an award winning OPF paper from the IEEE Power Engineering Society. David received his Ph.D. from the University of Texas, and MS/BS from the Rensselaer Polytechnic Institute. His current position is Chief Scientist for ALSTOM Energy Management Business in Washington, DC. Dr Sun is a Fellow of the IEEE.

Eric Toolson was formerly a Consultant at Pinnacle Consulting, and is now a consultant at Plexos Solutions, LLC.

Xing Wang received his B.S., from North China University of Electrical Power in 1991, his M.S. from China Electrical Power Research Institute (CEPRI) in 1996, and his Ph.D. from Brunel University, UK, in 2001, all in Electrical Engineering. He worked at CEPRI on Energy Management System development and project delivery from 1991 to 1998. He joined AREVA T&D Inc. in 2001. Since then he has been working on Market Management System development and implementation for the deregulated electricity markets. He is a senior member of the IEEE.

Frank A. Wolak, Ph.D., is the Holbrook Working Professor of Commodity Price Studies at Stanford University and is Chair of the CAISO MSC. His Ph.D. is in economics from Harvard University. His fields of specialization are Industrial Organization and Econometric Theory. He is a visiting scholar at University of California Energy Institute and a Research Associate of the National Bureau of Economic Research (NBER).

Jian Yao is a senior associate in the Berkeley research department of MSCI Barra. He was senior quantitative analyst on FX and government Treasury markets at Money Management Group, software engineer on advanced planning & scheduling products at Oracle Corporation, and research assistant on B2B supply chain integration at IBM. His main research interest is optimization, and his current research focuses on financial optimization and risk management. He holds a Ph.D. in Industrial Engineering and Operations Research from the University of California at Berkeley, an M.S. in Computer Science from the University of North Carolina at Charlotte, and an M.S. and a B.S in Mechanical Engineering from Shanghai Jiao Tong University, China.

Mingxia Zhang, Ph.D., was Lead Market Monitoring Specialist at the CAISO, Department of Market Monitoring, and is now an independent consultant. Her Ph.D. is in Economics from the University of California, Davis.

Xiao-Ping Zhang received the B.Eng., M.Sc., and Ph.D. degrees in electrical engineering from Southeast University, China in 1988, 1990, 1993, respectively. He worked at State Grid EPRI (Formerly NARI, Ministry of Electric Power), China on

EMS/DMS advanced application software research and development between 1993 and 1998. He was visiting UMIST from 1998 to 1999. He was an Alexander-von-Humboldt Research Fellow with the University of Dortmund, Germany from 1999 to 2000. He was a lecturer and then an associate professor at the University of Warwick, UK till early 2007. Currently he is a reader at the University of Birmingham, England, UK. He is also Director of the Institute for Energy Research and Policy. He is a coauthor of the monograph "Flexible AC Transmission Systems: Modelling and Control" published in 2006. He is a senior member of the IEEE and a member of CIGRE. He is an IEEE PES Distinguished Lecturer.

CHAPTER 1

FUNDAMENTALS OF ELECTRIC POWER SYSTEMS

Xiao-Ping Zhang

1.1 INTRODUCTION OF ELECTRIC POWER SYSTEMS

Commercial use of electricity began in the late 1870s when the inventive genius of Edison (Fig. 1.1) brought forth the electric incandescent light bulb. The first complete electric power system was the Pearl Street system in New York, which began operation in 1882 and was actually a DC system with a steam-driven DC generator. With the development of the transformer, polyphase systems, and AC transmission, the first three-phase line in North America was put into operation in 1893. It was then found that AC transmission with the help of transformers was preferable because DC transmission was impractical due to higher power losses.

With the development of electric power systems, interconnection of neighboring electric power systems leads to improved system security and economy. However, with the advent of interconnection of large-scale power systems, operation, control and planning of such systems become challenging tasks. With the development of digital computers and modern control techniques, automatic generation control (AGC) and voltage and reactive power control techniques have been introduced to operate and control modern large-scale power systems. Load flow solution has become the most frequently performed routine method of power network calculation, and can be used in power system planning, operational planning, operation control, and security analysis. With the advent of interconnected large-scale electric power systems, new dynamic phenomena, including transient stability, voltage stability, and low-frequency oscillations, have emerged. With the development of an electricity market, electricity companies engage in as many transactions in one hour as they once conducted in an entire day. Such increased load demand along with uncertainty of transactions will further strain electric power systems. Moreover, large amounts of decentralized renewable generation, in particular wind generation, connected with the network will result in further uncertainty of load and power flow distribution and impose additional strain on electric power system operation and control. It is a real challenge to ensure that the transmission system is flexible enough to meet new and less predictable supply and demand conditions in competitive

Restructured Electric Power Systems: Analysis of Electricity Markets with Equilibrium Models,
Edited by Xiao-Ping Zhang
Copyright © 2010 Institute of Electrical and Electronics Engineers

George Westinghouse   **Thomas Alva Edison**   **Nikola Tesla**

Figure 1.1 Pioneers of electric power systems

electricity markets. FACTS (flexible AC transmission systems) devices are considered low-environmental-impact technologies and are a proven enabling solution for rapidly enhancing reliability and upgrading transmission capacity on a long-term cost-effective basis. FACTS can provide voltage regulation, congestion management, enhancement of transfer capability, fast control of power oscillations, voltage stability control, and fault ride-through. The ever-increasing frequency of blackouts seen in developed countries has increased the need for new power system control technologies such as FACTS devices. With the development of advanced technologies and operation concepts such as FACTS, high voltage DC (HVDC), wide area measurements, microgrid systems, smart metering, and demand-side management, the development of smart grids is underway. It has been recognized that SCADA/ EMS Supervisory Control and Data Acquisition/Energy Management System plays a key role in the operating electricity networks and that state estimation is the key function of an EMS.

1.2 ELECTRIC POWER GENERATION

1.2.1 Conventional Power Plants

1.2.1.1 Fossil Fuel Power Plants Basically, fossil fuel power plants burn fossil fuels such as coal, natural gas, or petroleum (oil) to produce electricity. Traditionally, fossil fuel power plants are designed for continuous operation and large-scale production, and they are considered one of the major electricity production sources. The basic production process of a fossil fuel power plant is that the heat energy of combustion is converted into mechanical energy via a prime mover, a steam or gas turbine, then the mechanical energy is further converted into electrical energy via an AC generator, a synchronous generator.

It should be mentioned that the by-products of power plant production such as carbon dioxide, water vapor, nitrogen, nitrous oxides, and sulfur oxides need to be considered in both the design and operation of the power plant. Some of these by-products are harmful to the environment. In dealing with this, clean coal technology can remove sulfur dioxide and reburn it, which can enhance both the efficiency and the environmental acceptability of coal extraction, preparation, and use.

In addition to coal, natural gas is considered another major source of electricity generation, using gas and steam turbines. It is well known that high efficiencies can be achieved by combining gas turbines with a steam turbine in the so-called combined cycle mode. Basically, natural gas generation is cleaner than other fossil fuel power plants using oil and coal, and hence produces less carbon dioxide per unit energy generated. It is worth mentioning that fuel cell technology may provide cleaner options for converting natural gas into electricity, though such a generation technology is still not competitive in terms of generation costs.

1.2.1.2 CCGT Power Plants   The combined cycle gas turbine (CCGT) process utilizes rotational energy produced from gas turbines driving AC generators as well as the additional power made available from the waste heat contained in the gas turbine exhaust. The heat is passed through a heat recovery steam generator, one for each gas turbine, and the steam generated is then used to produce rotational energy in a steam turbine driving a second AC generator.

For a thermal power station, high-pressure steam requires strong, bulky components and high temperatures require expensive alloys made from nickel or cobalt. Due to the physical limitation of the alloys, practical steam temperatures do not exceed 655 °C, while the lower temperature of a steam plant is determined by the boiling point of water. Considering these constraints, the maximum efficiency of a steam plant is between 35% and 42%.

In contrast, for a combined cycle power plant, the heat of the gas turbine's exhaust can be used to generate steam driving a heat recovery steam generator with a steam temperature between 420 and 580 °C. This will in turn increase the CCGT plant thermal efficiency to 54%.

1.2.1.3 Nuclear Power Plants   Nuclear power technology extracts usable energy from atomic nuclei via controlled nuclear reactions and includes nuclear fission, nuclear fusion, and radioactive decay methods. Nuclear fission is the one most widely used for power generation today. The production process of a nuclear power plant is that nuclear reactors are used to heat water to produce steam, the steam is converted into mechanical energy via a turbine, and the mechanical energy can then be further converted into electrical energy via an AC generator, a synchronous generator. More than 15% of the world's electricity comes from nuclear power, where nuclear electricity generation is nearly carbon-free. It is estimated that replacing a coal-fired power plant with a 1 GW nuclear power plant can avoid emission of 6–7 million tons of CO_2 per year.

According to data from the International Energy Agency, existing nuclear power plants in operation worldwide have a total capacity of 370 GW. Most of them are second-generation light-water reactors (LWR) that were built in the 1970s and 1980s. Around 85% of the nuclear generation capacity is in US, France, Japan, Russia, the UK, Korea, and India. Third-generation nuclear power plant technology was developed in the 1990s to improve the safety and economics of nuclear power. However, due to the Chernobyl nuclear power accident in 1986, demand for constructing new nuclear power plants was much reduced and hence only a limited number of third-generation reactors have been built. The fourth generation of nuclear reactors has been developed within an international framework where safety and economic performance are improved, nuclear waste is minimized, and proliferation resistance is enhanced.

Nuclear power is a capital-intensive technology where the cost of electricity generated from new power plants depends on investment cost, discount rate, construction time (typically 5–7 years or even longer), and economic lifetime (say 25–40 years). It was estimated by the International Energy Agency in 2006 that, with an assumed carbon price of $25/tCO$_2$, the contribution of nuclear power generation to global electricity supply would increase to some 19–22% by 2050, where global nuclear power plant capacity would be at least doubled. Nuclear power could reduce global CO$_2$ emissions by 6% to 10%. With increasing oil prices and concerns about CO$_2$ emissions, there is growing interest in nuclear power generation. It has been recognized that nuclear power is one of the options to secure the supply of energy.

With the development of hydrogen technology, it will be important to find ways to produce hydrogen more efficiently. Hydrogen does not occur freely in nature in large quantities. Nuclear power could be used to generate hydrogen when load demand and electricity prices are low. The generated hydrogen could be stored to generate electricity and be fed back to the power grid when load demand and electricity prices are high. Alternatively, the stored hydrogen could be used to power hydrogen vehicles. Such a scenario would have a great impact on the operations of nuclear power plants, power grids, and electricity markets. The economics of a mixed portfolio of nuclear power and hydrogen energy need further research.

1.2.2 Renewable Power Generation Technologies

1.2.2.1 Wind Energy Generation Wind energy is a clean, renewable, and relatively inexpensive source of renewable energy. It is considered one of the most developed and cost-effective renewable energy technologies. Electricity from wind generation is generally competitive with electricity produced by conventional power plants. Wind turbines can be situated either onshore or offshore. According to the Global Wind Energy Council, global wind power capacity has continued to grow at an average cumulative rate of over 30%, and in 2008 there was more than 27 GW of new installations. By the end of 2008, the total installed global wind power capacity was over 120 GW. It has been recognized that the United States overtook Germany to become the number one market in wind power while China's total capacity doubled for the fourth year in a row. The ten countries with the highest installed wind power capacity are the US (25 GW), Germany (24 GW), Spain (16.8 GW), China (12 GW), India (9.6 GW), Italy (3.7 GW), France (3.4 GW), UK (3.2 GW), Denmark (3.2 GW), and Portugal (2.9 GW).

With implementation of European Commission targets on the promotion of electricity produced from renewable sources in the internal electricity market, wind power in Europe increased from 48 GW in 2006 to 65 GW in 2008. According to [9], by the end of 2008, there was 65 GW of wind power capacity (63.5 GW onshore and 1.5 GW offshore) installed in the EU-27, of which 63.9 GW was in the EU-15. Among the EU countries, Germany and Spain continue to be Europe's leaders with total installed wind energy capacities of 24 GW and 17 GW, respectively where 63% of the EU's installed wind energy capacity is located in these two countries.

With the development of wind power generation, there is growing penetration of wind energy into power grids. A wind power generation system normally consists of a wind turbine, a generator, and grid interface converters, if applicable, among which the generator is one of the core components. In the development of wind power generation techniques, synchronous generators, induction generators, and doubly fed induction generators have been employed to convert wind power to electrical power. Wind turbines usually rotate at a speed of 30–50 rev/min, and generators should rotate at the speed of 1000–1500 rev/min, so as to interface with power systems. Hence, a gearbox must connected between a wind turbine and a generator and requires regular maintenance; it also causes unpleasant noise and increases the loss of wind power generation. In order to overcome these problems, wind power generation with a direct-drive permanent magnet generator without a gearbox was developed. The permanent magnet generator driven directly by the wind turbine is a multi-pole and low-speed generator. Different types of direct-drive permanent magnet generators were developed for wind power generation, such as axial-flux and radial-flux machines.

Rapid technology development has enabled these prices and market growth. There are technical and economic challenges for large-scale deployment of wind power generation due to the intermittent nature of wind power and unpredictability in comparison to traditional generation technologies. Hence, increasing levels of wind power generation on the system will increase the costs of balancing the system and managing system frequency within statuary limits. With the increase of wind power penetration on the system, the impact on system operations as well as market operations should be examined. With the development of advanced energy storage technologies, it is expected that the intermittency of wind power generation can be handled in more effective ways.

1.2.2.2 Ocean Energy Generation The oceans cover more than 70% of the earth's surface and are the earth's largest collector and retainer of the sun's vast energy. Ocean energy includes tidal and wave energy.

Tidal power generation is nonpolluting, reliable, and predictable, and most modern tidal concepts use a dam approach with hydraulic turbines where tidal energy exploits the natural ebb and flow of coastal tidal waters due to the interaction of the gravitational fields of the earth, moon, and sun. Coastal water levels change twice daily, filling and emptying natural basins along the shoreline. In order to be practical for energy production, the height difference needs to be at least 5 meters. The tidal currents flowing in and out of these basins can be used to drive mechanical devices to generate electricity. The first large-scale tidal power plant in the world was built in 1966 at La Rance, France, and can generate 240 MW. There is another related tidal energy technology called tidal stream technology. Tidal streams are fast sea currents caused by the tides, often magnified by topographical features such as headlands, inlets, and straits that force water through narrow channels due to the shape of the sea bed. According to the World Offshore Renewable Energy Report 2002–2007 generated in the UK, worldwide, the potential capacity of tidal energy is around 3000 GW. It is estimated, however, that less than 3% is located in areas suitable for power generation.

Wave energy, with its high energy density, has been the focus of research since the 1970s and has attracted more interest recently. The wave energy resource is substantial, and the total resource around the world is 10 TW in the open sea, which is comparable to the world's total power consumption. Research and development projects with wave energy have been carried out around the world.

In the past, a number of wave power devices have been developed and demonstrated, each having its merits and limitations. Research is needed to predict the performance of wave power systems and to develop grid interface and control systems to collect and transmit wave power to the power grid. In the development of wave energy conversion techniques, different types of devices have been proposed, such as Archimedes Wave Swing (AWS), Oscillating Water Column (OWC), Pelamis, Wave Dragon, Mighty Whale, etc. Among these wave energy conversion techniques, AWS, invented in the early 1990s is the first wave energy conversion device to adopt the direct-drive power takeoff. AWS is completely submerged, which makes the device less vulnerable in a storm. Since an AWS is invisible, public acceptance is better than for wind farms.

1.2.2.3 *Photovoltaic Generation Systems*

One of the most attractive potential sources of energy able to meet future energy needs is solar energy, which has enormous long-term potential as a large-scale, carbon-neutral source of power. Installations of photovoltaic (PV) cells and modules around the world has grown rapidly in the past few years, at an average annual rate of more than 35%. However, for large-scale use of solar energy, a significant reduction in the cost-to-efficiency ratio is required. It has been indicated that, with the development of new technologies, the cost of PV systems is falling as the efficiency of solar panels increases and the cost of manufacturing decreases. It is estimated that solar PV could become cost-competitive for electricity generation by 2020–2030. In an ambitious scenario analysis, it is estimated that PV could cover 20% of global electricity consumption by 2040. With the increase of PV power penetration into power grids, the impact of PV power on power system operations and electricity market trade should be researched.

1.2.2.4 *Bioenergy*

Bioenergy and biomass resources include forestry and agriculture crops, biomass residues, and wastes, which provide about 14% of the world's primary energy supplies. As a carbon-neutral carrier, biomass can be used for production of heat, power, transport fuels, and bioproducts. Bioenergy offers cost-effective and sustainable solutions, has the potential to provide a large fraction of world energy demand over the next century, and will contribute to the requirements of reducing carbon emissions from fossil fuels.

1.2.2.5 Geothermal Energy   Geothermal energy is heat from within the earth, and the hot water or steam from geothermal energy can be used to produce electricity or applied directly for building heating and industrial processes. Geothermal energy can help offset the emission of carbon dioxide from conventional fossil-powered electricity generation, industrial processes, and building thermal energy supply systems, and it has great potential to contribute enormous quantities of clean, carbon-free energy.

1.2.2.6 Hydrogen   In the longer term, hydrogen is considered to be one of the most promising clean, sustainable energy supply technologies. Hydrogen can be used for all of the energy sectors—transportation, buildings, electricity supply, and industry—and can also provide storage options for intermittent renewable technologies such as solar and wind. Although hydrogen does not occur freely in nature in large quantities, it can be extracted from a variety of sources such as natural gas, biomass, coal, and water. The desirable feature is that end uses of hydrogen are basically clean and pollution-free. Hydrogen can be used in fuel cells for the clean production of electricity, powering vehicles and heating and cooling buildings while providing electricity.

Hydrogen is a versatile energy carrier and, unlike electricity, it is easy to store. Energy storage is a big challenge for the operation of electric power systems and electricity markets. When hydrogen is coupled with electric power systems, the impact on the operations of future electric power systems and electricity markets will be significant. This will affect the dependence of different energy carrier systems and further research is required in this area.

1.3   STRUCTURE OF ELECTRIC POWER SYSTEMS

1.3.1   Structure

An electric power system is used to generate, transmit, and distribute electrical energy in secure, reliable, and economic ways. Normally, electricity generation and transmission are using three phase AC systems while distribution of electrical energy may involve single-phase or two-phase systems. For interconnected AC systems in North America and some countries in Asia the system frequency is 60 Hz, while for the interconnected systems in Europe, Australia and Asia the system frequency is 50 Hz. In principle, systems with different system frequencies cannot be interconnected directly; rather they can be interconnected via HVDC links. Typically, a traditional electrical power system consists of three subsystems according to their different functionalities:

- Generation systems, which are used to generate the electrical energy from primal energy fuels;
- Transmission systems, which are used to transmit the electrical energy, over long distance, from where it is generated to either distribution systems or load centers;

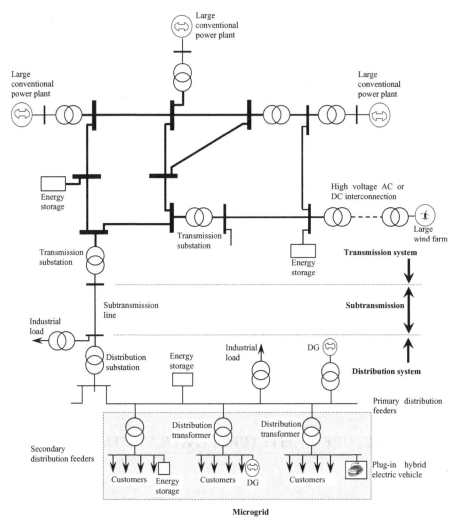

Figure 1.2 Structure of electric power systems

- Distribution systems, which are used to deliver the electrical energy to the end-use customers.

Structure of electric power systems is schematically shown in Figure 1.2. The highest transmission voltage level now in practical operation is 765 kV. The next voltage level is 1000 kV or above. Basically, high voltage (HV) transmission levels are between 100 and 275 kV, extra-high voltage transmission levels (EHV) are between 330 and 765 kV, and ultra-high voltage (UHV) levels are AC voltages higher than 765 kV and DC voltages higher than 600 kV. Subtransmission lines are used to interconnect high-voltage substations with distribution substations. The voltage levels of subtransmission are between 45 kV and 138 kV. Large synchronous generators are commonly used in conventional large generating power plants (including coal, gas, nuclear, and hydro) for electrical energy conversion where the output voltages of large generators are up to 25 kV.

According to IEC Standard 60038, low-voltage levels are between 100 and 1000 V including 230/400, 400/690, and 1000 V (at 50 Hz) and 120/208, 240, 277/480, 347/600, and 600 V (at 60 Hz). Medium voltage levels are between 1 and 35 kV, including 3.3, 6.6, 11, 22, and 33 kV. Low voltage and medium voltage levels belong to distribution voltage levels.

Distributed generation (DG) can either be connected to medium voltage levels or low voltage levels, depending on the size of the DG. Large decentralized power plants, such as wind farms and Combined Heat and Power (CHP) plants, are connected to high-voltage or extra-high voltage transmission system. Large offshore wind farms can be connected with transmission systems in three different ways: high-voltage AC interconnection, voltage sourced converter (VSC) based HVDC, or conventional line commutated HVDC.

In the UK, the transmission system consists of 275 kV and 400 kV, while 132 kV is considered as subtransmission voltage level and the distribution voltage levels include 0.4, 6.6, 11, and 33 kV. DG with a capacity up to 0.5 MW may be connected to a 400 V distribution network. It is a rule of thumb that DG with a capacity up to 5 MW may be connected to an 11 kV distribution network while DG with a capacity up to 20 MW may be connected to a 33 kV distribution network.

In addition, it can be anticipated that plug-in hybrid electric vehicles will be connected to distribution systems. With the development of energy storage technologies, various energy storage devices will be connected to electrical power systems at different voltage levels.

It has been pointed out that indiscriminant application of individual distributed generators can cause as many problems as it may solve [7]. It would be better to view distributed generation resources, energy storage devices, and associated loads as a subsystem. Such as subsystem of part of the distribution network is called a microgrid as indicated in Figure 1.2. The microgrid approach allows for local control of distributed generation in a coordinated ways and hence reduces the need for central dispatch or control. It is assumed that when there are disturbances in the distribution network, with which the microgrid is connected, the microgrid can separate from the distribution system to isolate the microgrid from the disturbance while the load can be supplied by the microgrid with coordination. It should be pointed out that the microgrid is a dynamic approach. From the viewpoint of distribution network operations and system developments, two more microgrids may be merged into one microgrid. A review of current worldwide R&D activities on microgrid technologies can be found in [8].

1.3.2   Benefits of System Interconnection

The benefits of an interconnected transmission system include:

- *Bulk Power Transfers*. There are a number of factors that influence the decision to site a power station at a particular location. These may include fuel availability, fuel price, fuel transport costs, financing, cooling water, land availability and the level of transmission system charges. Due to environmental concerns and other reasons, it may not be possible to site large power stations close to demand centers. This is also the case for renewable energy generation technologies such as wind or wave, which are location-based resources and will need to be connected to power grids and transmitted to load centers. One of benefits of an interconnected transmission system is that it can provide for the efficient bulk transfer of power from remote power plants where power is generated to demand centers where power is consumed. Transmission of electricity at high voltage is more efficient than transfer at lower voltage due to the lower capital cost per unit transmitted, the lower power losses, and the lower voltage drops.

- *Economic Operation and Electricity Trade.* An interconnected transmission system can also provide the main national electrical link between all market participants, including generation and demand, and it is then possible to optimize the generation portfolio available and hence provide the cheapest possible generation. In electricity market environments, for such an interconnected system, market participants can have the opportunity to choose to trade with the most competitive participants. The interconnection of different national transmission systems would provide the opportunity to optimize generation portfolio across region.

- *Security and Reliability of Supply.* Security and reliability of supply is a key issue for electricity supply. The system must be able to provide continuously an uninterrupted supply of electricity to customers under conditions of plant breakdown, transmission outages, or weather-induced failures for a wide range of demand conditions while the intact system is maintained and power quality standards in terms of voltage, waveform, and frequency are satisfied. Transmission circuits are more reliable than individual generating units, and hence an interconnected transmission system can enhance the security and reliability of supply in terms of breakdown of individual generating units.

- *Reduced Power Plant Capacity Margin.* In the operation of a transmission system, installed generation capacity should not be less than the forecast maximum demand. If a plant becomes unavailable due either to routine maintenance or a breakdown, extra demand under extreme weather conditions, or transmission line outages, additional capacity is required for security reasons to cover shortfalls. An interconnected transmission system enables surplus generating capacity in one zone to be used to cover shortfalls in other zones on the system. The requirement for additional installed generating capacity, to provide sufficient generation security for the whole system, is therefore smaller than the sum of individual zonal requirements of the subsystems when they are not interconnected.

- *Reduced Frequency Response and Active Power Reserve Requirements.* A transmission system operator has the responsibility to maintain frequency between certain specified limits, as large deviations in frequency can lead to widespread demand disconnections, generation disruptions, and even system splitting or collapse. If demand is greater than generation, frequency falls and, if generation is greater than demand, frequency rises. Basically, the frequency response of an interconnected transmission system should be less than the sum of each separate system.

1.4 ULTRA-HIGH VOLTAGE POWER TRANSMISSION

1.4.1 The Concept of Ultra-High Voltage Power Transmission

With the development of large-scale transmission systems, voltage levels have been increased continuously to achieve greater economies of scale. The highest transmission voltage level now in practical operation is 765 kV. The next voltage level is 1000 kV or above. High voltage (HV) transmission levels include 100 (110), 125, 132, 138, 161, 230 (220), an 275 kV; extra-high voltage levels (EHV) include 345 (330), 400, 500, and 765 (750) kV; and ultra-high voltage (UHV) levels are AC voltages higher than 765 kV and DC voltages higher than 600 kV.

The ultra-high voltage grid being considered in China and elsewhere is the transmission grid with voltages higher than 750 kV. It is mainly suitable for transferring bulk electricity energy over a very long distance.

It was recognized that a higher voltage grid should be introduced when the capacity of a power system is doubled. For this reason, the research on super-high voltage transmission grid became more important. Many countries, such as the US, Russia, Italy, France, Japan, and Sweden, have designed and planned at the development of ultra-high voltage grids in the past. However, these projects were not completed because of political and economic factors.

A ultra-high voltage grid research project in China has been conducted since 1986. China is a vast territory with great resources in hydro and coal, and less in oil and gas. Hydro resources and coal reserves are the most important energy resources in China. However, the distribution of electricity energy sources and electricity demand are seriously unbalanced. Nearly two thirds of hydro resources are distributed in the southwest and west of China, including Sichuan, Yunnan, and Tibet provinces. Two thirds of the coal reserves are distributed in the northwest and north of China. On the other hand, two thirds of electricity loads were mainly in the east areas of China, where there is a lack of electricity energy sources. The distances between the areas of energy resources and energy demand are up to 2000 km, which is a more reasonable transportation distance for an AC ultra-high voltage grid than for transportation of energy resources by train. Research has shown that it is more economical to transfer the electricity than to transport coal within 1500 km.

For the reasons mentioned above, the need for development of a super-ultra electricity transmission grid is clear. At the end of 2007, the installed generation capacity of 700 GW made China the second highest in the electricity production in the world. As China transforms into the manufacturing center of the world, the increase of electricity demand is tremendous. A new electric energy infrastructure is required to transfer bulk electricity energy over a very long distance. Although 500 kV AC and 500 kV DC transmission systems exist, such transmission networks are considered inadequate for bulk electricity energy transmission.

It is estimated that in China in 2020 the installed generation capability will be 1000 GW. Large amounts of hydro and coal resources are distributed in the southwest and northwest of China. In order to meet the energy consumption more generating plants will be constructed in these regions by 2020. On the other hand, load centers are situated in the east costal areas of China. According to the experience of the international power industry, when the capacity of power grid is expanded by 4 times, a higher voltage level should be introduced. The current 500 kV power grid is 30 years old, and the installed generation capability in 2004 was around 6 times that of the installed generation capability in 1982. Presently, the transmission of electricity across provincial power networks is mainly based on the 500 kV AC and ±500 kV DC transmission grids. Such transmission power grids will not meet the bulk power transmission requirements in 2020. Clearly, power grids of higher voltage levels are required. Long-distance bulk power transmission can satisfy future power growth, secure energy supply, and optimize energy resource allocation. Furthermore, overall social benefits will be increased by enhancing power grid security and reliability, saving right-of-way, coordinating the development of power plants and grids, alleviating coal transportation pressure, and boosting the harmonious development of regional economies.

Some general arguments for moving to higher voltages are:

- The higher the voltage of a transmission line, the higher the power that can be transported over a particular conductor of a transmission line, reducing line costs per power unit transferred.
- At the same time the space (right of way) required per transferred power unit decreases. The need for slightly higher insulation distances is compensated for by having to build fewer lines provided that the needed power transfer is beyond the capabilities of a single line.
- With the same transferred power, losses decrease with higher voltage levels due to the lower currents needed to sustain the power transfer.

Note that 1000 kV AC and ±800 kV DC transmission grids have different features. They are complementary. Historically, AC and DC have had clearly distinguishable application areas based on the following characteristics:

- AC connections can be more easily meshed and connected with the rest of the network. An AC/AC substation with breakers and transformers is less expensive and has lower losses than a converter station between AC and DC for the same power level. Reactive power compensation and insulation issues become more difficult to resolve with larger distances and higher AC voltages, so that scaling up has its limits. There is also a concern about the exposure to high electromagnetic fields at 50/60 Hz, particularly in populated areas.
- DC connections result in lower losses on the line and lower line costs per unit of transmitted power. When connecting two AC subnetworks with DC,

stability and oscillation problems on both AC subnetworks are effectively shielded from each other.

1.4.2 Economic Comparison of Extra-High Voltage and Ultra-High Voltage Power Transmission

The natural single circuit power transmission capacity of a 1000 kV AC transmission line is about 5 GW, and the maximum transmission power of a ±800 kV DC transmission line is at most 6.4 GW. The transmission capacity of a DC transmission line depends on the power system connection and is restricted by its maximum stability limit. Its capacity should be uplifted gradually. In contrast, the transmission capacity of a DC transmission line could achieve the design power capacity level as soon as an AC supply source is available. For the right of way, a transmission of 6 GW of power requires two parallel transmission lines, both for 500 kV DC and for 1000 kV AC, whereas only one transmission line is needed for 800 kV DC. The reasonable transmission distance of a 1000 kV AC transmission line is up to 2000 km. On the other hand, ±800 kV DC transmission is the most economical way of long distance power transmission when transmission distance is larger than 1000 km, for which the one-time conversion cost pays off through reduced line costs and losses.

A 1000 kV AC transmission grid, which has the general features of an AC transmission grid, can form a super backbone grid of a national power system. A "direct super high way," ±800 kV DC transmission, which is complementary to an AC transmission grid, can be used to transmit bulk power from large power generating bases to large load centers over long distances. However, ±800 kV DC transmission will not be suitable for forming a national backbone high-voltage power grid due to the limitation of interconnection.

Another option is to move primary energy, such as gas, and coal, closer to the place of consumption and convert it to electrical power there. However, this is not possible for hydropower. Besides the infrastructure costs for transportation (gas pipelines, railway infrastructure to transport coal), the issue of producing pollutants close to major population areas has to be taken into account.

Assume that 6 GW of power must be transported point to point over either 1000 km or 2000 km. In this scenario analysis, three transmission lines are necessary for 800 kV AC transmission while two lines are needed for 500 kV DC or 1000 kV AC and one line is sufficient for 800 kV DC. It has been found that, for the assumed power level, moving to higher voltages leads to reduced costs for both AC (moving from 800 kV to 1000 kV) and for DC (moving from 500 kV to 800 kV). For a bulk power transmission need of 6 GW, the losses for various transmission distances are compared. For this analysis, it is assumed that the conductor cross-section area is the same for the different transmission distances. It has been found that beyond a distance of 500 km, line losses dominate the station losses and thus DC solutions become more efficient than their AC counterparts. The comparison shows that for a distance of 1000 km (for which the various options are optimized in this example), losses for an 800 kV DC set-up are around 1% lower than for a 500 kV DC set-up.

1.4.3 Ultra-High Voltage AC Power Transmission Technology

The enabling technologies for establishing ultra-high voltage AC (U-HVAC) are breakers and gas insulated substations (GIS). Furthermore, current sensors, transformers, and FACTS devices should also be designed specifically. It has been already proven that ultra-high voltage level power transmission is technically feasible. Some manufacturers have gained experience that can be used in the actual development of ultra-high voltage level power transmission in China.

It should be mentioned that for building a 1000 kV AC network, a careful consideration of stability issues, compensation, and protection is required. Furthermore, the vulnerability to cascading blackouts in case of emergencies should be considered in the design and operations and special system supervision and emergency schemes need to be implemented.

1.4.4 Ultra-High Voltage DC Technology

There are a number of special R&D tasks to be executed to get a proper design of an 800 kV ultra-high voltage DC technology (U-HVDC):

- AC/DC converter transformer
- DC bushings
- External insulation
- Control and protection
- DC valves and thyristors
- System reliability and main circuit design

The last item is of great importance. Concentrating the power transfer on a smaller number of overhead lines obviously increases the operational risks posed by these lines being taken out of operation in an unplanned fashion (either due to random reasons, such as weather-related incidents or technical failures, or due to malicious attacks and sabotage). This would provide an argument against moving to 800 kV DC, for example, based on homeland security concerns. It would, however, also apply to some extent to 1000 kV AC links. On the other hand, this risk can be mitigated by a number of measures:

- The converter stations are designed in such a way as to employ redundancy (e.g., for protection and control) to reduce the likelihood that technical failures lead to a line trip.
- The two poles (+/−) of an HVDC line are designed to be as independent as feasible. Thus, even when one pole is tripped, a monopolar operation with the other pole remains possible.
- If a sufficient number of HVDC lines is available (the upper Jinsha River hydropower development shall be connected to East China with three 800 kV DC connections), when one of the lines is completely taken out of operation, the other lines are designed with an overload capability so that they can take up load from the lost line for a certain time period.

- The complete power system should be designed in such a way that the overall system stability would be maintained even if the power from one of the 800 kV DC links was suddenly completely lost. In the worst case, load shedding of noncritical loads may be employed.

1.4.5   Ultra-High Voltage Power Transmission in China

According to the strategic planning of State Grid Co., the national ultra-high voltage grid, which covers the majority of China, should be operated and dispatched through a united national control center. The national grid is mainly based upon AC interconnected networks with the assistance of DC interconnected networks. The backbone grid will be interconnected by 1000 kV AC transmission with the assistance of ±800 kV DC transmission. The northwest power grid will be interconnected by 750 kV AC transmission.

According to the design and planning of the national ultra-high voltage grid, two large capacity transmission channels should be constructed. One is the north-south channel from north via central to east, forming a huge UHV AC synchronizing power grid. The other is the west-east channel from Sichuan via central to east. The transmission of the electric energy in northwest coal bases should merge into this super UHV grid channel.

The 1000 kV UHV AC transmission grid should be constructed above 500 kV EHV AC transmission grid, and will be used to transfer the electricity from the coal bases in the northwest and north China to central and east China. To transfer the electric energy from the southwest hydro bases to east areas ±800 kV DC should be primarily used. Through the national super-high voltage backbone grid, coal transportation could be substituted for electricity transmission. In addition, the united national grid will promote bulk nationwide electricity trading between different regions.

As shown in Figure 1.3, the first UHV project connects North China and Central China. The UHV grid will expand to East China. A strong super UHV power grid will cover North, Central, and East China. The UHV grid therefore will cover major energy centers and load centers in China. It is estimated that the UHV capacity and regional transfer capacity will exceed 200 GW.

A feasibility study of the 1000 kV UHV AC pilot project has been carried out. The optimal option for the UHV AC pilot project has been identified. The project now enters the preliminary design stage. A 1000 kV UHV AC pilot demonstration project, from Jindongnan, southeast of Shanxi Province, via Nanyang, Henan Province, to Jinmen, Hubei Province, will be built. The total length of the pilot demonstration 1000 kV UHV AC line is a 640 km single line. Along with the 1000 kV UHV AC line, two 1000 kV UHV substations and one switching station have been built and the system has been put into operation since January 2009.

Taking the distribution of energy resources in China into consideration, the transportation expenses will rise because the quantity of coal transportation will greatly increase as demands increase with the development of national economy. It is much more economical to transfer electricity than to transport coal from north to

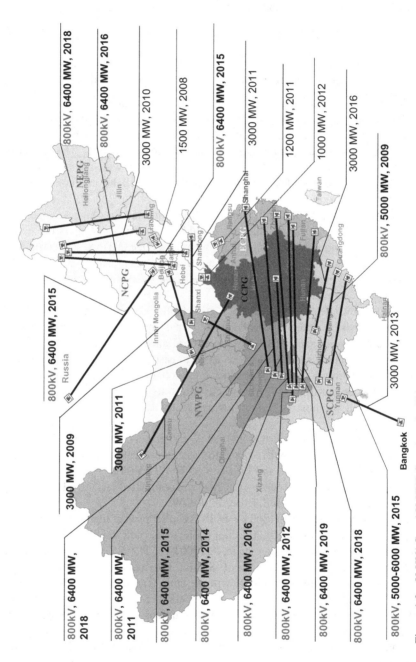

800kV, 6400 MW, 2018

800kV, 6400 MW, 2016

3000 MW, 2010

1500 MW, 2008

800kV, 6400 MW, 2015

3000 MW, 2011

1200 MW, 2011

1000 MW, 2012

3000 MW, 2016

800kV, 5000 MW, 2009

800kV, 6400 MW, 2015
Russia

3000 MW, 2009

3000 MW, 2011

800kV, 6400 MW, 2018

800kV, 6400 MW, 2011

800kV, 6400 MW, 2015

800kV, 6400 MW, 2014

800kV, 6400 MW, 2016

800kV, 6400 MW, 2012

800kV, 6400 MW, 2019

800kV, 6400 MW, 2018

800kV, 5000-6000 MW, 2015

3000 MW, 2013

Bangkok

NEPG
Heilongjiang
Jilin
Liaoning
NCPG
Beijing
Hebei
Inner Mongolia
Shanxi
Shandong
Shaanxi
Shanghai
Jiangsu
ECPG
CCPG
Hunan
Fujian
Taiwan
Guizhou
Guangdong
SCPG
Yunnan
Hainan
NWPG
Gansu
Qinghai
Xizang

Figure 1.3 U-HVAC and U-HVDC projects in China

south in China. Therefore, the construction of the national super-high voltage transmission grid in China is not only necessary but also timely.

The national ultra-high voltage grid should have the basic function of transmitting bulk capacity of electricity through very long distances and across different regions with lower power losses. The present imbalance between the electricity supply and demand would be remedied with the national ultra-high voltage grid. Furthermore, the national super-high voltage transmission grid will promote electricity trading between different regions over different scales, optimizing the utilization of energy sources within larger areas, meeting the requirements of economic development and creating competitive national energy markets.

1.4.6 Ultra-High Voltage Power Transmission in the World

There has been increasing interest worldwide in HVDC applications involving voltage levels above 500 kV. In Asia, in addition to the U-HVDC projects in China, India has announced an intention to go for ±800 kV. In Brazil, U-HVDC is also under consideration and Africa has also shown an interest in U-HVDC power transmission.

In Europe, the idea of a super power grid, which would link Africa and Europe, is being discussed. The super grid would connect geothermal energy in Iceland, biomass power in Poland, and solar power in the Sahara. For the super power grid, U-HVDC technologies would be one of the options.

In the US, a continental super grid has been discussed that envisions the use of underground, superconducting direct current cables for long-distance power transmission at levels of perhaps 5 to 10 GW. It has been suggested that, in addition to carrying electricity, the superconducting cables could also be used as a multiple energy carrier to transport hydrogen for use both as a cryogen and for end-use energy consumption with a significant development in hydrogen energy market in the future.

1.5 MODELING OF ELECTRIC POWER SYSTEMS

In power system analyses such as load flow, power system stability studies, power system components such as transmission lines, transformers, and static loads may be represented by algebraic equations. Synchronous generators are the most important components in power system analysis. They are usually represented by algebraic and differential equations in stability studies [1, 2]. The modeling philosophy of synchronous generators is applicable to modeling of HVDC and flexible AC transmission systems [3–6].

1.5.1 Transmission Lines

If it is assumed that angular frequency of the system is nearly constant and the three-phase transmission line is balanced in parameters, then the transmission line can be represented by a single-phase π section equivalent circuit as show in Figure 1.4.

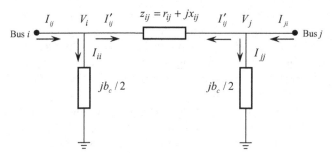

Figure 1.4   Equivalent π circuit model of transmission line

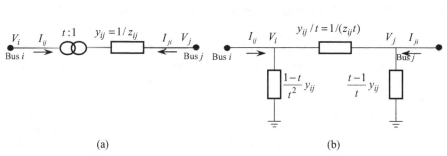

(a)   (b)

Figure 1.5   Transformer equivalent circuit with off-nominal tap ratio

In Figure 1.4, $z_{ij} = r_{ij} + jx_{ij}$, jb_c are the series impedance and shunt admittance of the transmission lines.

According to Kirchoff's current law, we have

$$I_{ij} = I'_{ij} + I_{ii} = (V_i - V_j)/z_{ij} + V_i(jb_c/2) \tag{1.1}$$
$$I_{ji} = I'_{ji} + I_{jj} = (V_j - V_i)/z_{ij} + V_j(jb_c/2) \tag{1.2}$$

Equations (1.1) and (1.2) may be written in a compact form as follows

$$\begin{bmatrix} y_{ij} + jb_c/2 & -y_{ij} \\ -y_{ij} & y_{ij} + jb_c/2 \end{bmatrix} \begin{bmatrix} V_i \\ V_j \end{bmatrix} = \begin{bmatrix} I_{ij} \\ I_{ji} \end{bmatrix} \tag{1.3}$$

where y_{ij} is the series branch admittance and given by $y_{ij} = 1/z_{ij}$. Equation (1.3) is bus voltage equation of the transmission line, which can be directly incorporated into the network voltage equation for system analysis.

1.5.2   Transformers

Similar to the modeling of transmission lines, transformers can also be represented by the equivalent circuit and bus voltage equation. A transformer represented by an ideal transformer $t:1$ in series with an impedance z_{ij} is shown in Figure 1.5 (a). The equivalent circuit in Figure 1.5 (a) can be transformed into Figure 1.5 (b). In Figure 1.5, t is the off nominal tap ratio, y_{ij} is the short-circuit or leakage admittance.

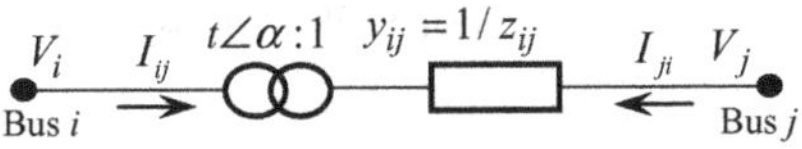

Figure 1.6   Transformer equivalent circuit with off-nominal tap ratio

The bus voltage equation of the transformer is given by

$$\begin{bmatrix} \dfrac{y_{ij}}{t^2} & -y_{ij}/t \\ -y_{ij}/t & y_{ij} \end{bmatrix} \begin{bmatrix} V_i \\ V_j \end{bmatrix} = \begin{bmatrix} I_{ij} \\ I_{ji} \end{bmatrix} \tag{1.4}$$

Basically, phase-shifting transformers (PST) are used to control the real power flow of in transmission lines by regulating the phase angle difference. Both the magnitude and the direction of the power flow can be controlled by varying the phase shift. Under electricity market environments, transmission congestion can be managed by PST in a very economical way.

A phase shifting transformer represented by an ideal transformer in series with an impedance z_{ij} is shown in Figure 1.6. In Figure 1.6, $t\angle\alpha:1$ is the off nominal tap ratio, which is a complex number. y_{ij} is the short-circuit or leakage admittance. Then we have the following equations:

$$I_{ij}\, t\angle -\alpha = (V_i/t\angle\alpha - V_j)y_{ij} \tag{1.5}$$
$$I_{ji} = (V_j - V_i/t\angle\alpha)y_{ij} \tag{1.6}$$

The bus voltage equation of the PST is given by

$$\begin{bmatrix} \dfrac{y_{ij}}{t^2} & -y_{ij}/t\angle -\alpha \\ -y_{ij}/t\angle\alpha & y_{ij} \end{bmatrix} \begin{bmatrix} V_i \\ V_j \end{bmatrix} = \begin{bmatrix} I_{ij} \\ I_{ji} \end{bmatrix} \tag{1.7}$$

In comparison to (1.4), the system admittance matrix of (1.7) is unsymmetrical due to the complex off-normal tap ratio, and in this situation, the model shown in (1.7) cannot be represented by an equivalent circuit.

1.5.3 Loads

The static loads may be classified into three categories:

1. Constant power:

$$P = P_0(V)^0, \quad Q = Q_0(V)^0 \tag{1.8}$$

where P_0, Q_0 are constant powers at nominal voltage.

2. Constant current:

$$P = P_0(V)^1, \quad Q = Q_0(V)^1 \tag{1.9}$$

3. Constant impedance:

$$P = P_0(V)^2, \quad Q = Q_0(V)^2 \tag{1.10}$$

A general representation of the static loads as functions of voltage magnitude and frequency deviation may be given by

$$P = P_0 \left[a_0 (V)^0 + a_1 (V)^1 + a_2 (V)^2 \right] (1 + K_P \Delta f) \tag{1.11}$$

$$P = P_0 \left[b_0 (V)^0 + b_1 (V)^1 + b_2 (V)^2 \right] (1 + K_Q \Delta f) \tag{1.12}$$

where a_0, a_1, a_2, b_0, b_1, and b_2 are voltage coefficients while K_p and K_Q are frequency coefficients.

1.5.4 Synchronous Generators

In load flow analysis, a synchronous generator is simply represented by algebraic constraints. For instance, a slack generator is represented by a constant voltage source. A PV generator is represented by constant active power injection and controlled voltage bus.

In power system stability studies, synchronous generators are usually represented by equivalent circuits and algebraic and differential equations [1, 2]. Similarly, dynamic motors can also be represented by algebraic and differential equations. In addition to the modeling of synchronous generators, excitation systems and speed-governing systems, which are usually described by differential equations, are also need to be represented.

1.5.5 HVDC Systems and Flexible AC Transmission Systems (FACTS)

In load flow analysis, HVDC systems and Flexible AC Transmission Systems are represented by algebraic equations while in stability studies they are represented by algebraic and differential equations [6].

1.6 POWER FLOW ANALYSIS

It is well known that load flow solution is the most frequently performed routine power network calculations, which can be used in power system planning, operational planning, and operation/control. It is also considered as the fundamental of power system network calculations. From a load flow solution, the voltage magnitude and angle at each bus and active and reactive power flows and power losses in each line can be obtained. In the following, classifications of buses for load flow analysis are introduced first. Then load flow solution methods are presented. Numerical examples on a simple system are used to show the principles of load flow analysis.

1.6.1 Classifications of Buses for Power Flow Analysis

1.6.1.1 Slack Bus For load flow analysis, a slack bus should be selected where the voltage magnitude and angle are given and kept constant while the active and reactive power injections are not known before a load flow solution is found. The slack bus basically performs two functions:

- the reference of the system;
- balancing the active and reactive powers of the system.

In load flow analysis, usually there is only one slack bus in the system. For a slack bus, the active and reactive power injections need to be determined once a load flow solution has been found.

In some situations, a distributed slack bus concept may be used for large-scale power systems where a number of buses can be selected as a slack bus group and the system active and reactive power can be balanced by a group of buses rather a single bus.

1.6.1.2 *PV Buses*   A PV bus, also called a voltage-controlled bus, is a bus where the voltage magnitude is given and kept constant and the active power injection is specified while the voltage angle and the reactive power injection need to be deter-mined by load flow analysis. A generator bus may be considered as a PV bus if the voltage of the bus and the active power output from the generator are controlled to the specified values. Sometimes a bus, which a reactive control resource like a syn-chronous condenser is connected with, may also be taken as a PV bus. For a practical interconnected power system, there may be one or more PV buses.

1.6.1.3 *PQ Buses*   A PQ bus is a bus where the active and reactive power injections are given and kept constant. Usually a nongenerator bus is considered as a PQ bus.

1.6.2 Formulation of Load Flow Solution

The features of load flow solution will be shown on a three-bus power system as shown in Figure 1.7. In the system, there are two generators, which are connected to bus 1 and 2, respectively. A load is connected with bus 3 of the system. In the figure, we assume that the generator bus 1 is taken as the slack bus; the generator

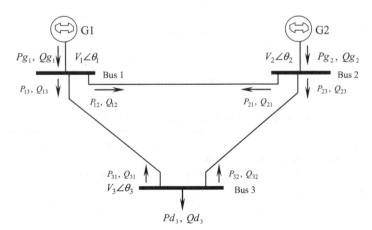

Figure 1.7   A three-bus system

bus 2 is taken as a PV bus; and the load bus 3 is considered as a PQ bus. For the three-bus system, we assume that the transmission lines are represented by series impedances only.

According to Kirchoff's voltage law, the relationship between the bus voltages and current injections is given by

$$
\begin{bmatrix} \mathbf{I}_1 \\ \mathbf{I}_2 \\ \mathbf{I}_3 \end{bmatrix} = \begin{bmatrix} Y_{11} & Y_{12} & Y_{13} \\ Y_{21} & Y_{22} & Y_{23} \\ Y_{31} & Y_{32} & Y_{33} \end{bmatrix} \begin{bmatrix} \mathbf{V}_1 \\ \mathbf{V}_2 \\ \mathbf{V}_3 \end{bmatrix}
\tag{1.13}
$$

where, $Y_{11} = 1/z_{12} + 1/z_{13}$, $Y_{22} = 1/z_{12} + 1/z_{23}$, $Y_{33} = 1/z_{13} + 1/z_{23}$, $Y_{12} = Y_{21} = -1/z_{12}$, $Y_{13} = Y_{31} = -1/z_{13}$, $Y_{23} = Y_{32} = -1/z_{23}$. z_{12}, z_{13}, and z_{23} are the impedances of transmission lines 1–2, 1–3, and 2–3, respectively. The bus voltages and current injections are phasors. However, in load flow analysis, usually active and reactive power injections rather than current injections are given. For this reason, equation (1.13) needs to be transformed into the following power equation:

$$
P_i + jQ_i = V_i I_i^* = V_i \sum_{j=1}^{N} (Y_{ij} V_j)^*
\tag{1.14}
$$

where * represents the conjugate operation. N is the total number of buses. For the three-bus system, N equals 3. In the polar coordinates, assuming that $V_i = V_i \angle \theta_i$ ($i = 1, 2, \ldots, N$) and $Y_{ij} = G_{ij} + jB_{ij}$, the active and reactive power injection equation (1.14) may be represented separately as

$$
P_i = V_i \sum_{j=1}^{N} V_j (G_{ij} \cos \theta_{ij} + B_{ij} \sin \theta_{ij})
\tag{1.15}
$$

$$
Q_i = V_i \sum_{j=1}^{N} V_j (G_{ij} \sin \theta_{ij} - B_{ij} \cos \theta_{ij})
\tag{1.16}
$$

As pointed out above, in load flow analysis, active and reactive power injections are known, then the following power mismatch equations can be obtained:

$$
\Delta P_i = P_i^{Spec} - P_i
\tag{1.17}
$$

$$
\Delta Q_i = Q_i^{Spec} - Q_i
\tag{1.18}
$$

where P_i^{Spec} and Q_i^{Spec} are the specified active and reactive power injections, respectively. $P_2^{Spec} = Pg_2$, $P_3^{Spec} = -Pd_3$, and $Q_3^{Spec} = -Qd_3$. The objective of a load flow solution is to find a solution to equations (1.17) and (1.18) while the voltage of bus 1 and the voltage magnitude of bus 2 are given.

1.6.3 Power Flow Solution by Newton-Raphson Method

Equations (1.17) and (1.18) represent a set of nonlinear equations of the system studied. The most efficient method for solving nonlinear equations may be the well-known Newton-Raphson method. For the following nonlinear equations with three unknown variables:

$$
f_1(x_1, x_2, x_3) = 0
\tag{1.19}
$$

$$
f_2(x_1, x_2, x_3) = 0
\tag{1.20}
$$

$$
f_3(x_1, x_2, x_3) = 0
\tag{1.21}
$$

we use a Taylor expansion of the functions about x_1^0, x_2^0, x_3^0.

$$f_1\left(x_1^0, x_2^0, x_3^0\right) + \frac{\partial f_1\left(x_1^0, x_2^0, x_3^0\right)}{\partial x_1} \Delta x_1 + \frac{\partial f_1\left(x_1^0, x_2^0, x_3^0\right)}{\partial x_2} \Delta x_2 + \frac{\partial f_1\left(x_1^0, x_2^0, x_3^0\right)}{\partial x_3} \Delta x_3 = 0$$

(1.22)

$$f_2\left(x_1^0, x_2^0, x_3^0\right) + \frac{\partial f_2\left(x_1^0, x_2^0, x_3^0\right)}{\partial x_1} \Delta x_1 + \frac{\partial f_2\left(x_1^0, x_2^0, x_3^0\right)}{\partial x_2} \Delta x_2 + \frac{\partial f_2\left(x_1^0, x_2^0, x_3^0\right)}{\partial x_3} \Delta x_3 = 0$$

(1.23)

$$f_3\left(x_1^0, x_2^0, x_3^0\right) + \frac{\partial f_3\left(x_1^0, x_2^0, x_3^0\right)}{\partial x_1} \Delta x_1 + \frac{\partial f_3\left(x_1^0, x_2^0, x_3^0\right)}{\partial x_2} \Delta x_2 + \frac{\partial f_3\left(x_1^0, x_2^0, x_3^0\right)}{\partial x_3} \Delta x_3 = 0$$

(1.24)

The above equations may be written as the compact form

$$\begin{bmatrix} \dfrac{\partial f_1}{\partial x_1} & \dfrac{\partial f_1}{\partial x_2} & \dfrac{\partial f_1}{\partial x_3} \\[2mm] \dfrac{\partial f_2}{\partial x_1} & \dfrac{\partial f_2}{\partial x_2} & \dfrac{\partial f_2}{\partial x_3} \\[2mm] \dfrac{\partial f_3}{\partial x_1} & \dfrac{\partial f_3}{\partial x_2} & \dfrac{\partial f_3}{\partial x_3} \end{bmatrix} \begin{bmatrix} \Delta x_1 \\ \Delta x_2 \\ \Delta x_3 \end{bmatrix} = - \begin{bmatrix} f_1\left(x_1^0, x_2^0, x_3^0\right) \\ f_2\left(x_1^0, x_2^0, x_3^0\right) \\ f_3\left(x_1^0, x_2^0, x_3^0\right) \end{bmatrix}$$

(1.25)

With the initial point x_1^0, x_2^0, x_3^0, the incremental changes of Δx_1, Δx_2, and Δx_3 can be obtained, then a new point $x_1 = x_1^0 + \Delta x_1$, $x_2 = x_2^0 + \Delta x_2$, and $x_3 = x_3^0 + \Delta x_3$ can be found. With this new point, a new Newton-Raphson equation like (1.24) can be reformulated, then incremental changes of Δx_1, Δx_2, and Δx_3 can be obtained and a new point can be found. The iterating process continues until $f_1(x_1^0, x_2^0, x_3^0)$, $f_2(x_1^0, x_2^0, x_3^0)$, and $f_3(x_1^0, x_2^0, x_3^0)$ are very close to zero.

Applying the Newton-Raphson method to the load flow problem: (1.17) and (1.18), we have the following Newton iterative equation:

$$\begin{bmatrix} \dfrac{\partial \Delta P_1}{\partial \theta_1} & \dfrac{\partial \Delta P_1}{\partial V_1} & \dfrac{\partial \Delta P_1}{\partial \theta_2} & \dfrac{\partial \Delta P_1}{\partial V_2} & \dfrac{\partial \Delta P_1}{\partial \theta_3} & \dfrac{\partial \Delta P_1}{\partial V_3} \\[2mm] \dfrac{\partial \Delta Q_1}{\partial \theta_1} & \dfrac{\partial \Delta Q_1}{\partial V_1} & \dfrac{\partial \Delta Q_1}{\partial \theta_2} & \dfrac{\partial \Delta Q_1}{\partial V_2} & \dfrac{\partial \Delta Q_1}{\partial \theta_3} & \dfrac{\partial \Delta Q_1}{\partial V_3} \\[2mm] \dfrac{\partial \Delta P_2}{\partial \theta_1} & \dfrac{\partial \Delta P_2}{\partial V_1} & \dfrac{\partial \Delta P_2}{\partial \theta_2} & \dfrac{\partial \Delta P_2}{\partial V_2} & \dfrac{\partial \Delta P_2}{\partial \theta_3} & \dfrac{\partial \Delta P_2}{\partial V_3} \\[2mm] \dfrac{\partial \Delta Q_2}{\partial \theta_1} & \dfrac{\partial \Delta Q_2}{\partial V_1} & \dfrac{\partial \Delta Q_2}{\partial \theta_2} & \dfrac{\partial \Delta Q_2}{\partial V_2} & \dfrac{\partial \Delta Q_2}{\partial \theta_3} & \dfrac{\partial \Delta Q_2}{\partial V_3} \\[2mm] \dfrac{\partial \Delta P_3}{\partial \theta_1} & \dfrac{\partial \Delta P_3}{\partial V_1} & \dfrac{\partial \Delta P_3}{\partial \theta_2} & \dfrac{\partial \Delta P_3}{\partial V_2} & \dfrac{\partial \Delta P_3}{\partial \theta_3} & \dfrac{\partial \Delta P_3}{\partial V_3} \\[2mm] \dfrac{\partial \Delta Q_3}{\partial \theta_1} & \dfrac{\partial \Delta Q_3}{\partial V_1} & \dfrac{\partial \Delta Q_3}{\partial \theta_2} & \dfrac{\partial \Delta Q_3}{\partial V_2} & \dfrac{\partial \Delta Q_3}{\partial \theta_3} & \dfrac{\partial \Delta Q_3}{\partial V_3} \end{bmatrix} \begin{bmatrix} \Delta \theta_1 \\ \Delta V_1 \\ \Delta \theta_2 \\ \Delta V_2 \\ \Delta \theta_3 \\ \Delta V_3 \end{bmatrix} = - \begin{bmatrix} \Delta P_1 \\ \Delta Q_1 \\ \Delta P_2 \\ \Delta Q_2 \\ \Delta P_3 \\ \Delta Q_3 \end{bmatrix}$$

(1.26)

For the slack bus, θ_1, and V_1 are given and kept constant. In the above equations, the differentials with respect to these variables are zero and the first two equations at the slack bus are not needed and should be removed. For the PV bus, the bus voltage magnitude V_2 is given and kept constant. The differentials with respect to this variable should be zero and the fourth equation (reactive power mismatch equation at the PV bus) in (1.18) should be removed. Then we have the following reduced order Newton equation

$$
\begin{bmatrix}
\dfrac{\partial \Delta P_2}{\partial \theta_2} & \dfrac{\partial \Delta P_2}{\partial \theta_3} & \dfrac{\partial \Delta P_2}{\partial V_3} \\[2mm]
\dfrac{\partial \Delta P_3}{\partial \theta_2} & \dfrac{\partial \Delta P_3}{\partial \theta_3} & \dfrac{\partial \Delta P_3}{\partial V_3} \\[2mm]
\dfrac{\partial \Delta Q_3}{\partial \theta_2} & \dfrac{\partial \Delta Q_3}{\partial \theta_3} & \dfrac{\partial \Delta Q_3}{\partial V_3}
\end{bmatrix}
\begin{bmatrix}
\Delta \theta_2 \\ \Delta \theta_3 \\ \Delta V_3
\end{bmatrix}
= -
\begin{bmatrix}
\Delta P_2 \\ \Delta P_3 \\ \Delta Q_3
\end{bmatrix}
\qquad (1.27)
$$

The load flow solution of the three-bus system can be found by iteratively solving (1.27). For a system with N buses, the Newton load flow model may be given by the following compact form:

$$
\begin{bmatrix}
\dfrac{\partial \mathbf{P}}{\partial \boldsymbol{\theta}} & \dfrac{\partial \mathbf{P}}{\partial \mathbf{V}} \\[2mm]
\dfrac{\partial \mathbf{Q}}{\partial \boldsymbol{\theta}} & \dfrac{\partial \mathbf{Q}}{\partial \mathbf{V}}
\end{bmatrix}
\begin{bmatrix}
\Delta \boldsymbol{\theta} \\ \Delta \mathbf{V}
\end{bmatrix}
= -
\begin{bmatrix}
\Delta \mathbf{P} \\ \Delta \mathbf{Q}
\end{bmatrix}
\qquad (1.28)
$$

1.6.4 Fast Decoupled Load Flow Method

Noting the physical coupling between P and V and between Q and θ in (1.28), the differentials $\dfrac{\partial \mathbf{P}}{\partial \mathbf{V}}$, $\dfrac{\partial \mathbf{Q}}{\partial \boldsymbol{\theta}}$ may be simply set to zero. Then we have the decoupled load flow model

$$
\left[\dfrac{\partial P}{\partial \boldsymbol{\theta}} \right] [\Delta \boldsymbol{\theta}] = -[\Delta P] \qquad (1.29)
$$

$$
\left[\dfrac{\partial Q}{\partial V} \right] [\Delta V] = -[\Delta Q] \qquad (1.30)
$$

The above two equations can be further simplified such that the system matrix becomes constant. Then we have

$$
[B'][\Delta \boldsymbol{\theta}] = -[\Delta P / V] \qquad (1.31)
$$

$$
[B''][\Delta V] = -[\Delta Q / V] \qquad (1.32)
$$

The model in (1.31) and (1.32) is the well-known fast decoupled load flow. The fast decoupled load flow method is much faster than the standard Newton-Raphson load flow method. The fast decoupled load flow method can be used in security analysis in the real-time environment of energy management systems due to its superior computational performance. However, the fast decoupled load flow method may have difficulty modeling novel power system controllers like flexible AC transmission systems (FACTS). Most production-grade load flow programs usually include both the load flow solution algorithms.

1.6.5   DC Load Flow Method

With certain assumptions, the AC load flow models discussed in Sections 1.6.2–1.6.4 can be simplified. The basic assumptions for the DC load flow model are as follows:

- Only the angles of the complex bus voltages vary, and the angle differences of transmission lines are small.
- Voltage magnitudes are assumed to be constant (usually set to 1.0 in per unit).
- Transmission lines are assumed to have no resistance, and therefore no losses.
- Transformer tap ratio control is not considered (usually tap ratio is set to 1.0), though the transformer shifting can be modeled if applicable.

With the above assumptions, a DC load flow model, which has advantages for speed of computation, can provide a reasonable approximation of the real power system. In addition, such a model has the following properties:

- *Linearity*: The power flow of a particular transmission line is the linear combination of the power injections of the system.
- *Superposition*: The power flows can be broken down into a sum of power flow components of different transactions.

With the DC load flow model assumptions, the power flows of transmission line *i-j* are given by the following linear function of the angles of the transmission line:

$$P_{ij} = \frac{\theta_i - \theta_j}{x_{ij}} \tag{1.33}$$

$$P_{ji} = \frac{\theta_j - \theta_i}{x_{ij}} \tag{1.34}$$

Taking the three-bus system shown in Figure 1.3 as an example, we have the following power flow equations:

$$P_{12} = \frac{\theta_1 - \theta_2}{x_{12}}, P_{21} = \frac{\theta_2 - \theta_1}{x_{12}}, P_{13} = \frac{\theta_1 - \theta_3}{x_{13}}, P_{31} = \frac{\theta_3 - \theta_1}{x_{13}}, P_{23} = \frac{\theta_2 - \theta_3}{x_{23}}, P_{32} = \frac{\theta_3 - \theta_2}{x_{32}}$$

Considering the power balance at each bus of the system, we have

$$Pg_1 = P_{12} + P_{13}, Pg_2 = P_{23} + P_{21}, -Pd_3 = P_{31} + P_{32}$$

hence we have the following compact format:

$$B'\boldsymbol{\theta} = \boldsymbol{P} \tag{1.35}$$

where diagonal elements of **B'** are given by $B_{ii} = \sum_{i=1}^{N} 1/x_{ij}$ and off-diagonal elements are given by $B_{ij} = -1/x_{ij}$. $P = Pg - Pd$. In (1.29), if we take bus 1 as the reference bus, the row and column related to bus 1 should be removed and then the dimension of equation (1.29) becomes $N - 1$. Equation (1.29) now gives the DC load flow model. Unlike the solving of AC load flow problems, direct solution of DC load flow problem (1.29) can be obtained without any iterations. However, the deficits of the DC load flow model are obvious: a) in the model, power loss is not considered; b) reactive power and control is excluded. For heavily loaded system conditions, the DC load flow analysis may bring significant errors.

1.7 OPTIMAL OPERATION OF ELECTRIC POWER SYSTEMS

1.7.1 Security-Constrained Economic Dispatch

Economic dispatch determines the optimal power output of each generating unit while minimizing the overall cost of fuel to serve the system load. Normally any operational limits of generation and transmission facilities should be recognized. In the next section, we will start with classic economic dispatch with transmission network power loss. We will then introduce the security-constrained economic dispatch (SCED).

1.7.1.1 Classic Economic Dispatch Without Transmission Network Power Loss The classic economic dispatch can be formulated as:

$$\text{Minimize } f(Pg) = \sum_{i}^{Ng} f_i(Pg_i) = \sum_{i}^{Ng} \left(\alpha_i * Pg_i^2 + \beta_i * Pg_i + \gamma_i \right) \quad (1.36)$$

while subject to the following constraints:

Equality constraint:

$$\sum_{i=1}^{Ng} Pg_i - Pd - P_L = 0 \quad (1.37)$$

Inequality constraints

$$Pg_i^{\min} \leq Pg_i \leq Pg_i^{\max} \quad (1.38)$$

where Ng is the total number of generators. $f_i(Pg_i)$ is the fuel cost of generator i. P_L is the total transmission network power loss. Pd is the total system demand and it is assumed that this is constant. Now the key issue is to represent the total transmission network power loss P_L. Assume that the inequality constraints above are not binding, the Lagrange function of the above problem will be:

$$L(Pg) = \sum_{i}^{Ng} f_i(Pg_i) - \lambda \left(\sum_{i=1}^{Ng} Pg_i - Pd - P_L \right) \quad (1.39)$$

where λ is the incremental fuel cost of the system, which is called the Lagrange multiplier. The necessary optimization conditions for (1.39) are:

$$\frac{\partial L(Pg_i)}{\partial Pg_i} = \frac{\partial f_i(Pg_i)}{\partial Pg_i} - \lambda \left(1 - \frac{\partial P_L}{\partial Pg_i} \right) = 0 \quad (1.40)$$

where these are called the classic coordination equations.

Then we have:

$$\lambda = \left(\frac{1}{1 - \dfrac{\partial P_L}{\partial Pg_i}} \right) \frac{\partial f_i(Pg_i)}{\partial Pg_i} \tag{1.41}$$

where $\dfrac{\partial P_L}{\partial Pg_i}$ are the incremental losses. The above equation can be rewritten as:

$$\lambda = PF_i \frac{\partial f_i(Pg_i)}{\partial Pg_i} \tag{1.42}$$

where PF_i is called the penalty factor of generator i, and is given by

$$PF_i = \left(\frac{1}{1 - \dfrac{\partial P_L}{\partial Pg_i}} \right) \tag{1.43}$$

The penalty factor can be determined in two different ways. The first approach is based on the so-called B coefficient model [30–32]. The calculations of B coefficients have been improved upon in [33] for efficient implementation by using sparsity techniques. The B coefficients calculations are based on an approximation of the system losses as a quadratic function of the generation powers:

$$P_L = Pg^T BPg + Pg^T B_0 + B_{00} \tag{1.44}$$

where $Pg = [Pg_1, Pg_2, \ldots, Pg_{Ng}]^T$. B is a square matrix, B_0 is a vector.

The second approach is based on Newton's method [34]. Assume bus 1 is a slack bus or a reference bus and Pg_1 is the active power of the generator at bus 1, we have:

$$\begin{bmatrix} \dfrac{\partial P}{\partial \theta} & \dfrac{\partial P}{\partial V} \\[2ex] \dfrac{\partial Q}{\partial \theta} & \dfrac{\partial Q}{\partial V} \end{bmatrix}^T \begin{bmatrix} \dfrac{\partial Pg_1}{\partial Pg} \\[2ex] \dfrac{\partial Pg_1}{\partial Qg} \end{bmatrix} = - \begin{bmatrix} \dfrac{\partial Pg_1}{\partial \theta} \\[2ex] \dfrac{\partial Pg_1}{\partial V} \end{bmatrix} \tag{1.45}$$

Note in the above equation, the system matrix is the transpose of the Newton power flow Jacobian Matrix in (1.28). The equation (1.45) can be solved very efficiently using sparsity matrix techniques.

We know $P_L = \displaystyle\sum_{i=1}^{Ng} Pg_i - Pd$, then we can get

$$\frac{\partial P_L}{\partial Pg_i} = \frac{\partial Pg_1}{\partial Pg_i} + \frac{\partial \displaystyle\sum_{k=2}^{Ng} Pg_k}{\partial Pg_i} = \frac{\partial Pg_1}{\partial Pg_i} + 1 \tag{1.46}$$

Rewriting the above equation, we have

$$-\frac{\partial Pg_1}{\partial Pg_i} = 1 - \frac{\partial P_L}{\partial Pg_i} \tag{1.47}$$

Then the penalty factors are given by

$$PF_i = \frac{1}{-\dfrac{\partial Pg_1}{\partial Pg_i}} \tag{1.48}$$

where $-\dfrac{\partial Pg_1}{\partial Pg_i}$ can be found by solving (1.45).

1.7.1.2 Security Constrained Economic Dispatch In comparison with the classic economic dispatch problem, a security constrained economic dispatch problem can consider transmission line thermal limits, which are also called security constraints. The security-constrained economic dispatch can be formulated as follows:

$$\text{Minimize } f(Pg) = \sum_{i}^{Ng} f_i(Pg_i) = \sum_{i}^{Ng} \left(\alpha_i * Pg_i^2 + \beta_i * Pg_i + \gamma_i\right) \tag{1.49}$$

while subject to the following constraints:

Equality constraint:

$$\sum_{i=1}^{Ng} Pg_i - PL - P_{Loss} = 0 \tag{1.50}$$

Inequality constraints:

$$Pg_i^{\min} \leq Pg_i \leq Pg_i^{\max} \tag{1.51}$$

$$P_{ij}^{\min} \leq P_{ij}(\mathbf{Pg}) \leq P_{ij}^{\max} \tag{1.52}$$

where P_{ij} is the power flow of line ij and given by (1.33). From (1.35), the angles in (1.33) can be represented as a function of **Pg** by solving equation (1.35). Survey papers on economic dispatch can be found in [36–39]. In comparison to an economic dispatch problem, an optimal power flow problem is a more general optimization problem, which can include detailed network representation and various operating, control, and contingency constraints. Consider variables of various control devices such as generators, transformers, reactive compensation devices, FACTS devices, load shedding actions, DC lines, and network switching, and adopt various objective functions concerning economy and security. In the next section, optimal power flow techniques are reviewed, then detailed formulations and solutions of optimal power flow problems are introduced.

1.7.2 Optimal Power Flow Techniques

1.7.2.1 Development of Optimization Techniques in OPF Solutions The optimal power flow (OPF) problem was initiated by the desire to minimize the operating cost of the supply of electric power when load is given [31, 32]. In 1962 a generalized nonlinear programming formulation of the economic dispatch problem including voltage and other operating constraints was proposed by Carpentier [35]. The OPF problem was defined in early 1960s as an expansion of conventional economic dispatch to determine the optimal settings for control variables in a power network considering various operating and control constraints [40–44]. The OPF method proposed in [40] has been known as the reduced gradient method, which can be formulated by eliminating the dependent variables based on a solved load flow. Since the concept of the reduced gradient method for the solution of the OPF problem was proposed, continuous efforts in the development of new OPF methods have been found. Several review papers were published [39, 45–47]. Among the various OPF methods proposed, it has been recognized that the main techniques for solving the OPF problems are the gradient method [40], linear programming (LP) method [49, 50], successive sparse quadratic programming (QP) method [52], successive nonsparse QP method [54], Newton's method [55], and Interior Point Methods [61–66]. Each method has its own advantages and disadvantages. These algorithms have been employed with varied success.

OPF problems are very complex mathematical programming problems. Numerous papers on the numerical solution of the OPF problems have been published [39, 42, 45–47]. In this section, a review of several OPF methods is given.

The widely used gradient methods for the OPF problems include the reduced gradient method [40] and the generalized gradient method [48]. Gradient methods exhibit slow convergence characteristics near the optimal solution. In addition, the methods are difficult to solve in the presence of inequality constraints.

LP methods have been widely used in the OPF problems. The main strengths of LP-based OPF methods are summarized as follows: a) efficient handling of inequalities and detection of infeasible solutions; b) dealing with local controls; and c) incorporation of contingencies.

It is quite common in OPF problems that nonlinear equalities and inequalities and objective function need to be handled. In this situation, all the nonlinear constraints and objective functions should be linearized around the current operating point such that LP methods can be applied to solve the linear optimal problems. For a typical LP-based OPF, the solution can be found through the iterations between load flow and linearized LP subproblem. The LP-based OPF methods have been shown to be effective for problems where the objectives are separable and convex. However, the LP-based OPF methods may not be effective where the objective functions are nonseparable, for instance in the minimization of transmission losses.

QP based OPF methods [51–54] are efficient for some OPF problems, especially for the minimization of power network losses. In [54], the nonsparse implementation of the QP-based OPF was proposed while in [51–53], the sparse implementation of the QP-based OPF algorithm for large-scale power systems was presented. In [51, 52], the successive QP-based OPF problems are solved through a sequence of linearly constrained subproblems using a quasi-Newton search direction. The QP formulation can always find a feasible solution by adding extra shunt compensation. In [53], the QP method, which is a direct solution method, solves a set of linear equations involving the Hessian matrix and the Jacobian matrix by converting the inequality constrained quadratic program (IQP) into the equality constrained quadratic program (EQP) with an initial guess at the correct active set. The computational speed of the QP method in [53] has been much improved in comparison to those in [51, 52]. The QP methods in [51–53] are solved using MINOS, developed at Stanford University.

The development of the OPF algorithm by Newton's method [55, 57, 58], is based on the success of the Newton's method for the power flow calculations. Sparse matrix techniques applied to the Newton power flow calculations are directly applicable to the Newton OPF calculations. The major idea is that the OPF problems are solved by the sequence of the linearized Newton equations where inequalities are treated as equalities when they are binding. However, the most critical aspect of Newton's algorithm is that the active inequalities are not known prior to the solution and the efficient implementations of the Newton's method usually adopt the so-called trial iteration scheme where heuristic constraint enforcement/release is iteratively performed until acceptable convergence is achieved. In [56, 59], alternative approaches using linear programming techniques have been proposed to identify the active set efficiently in the Newton's OPF. In principle, the successive QP methods and Newton's method both using the second derivatives, considered a second-order optimization method, are theoretically equivalent.

Since Karmarkar published his paper on an interior point method for linear programming in 1984 [60], a great interest on the subject has arisen. Interior point methods have proven to be a promising alternative for the solution of power system optimization problems. In [61, 62], a security-constrained economic dispatch (SCED) is solved by sequential linear programming and the IP dual-affine scaling (DAS). In [63], a modified IP DAS algorithm was proposed. In [64], an interior point method was proposed for linear and convex quadratic programming. It is used to solve power system optimization problems such as economic dispatch and reactive power planning. In [65–70], nonlinear primal-dual interior point methods for power system optimization problems were developed. The nonlinear primal-dual methods proposed can be used to solve the nonlinear power system OPF problems efficiently. The theory of nonlinear primal-dual interior point methods has been established based on three achievements: Fiacco and McCormick's barrier method for optimization with inequalities, Lagrange's method for optimization with equalities, and Newton's method for solving nonlinear equations [71]. Experience with application of interior point methods to power system optimization problems has been quite positive.

1.7.2.3 OPF Formulation

The OPF problem may be formulated as follows:

$$\text{Minimize: } f(\mathbf{x}, \mathbf{u}) \tag{1.53}$$

subject to:

$$\mathbf{g}(\mathbf{x}, \mathbf{u}) = 0 \tag{1.54}$$

$$\mathbf{h}_{min} \leq \mathbf{h}(\mathbf{x}, \mathbf{u}) \leq \mathbf{h}_{max} \tag{1.55}$$

where

  u-the set of control variables

  x-the set of dependent variables

  $f(\mathbf{x}, \mathbf{u})$-a scalar objective function

  $\mathbf{g}(\mathbf{x}, \mathbf{u})$-the power flow equations

  $\mathbf{h}(\mathbf{x}, \mathbf{u})$-the limits of the control variables and operating limits of power system components.

TABLE 1.1   Objectives, constraints, and control variables of the OPF problems

Objectives	• Minimum cost of generation and transactions
	• Minimum transmission losses
	• Minimum shift of controls
	• Minimum number of controls shifted
	• Mininum number of controls rescheduled
	• Minimum cost of VAr investment
Equality constraints	• Power flow constraints
	• Other balance constraints
Inequality constraints	• Limits on all control variables
	• Branch flow limits (amps, MVA, MW, MVAr)
	• Bus voltage variables
	• Transmission interface limits
	• Active/reactive power reserve limits
Controls	• Real and reactive power generation
	• Transformer taps
	• Generator voltage or reactive control settings
	• MW interchange transactions
	• HVDC link MW controls
	• FACTS voltage and power flow controls
	• Load shedding

The objectives, controls and constraints of the OPF problems are summarized in Table 1.1. The limits of the inequalities in Table 1.1 can be classified into two categories: a) physical limits of control variables; b) operating limits of power system. In principle, physical limits on control variables cannot be violated while operating limits representing security requirements can be violated or relaxed temporarily.

In addition to the steady state power flow constraints, for the OPF formulation, stability constraints, which are described by differential equations, may be considered and incorporated into the OPF. In recent years, stability constrained OPF problems have been proposed [72–74].

1.7.2.4   Optimal Power Flow Solution by Nonlinear Interior Point Methods

1.7.2.4.1   Power Mismatch Equations   The power mismatch equations in rectangular coordinates at a bus are given by:

$$\Delta P_i = Pg_i - Pd_i - P_i \tag{1.56}$$

$$\Delta Q_i = Qg_i - Qd_i - Q_i \tag{1.57}$$

where Pg_i and Qg_i are real and reactive powers of generator at bus i, respectively; Pd_i and Qd_i the real and reactive load powers, respectively; P_i and Q_i the power injections at the node and are given by:

$$P_i = V_i \sum_{j=1}^{N} V_j \left(G_{ij} \cos\theta_{ij} + B_{ij} \sin\theta_{ij} \right) \tag{1.58}$$

$$Q_i = V_i \sum_{j=1}^{N} V_j \left(G_{ij} \sin\theta_{ij} - B_{ij} \cos\theta_{ij} \right) \tag{1.59}$$

where V_i and θ_i are the magnitude and angle of the voltage at bus i, respectively; $Y_{ij} = G_{ij} + jB_{ij}$ is the system admittance element while $\theta_{ij} = \theta_I - \theta_j$. N is the total number of system buses.

1.7.2.4.2   Transmission Line Limits   The transmission MVA limit may be represented by:

$$\left(P_{ij} \right)^2 + \left(Q_{ij} \right)^2 \le \left(S_{ij}^{\max} \right)^2 \tag{1.60}$$

where $S_{ij}^{\max}$ is the MVA limit of the transmission line ij. P_{ij} and Q_{ij} are given by:

$$P_{ij} = -V_i^2 G_{ij} + V_i V_j \left(G_{ij} \cos\theta_{ij} + B_{ij} \sin\theta_{ij} \right) \tag{1.61}$$

$$Q_{ij} = V_i^2 b_{ii} + V_i V_j \left(G_{ij} \sin\theta_{ij} - B_{ij} \cos\theta_{ij} \right) \tag{1.62}$$

where $b_{ii} = -B_{ii} + bc_{ij}/2$. bc_{ij} is the shunt admittance of transmission line ij.

1.7.2.4.3   Formulation of the Nonlinear Interior Point OPF   Mathematically, as an example, the objective function of an OPF may minimize the total operating cost as follows:

$$\text{Minimize } f(x) = \sum_{i}^{Ng} \left(\alpha_i * Pg_i^2 + \beta_i * Pg_i + \gamma_i \right) \tag{1.63}$$

while being subject to the following constraints:
Nonlinear equality constraints:

$$\Delta P_i(x) = Pg_i - Pd_i - P_i(t, e, f) = 0 \tag{1.64}$$

$$\Delta Q_i(x) = Qg_i - Qd_i - Q_i(t, e, f) = 0 \tag{1.65}$$

Nonlinear inequality constraints:

$$h_j^{\min} \le h_j(x) \le h_j^{\max} \tag{1.66}$$

where

$x = [Pg, Qg, t, \theta, V]^T$ is the vector of variables

$\alpha_i, \beta_i, \gamma_i$ coefficients of production cost functions of generator

$\Delta P(x)$ bus active power mismatch equations

$\Delta Q(x)$ bus reactive power mismatch equations

$h(x)$ functional inequality constraints including line flow and voltage magnitude constraints, simple inequality constraints of variables such as generator active power, generator reactive power, and transformer tap ratio

Pg the vector of active power generation

Qg the vector of reactive power generation

t the vector of transformer tap ratios

θ the vector of bus voltage magnitude

V the vector of bus voltage angle

Ng the number of generators

By applying Fiacco and McCormick's barrier method, the OPF problem equations (1.63)–(1.67) can be transformed into the following equivalent OPF problem:

$$\text{Objective:} \quad Min\left\{ f(x)-\mu\sum_{j=1}^{M}\ln(sl_j)-\mu\sum_{j=1}^{M}\ln(su_j)\right\} \tag{1.67}$$

subject to the following constraints:

$$\Delta P_i = 0 \tag{1.68}$$
$$\Delta Q_i = 0 \tag{1.69}$$
$$h_j - sl_j - h_j^{\min} = 0 \tag{1.70}$$
$$h_j + su_j - h_j^{\max} = 0 \tag{1.71}$$

where $sl > 0$ and $su > 0$.

Thus, the Lagrangian function for equalities optimization of equations (1.67)–(1.71) is given by:

$$L = f(x)-\mu\sum_{j=1}^{M}\ln(sl_j)-\mu\sum_{j=1}^{M}\ln(su_j)-\sum_{i=1}^{N}\lambda p_i\Delta P_i-\sum_{i=1}^{N}\lambda q_i\Delta Q_i$$
$$-\sum_{j=1}^{M}\pi l_j\left(h_j-sl_j-h_j^{\min}\right)-\sum_{j=1}^{M}\pi u_j\left(h_j+su_j-h_j^{\max}\right) \tag{1.72}$$

where λp_i, λq_i, πl_j, πu_j are Langrage multipliers for the constraints of equations (1.68)–(1.71), respectively. N represents the number of buses and M the number of inequality constraints. Note that $\mu > 0$. The Karush-Kuhn-Tucker (KKT) first order conditions for the Lagrangian function shown in equation (1.72) are as follows:

$$\nabla_x L_\mu = \nabla f(x)-\nabla\Delta P^T\lambda p-\nabla\Delta Q^T\lambda q-\nabla h^T\pi l-\nabla h^T\pi u = 0 \tag{1.73}$$
$$\nabla_{\lambda p}L_\mu = -\Delta P = 0 \tag{1.74}$$
$$\nabla_{\lambda q}L_\mu = -\Delta Q = 0 \tag{1.75}$$
$$\nabla_{\pi l}L_\mu = -\left(h-sl-h^{\min}\right)=0 \tag{1.76}$$
$$\nabla_{\pi u}L_\mu = -\left(h+su-h^{\max}\right)=0 \tag{1.77}$$
$$\nabla_{sl}L_\mu = \mu - Sl\Pi l = 0 \tag{1.78}$$
$$\nabla_{su}L_\mu = \mu + Su\Pi u = 0 \tag{1.79}$$
$$\nabla_{\lambda q}L_\mu = -\Delta Q = 0 \tag{1.80}$$

where $Sl = diag(sl_j)$, $Su = diag(su_j)$, $\Pi l = diag(\pi l_j)$, $\Pi u = diag(\pi u_j)$. As suggested in [65], the above equations can be decomposed into the following three sets of equations:

$$\begin{bmatrix} -\Pi l^{-1}Sl & 0 & -\nabla h & 0 \\ 0 & \Pi u^{-1}Su & -\nabla h & 0 \\ -\nabla h^T & -\nabla h^T & H & -J^T \\ 0 & 0 & -J & 0 \end{bmatrix}\begin{bmatrix} \Delta\pi l \\ \Delta\pi u \\ \Delta x \\ \Delta\lambda \end{bmatrix}=\begin{bmatrix} -\nabla_{\pi l}L_\mu-\Pi l^{-1}\nabla_{sl}L_\mu \\ -\nabla_{\pi u}L_\mu-\Pi u^{-1}\nabla_{su}L_\mu \\ -\nabla_x L_\mu \\ -\nabla_\lambda L_\mu \end{bmatrix} \tag{1.81}$$

$$\Delta sl = \Pi l^{-1}\left(\nabla_{sl}L_\mu - Sl\Delta\pi l\right) \qquad (1.82)$$

$$\Delta su = \Pi u^{-1}\left(-\nabla_{su}L_\mu - Su\Delta\pi u\right) \qquad (1.83)$$

where $H(x, \lambda, \pi l, \pi u) = \nabla^2 f(x) - \sum \lambda\nabla^2 g(x) - \sum(\pi l + \pi u)\nabla^2 h(x),$

$$J(x) = \left[\frac{\partial\Delta P(x)}{\partial x}, \frac{\partial\Delta Q(x)}{\partial x}\right], g(x) = \left[\begin{matrix}\Delta P(x)\\ \Delta Q(x)\end{matrix}\right], \quad \text{and} \quad \lambda = \left[\begin{matrix}\lambda_p\\ \lambda_q\end{matrix}\right].$$

The elements corresponding to the slack variables sl and su have been eliminated from equation (1.81) using analytical Gaussian elimination. By solving equation (1.81), $\Delta\pi l$, $\Delta\pi u$, Δx, $\Delta\lambda$ can be obtained, then by solving equations (1.82) and (1.83), respectively, Δsl, Δsu can be obtained. With $\Delta\pi l$, $\Delta\pi u$, Δx, $\Delta\lambda$, Δsl, Δsu known, the OPF solution can be updated using the following equations:

$$sl^{(k+1)} = sl^{(k)} + \sigma\alpha_p\Delta sl \qquad (1.84)$$

$$su^{(k+1)} = su^{(k)} + \sigma\alpha_p\Delta su \qquad (1.85)$$

$$x^{(k+1)} = x^{(k)} + \sigma\alpha_p\Delta x \qquad (1.86)$$

$$\pi l^{(k+1)} = \pi l^{(k)} + \sigma\alpha_d\Delta\pi l \qquad (1.87)$$

$$\pi u^{(k+1)} = \pi u^{(k)} + \sigma\alpha_d\Delta\pi u \qquad (1.88)$$

$$\lambda^{(k+1)} = \lambda^{(k)} + \sigma\alpha_d\Delta\lambda \qquad (1.89)$$

where k is the iteration count, parameter $\sigma \in [0.995 - 0.99995]$ and αp and αd are the primal and dual step-length parameters, respectively. The step-lengths are determined as follows:

$$\alpha_p = \min\left[\min\left(\frac{sl}{-\Delta sl}\right), \min\left(\frac{su}{-\Delta su}\right), 1.00\right] \qquad (1.90)$$

$$\alpha_d = \min\left[\min\left(\frac{\pi l}{-\Delta\pi l}\right), \min\left(\frac{\pi u}{-\Delta\pi u}\right), 1.00\right] \qquad (1.91)$$

for those $sl < 0$, $\Delta su < 0$, $\Delta\pi l < 0$, and $\Delta\pi u > 0$.
The barrier parameter μ can be evaluated by:

$$\mu = \frac{\beta \times Cgap}{2 \times M} \qquad (1.92)$$

where $\beta \in [0.01 - 0.2]$ and Cgap is the complementary gap for the nonlinear interior point OPF and can be determined using:

$$Cgap = \sum_{j=1}^{M}\left(sl_j\pi l_j - su_j\pi u_j\right) \qquad (1.93)$$

1.8 OPERATION AND CONTROL OF ELECTRIC POWER SYSTEMS—SCADA/EMS

1.8.1 Introduction of SCADA/EMS

SCADA/EMS (Supervisory Control and Data Acquisition/Energy Management System) is a computer monitoring and control system that can be used to supervise,

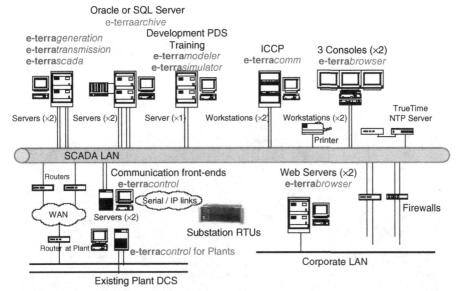

Figure 1.8 Schematic diagram of SCADA/EMS of energy control centers (courtesy of AREVA T&D)

control, optimize, and manage interconnected transmission systems. Electric power networks are complex systems that cannot be efficiently and securely operated without a SCADA/EMS. In contrast to SCADA/EMS for electric transmission networks, SCADA/DMS (Distribution Management System) is applied for electric distribution networks performing the very similar functions. A schematic diagram of SCADA/EMS of energy control centers is shown in Figure 1.8.

In Figure 1.8, e-terra*scada* is a distributed, scalable SCADA system that gathers real-time data from remote terminal units (RTUs) and other communication sources in the field and enables control of field devices from consoles. e-terra*generation* includes a suite of software applications such as real-time automatic generation control (AGC), transaction and unit scheduling, unit commitment, product costing, and load forecast. e-terra*simulator* provides training functionalities by helping operators acquire more knowledge, skills, and experience to operate real-time systems with the highest reliability standards. e-terra*transmission* is a suite of integrated network analysis applications designed for the support of real-time operation of a large electric power transmission network: network topology, state estimator, contingency analysis, dynamic stability, and short-circuit analysis as well as optimization and security constrained dispatch. A secure intercontrol center communication (ICCP) gateway facilitates the open exchange of data between interconnected and interdependent electric power utilities while e-terra*control* is a network-based distributed system that implements SCADA across a wide area network (WAN) while e-terra*control* for substations provides an access gateway into substations.

The energy control center is considered the central nerve system of the power system, which senses the status and measures the power, voltage, and current of the

Figure 1.9   The energy control center of Midwest Independent System Operator (MISO)
(© Midwest ISO, all rights reserved)

power system, adjusts its operating conditions, coordinates its controls, and provides defense against abnormal events. After the 1965 power blackout in the US, great efforts in the development of energy control centers occurred in improving power system monitoring, operation, control, and planning using advanced computer techniques. Basically, the power system operation, control, planning, and intelligent management functions form the Energy Management System (EMS), which has the nature of centralized control. An energy control center is shown in Figure 1.9.

1.8.2   SCADA/EMS of Conventional Energy Control Centers

SCADA/EMS of energy control centers have evolved over the years into a complex computer based information and control system. For a SCADA system, the functions may include:

- Data acquisition
- Device and sequence control
- Events management
- Dynamic network coloring
- Intercontrol center communications

A SCADA/EMS system in an energy control center plays a very important role in the operation of a power system. The functions of SCADA/EMS are usually called applications. The major functions of EMS may include the following categories:

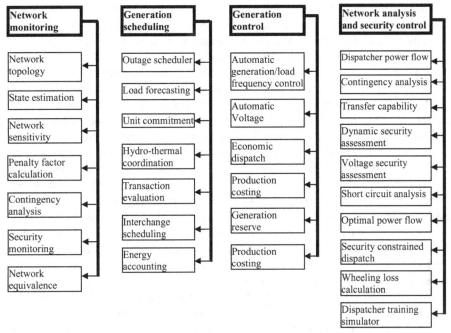

Figure 1.10   Functions of EMS

- Network monitoring
- Generation scheduling
- Generation control
- Network analysis and security control

The detailed functionalities are shown in Figure 1.10.

1.8.3 New Development Trends of SCADA/EMS of Energy Control Centers

1.8.3.1 New Environments   Early SCADA and EMS of energy control centers were developed for the industry of a vertically integrated monopoly. Significant changes have taken place in the deregulated industry environment. The functionalities of energy control centers need to be reshaped with increased complexities, flexibility, and openness in response to the fasting-changing of rules for market operations and regulatory framework and the increasing information exchange requirements between energy control centers. Advances in technology provide great opportunities to integrate different automation and information products more efficiently and reliably. For instance, the integration of various automation products, such as SCADA, EMS, DA (distribution automation), AM/FM/GIS (automated mapping and facility management, geographic information system), and MIS (management information system), has been implemented.

1.8.3.2 Advanced Software Technologies With the development of information and control technologies, along with the development of worldwide electricity markets, the current EMS may need to migrate into a more decentralized environment while new emerging technologies would speed up this migration process.

Communications protocols play a key role in the development of SCADA/ EMS of energy control centers. With standard protocols, computer communications can be implemented either via the Internet or in LANs. Standard protocols are based on the so-called open system interconnection (OSI) layered model, which is an industry standard framework including the functions of networking divided into seven distinct layers: physical, data link, network, transport, session, presentation, and application layers. The TCP/IP suite of protocol is the dominant standard for internetworking, which specifies how packets of information are exchanged between computers over one or more networks. The standard IP protocol provides a high degree of interoperability. Within the power industry, the intercontrol center protocol (ICCP) for intercontrol center communications based on the OSI model has been developed and is an IEC standard. For communications, electricity market operations use e-commerce standards like XML (eXtensible Markup Language) for documents containing structured information.

In the past 20 years or so, distributed system technologies, including distributed file systems, distributed memory storage systems, network operating systems, and middleware systems, have been developed, along with the recent advances in high-speed networking. Object-oriented concepts and methodology were developed along with the development of object-oriented programming in the late 1980s as a revolutionary attempt to change the paradigm of software design and engineering. Object-oriented programming provides a modular approach for software design. Each object combines data and procedures where object-oriented languages can provide a well-defined interface to their objects through classes and encapsulation, leading to more self-contained verifiable, modifiable, and maintainable software. By reusing developed classes, new applications can be built up much faster than in the traditional paradigm of software design in terms reliability, efficiency, and consistency of design. Based on the object design paradigm, C++, Java, and UML (the Unified Modeling Language) have been developed.

Another software technology is called component technology, where components consist of multiple objects and functionalities of the objects can be combined to offer a single software building block that can be adapted and reused. A component should have a standard interface, via which other components of the application can invoke its functions and access and manipulate the data within the component. The benefit is that software components can be independently developed, and they can be assembled and deployed, very much like standard reconfigurable hardware. The well-recognized component models include Enterprise JavaBeans, CORBA Components, and Microsoft COM/DCOM. In distributed object design technology, complex applications are usually decomposed into software components. Middleware technology was developed based on distributed object technology. Middleware is computer software that connects software components or applications.

Online power system security analysis and control for future smart grids requires a significant amount of computational power. Grid computing technology is an emerging technology for providing high-performance computing capability and a collaboration mechanism for solving complex problems while using the existing distributed resources. It can be used in the computation of intensive power system operation and control problems.

There is a need for synchronizing fast measurements across wide areas for online security analysis and stability control and coordination using phasor measurement units (PMUs). Industry research and development is working towards the systematic deployment of these PMUs and their integration into the existing energy control center design.

An energy control center with SCADA and EMS provides real-time data acquisition, monitoring, security, and control functions so as to support the operation of a power system and ensure the reliability and security of power system operations and the efficiency of electricity market operations. Systematic integration of enabling and emerging technologies such as PMUs, FACTS, and HVDC technologies, advanced energy storage technologies, advanced information technologies, and demand side management along with advanced operating concepts such as virtual power plants and microgrids will shape the way towards smart grids or intelligent grids. Such a trend of integration will facilitate the advanced energy control centers with the features of decentralization, flexibility, and openness in a more complicated industry environment.

1.9   ACTIVE POWER AND FREQUENCY CONTROL

1.9.1   Frequency Control and Active Power Reserve

A transmission system operator has the responsibility to maintain frequency between certain specified limits as large deviations in frequency can lead to widespread demand disconnections, generation disruptions, and even system splitting or collapse. If demand is greater than generation, frequency falls, and, if generation is greater than demand, frequency rises. The frequency control requirement of an interconnected transmission system should be less than the sum of that of each separate system. In order to fulfill the frequency control requirement, the system should have sufficient active power reserve. The active power reserve may be classified into the following categories in terms of system contingencies [10, 11]:

- *Spinning reserves:* Power generating units, synchronized to the grid, that can increase output immediately in response to a major generator or transmission outage and can reach full output, for instance, within 10 minutes to comply with NERC's Disturbance Control Standard.

- *Supplemental reserve:* Same as spinning reserve, but need not respond immediately; units can be offline but still must be capable of reaching full output within the required 10 minutes.

- *Replacement reserve:* Same as supplemental reserve, but with a 30-minute response time; used to restore spinning and supplemental reserves to their precontingency status.

The definition of the various reserves may vary for different system operators. In electricity market environments, the above reserve provisions are available in the framework of ancillary services. In addition to these active power reserve services for frequency control in terms of system contingencies, there is also the so-called "regulation and load following" service, which is provided by the real-time energy market and is used to continuously balance generation and load under normal conditions. Load following and regulation can be implemented in AGC.

1.9.2 Objectives of Automatic Generation Control

For the normal operation of an interconnected power system, frequency of the system and bus voltages should be controlled within limits in order to provide satisfactory service to the customers and to ensure the security operation of equipment. Due to the different physical characteristics, frequency and voltage can controlled in decomposed manner and in different time scales. The active power and reactive power in a transmission network are relatively independent of each other and their control can be implemented separately. System frequency control is more relevant to active power control while voltage control is more related to reactive power control.

As system load is continuously changing, the output of generators should be changed automatically. Automatic generation control (AGC) is a control system used in active power and frequency control. AGC has three major objectives:

- To maintain system frequency at or very close to a specified nominal value (e.g., 50 Hz).
- To maintain the scheduled tie-line loading of interchange power between control areas.
- To maintain each generator's output at the most economic value.

AGC systems have many advantages over governor speed control systems. The AGC systems transmit control signals to the governor systems and operate the control valves to decrease or increase the input to turbines to restore and maintain correct frequency if required. AGC systems have the capability to allocate the governing responses among generators of power plants. AGC systems have more powerful control ability to keep the system frequency within limits than that of conventional governing control systems.

1.9.3 Turbine-Generator-Governor System Model

A diagram that integrates the turbine, generator and governor into a system is shown in Figure 1.11. The characteristics of the turbine and generator as well as the primary control of governor have been included in the diagram. The AGC is related to the supplementary control and tie-line control that will be discussed later.

In Figure 1.12, P_{mech} is mechanical power of the generator, P_{elec} is electrical power output of the generator, and ω is rotational speed of the generator. The relationship between mechanical power, electrical power, and speed change may be given by

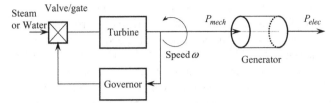

Figure 1.11   Conceptual description of a generator and its associated control

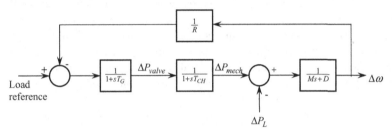

Figure 1.12   Turbine-generorat-governor system model

$$\Delta P_{mech} - \Delta P_{elec} = M \frac{d}{dt}(\Delta \omega) \qquad (1.94)$$

where ΔP_{mech}, ΔP_{elec}, and $\Delta \omega$ are mechanical power, electrical power output, and rotational speed change, respectively. The electrical power change ΔP_{elec} can be further represented by nonfrequency sensitive load change and frequency-sensitive load change

$$\Delta P_{elec} = \Delta P_L + D\Delta \omega \qquad (1.95)$$

Usually a generator is driven by a steam turbine or a hydro turbine. A non-reheat turbine model, which describes the relationship between the mechanical power and valve position, is shown in Figure 1.12. T_{CH} is the time constant of the turbine system. In the a generator governor model shown in Figure 1.12, the governor has a gain $\frac{1}{R}$ and a time constant T_G. R determines the change of generator's output for a change in frequency.

For normal operation of a power system, system frequency should be controlled within its limits. Because system frequency is common to all parts of the system and is easily measured, it was probably the first quantity applied to system control. The governors of generating units make use of rotating flyballs. These actuate a hydraulic system to open or close the throttle valves of the prime moves of the machines. This action increases or decreases energy input (for instance, fuel in a thermal plant or water in a hydro plant) to maintain speed (and hence frequency) at the desired value. Modern electronic governors sense frequency and actuate hydraulic devices to control gate or throttle position without the use of flyballs.

In order to operate machines stably in parallel with the system, it is necessary that the governors have drooping characteristics. That is, as load increases, speed decreases. Governor droops are expressed in percentage of speed change from no load to full load. If governors had zero droop, or if they were adjusted so that the speed characteristics increased with load, operation would be unstable. If one machine has a lower governor droop setting than others, when two or more generating units are operated in parallel on an AC system, on a system frequency drop the machine with the lower droop characteristic will pick up proportionally more load. Since generators operated in parallel cannot be separated to adjust the governor, when there is a load change each time, the governor droop characteristic is adjusted during a series of tests and is then fixed. Because governors are combination of hydraulic and mechanical components, an appreciable change in system speed is required before the governor can sense it and take corrective action. Consequently, the correction is delayed by a discrete time interval from the time the speed or frequency change occurred. As a result, machines or systems controlled only by governors have a dead band on the order of 0.02 Hz.

1.9.4 AGC for a Single-Generator System

AGC for a single-generator system is shown Figure 1.13. Once a load change has occurred, the generator governing system as shown in Figure 1.12 can reduce frequency deviation, however, it can not restore system frequency back to nominal value. In this situation, a supplementary control by AGC is required to restore frequency to nominal value. The principle of the supplementary control by AGC is to reset the load reference point as shown in Figure 1.13 and to force the frequency error to zero.

The AGC for a single generator system can be very easily extended to multimachine systems by modifying the output of the supplementary control block based on generator's participation factor determined by economic dispatching algorithm. Their allocated load reference P_{ref} is input to corresponding generator.

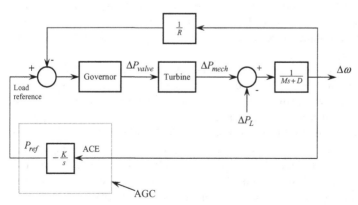

Figure 1.13   AGC for a single-generator system

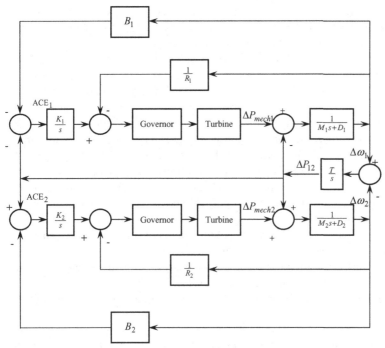

Figure 1.14   AGC for two-area system

1.9.5   AGC for Two-Area Systems

For two-area systems, the basic objective of supplementary control is to restore balance between each area load and generation (Figure 1.14). This is usually done by the following control actions:

1. Maintaining frequency at the scheduled value;
2. Maintaining net interchange power at the scheduled value.

For two-area systems, area control error signals for the two systems are:

$$ACE_1 = -\Delta P_{12} - B_1 \Delta \omega \qquad (1.96)$$
$$ACE_2 = \Delta P_{12} - B_2 \Delta \omega \qquad (1.97)$$

where ΔP_{12} is the net interchange power change. B_1 and B_2 bias factors.

A thorough review of some key issues of AGC is found in [12]. AGC logic based on NERC (North American Electric Reliability Corporation)'s new control performance standard and disturbance control standard is discussed in [13, 14].

1.9.6   Frequency Control and AGC in Electricity Markets

It has been reported [10] that hourly markets for regulation and the contingency reserves (spinning, supplemental, and sometimes replacement reserves) are either in operation or are being established in ISO regions, such as New England, New York,

PJM (Pennsylvania-New Jersey-Maryland Interconnection), the Electric ERCOT (Electric Reliability Council of Texas), California, Ontario and Alberta, Canada. Design of such ancillary markets is referred to in Chapter 2.

Further discussions on frequency control and active power reserve and AGC in electricity market environments are found in [15–23].

In the UK transmission system, mandatory frequency response, which is an automatic change in active power output in response to a frequency change, is provided by all generators in the light of the requirements of the grid code. The capability to provide mandatory frequency response service is a condition of connection of generators to the system. The purpose of the mandatory frequency response is to fulfill the national grid's obligation to ensure that sufficient generation and/or demand is held in terms of all credible frequency change contingencies. According to the service delivered, a generating unit will be paid according to the Connection and Use of System Code (CUSC) with two types of payment, holding payment (£/h) and response energy payment (£/MWh). The holding payment is made for the capability of the unit to provide the frequency response in responsive mode. Response energy payment (£/MWh) is made for the amount of energy delivered to and from the system in providing the frequency response service.

The holding payment for a generating unit is calculated based on the following:

$$HP_M = P_M + H_M + S_M \qquad (1.98)$$

where HP_M is the holding payment in pounds sterling per minute. P_M is the payment per minute for the ancillary service of primary response. H_M is the payment per minute for the ancillary service of high frequency. S_M is the payment per minute for the ancillary service of secondary response.

Response energy payment (£/MWh) is calculated as follows:

$$REP_{ij} = RE_{ij} \times \text{Reference Price} \qquad (1.99)$$

where RE_{ij} is the expected response energy for generating unit i in settlement period j, which is given by.

$$RE_{ij} = \int_0^{SPD} [\max(FR_{ij}(t), 0) \times (1 - SF_{LF}) + \min(FR_{ij}(t), 0) \times (1 - SF_H)] \times K_T \times K_{GRC} dt \qquad (1.100)$$

where the integral is over the settlement period duration. $FR_{ij}(t)$ is the expected change in active power output for generating unit i at time t. SF_{LF} indicates the provision of primary and secondary services. K_T and K_{GRC} are adjustment factors.

1.10   VOLTAGE CONTROL AND REACTIVE POWER MANAGEMENT

1.10.1   Introduction of Voltage Control and Reactive Power Management

For efficient, secure, and reliable operation of electric power systems, it has been recognized that the following operating objectives should be satisfied:

- *Voltage limits*: Bus voltage magnitudes should be within acceptable limits since electric power equipment and customer equipment can only be safely operated at a voltage very close to its rating. Equipment being operated outside of voltage limits may either not perform properly or be damaged.

- *System stability enhancement*: System transient stability and voltage stability can usually be enhanced by proper voltage control and reactive power (VAR) management. Subsequently the use of the transmission system asset can be maximized.

- *Minimized reactive power flows*: Reactive power flows should be minimized such that the active and reactive power losses can be reduced. In addition, the by-product of the minimized reactive power flows can actually reduce the voltage drop across transmission lines and transformers.

1.10.2   Reactive Power Characteristics of Power System Components

In order to understand the voltage control and VAR management problems of electric power systems, the reactive power characteristics of power system components will be reviewed first.

- *Synchronous generators:* Synchronous generators are very important reactive sources that can generate or absorb reactive power depending on excitation control. Equipped with modern excitation control systems, synchronous generators can provide both static but also dynamic voltage control and reactive power support.

- *Overhead transmission lines:* When load levels are less than the natural load, transmission lines can produce reactive power; when load levels are higher than the natural load, transmission lines absorb reactive power.

- *Underground cables:* Since underground cables usually have high capacitance and high natural load, cables generate reactive power.

- *Transformers:* In transformers, capacitance is usually very small in comparison to their reactance. Transformers always absorb reactive power.

- *Loads:* Most loads absorb reactive power. Usually it is required that larger industry customers maintain high lagging power factors.

- *Conventional HVDC:* Conventional HVDC systems for large power transfer over long distance usually absorb reactive power. Reactive shunt compensating devices must be installed at the terminals of the HVDC system.

1.10.3   Devices for Voltage and Reactive Power Control

In electrical power systems, voltage control and reactive power management requires various voltage control devices installed at different locations of the systems. In addition to the voltage control devices, suitable control algorithms and software tools are needed to determine control settings and coordinate the control actions of the voltage control devices sited at different locations of the systems. The voltage control devices will be briefly introduced and the characteristics of these devices will be discussed.

- *Shunt reactors:* Shunt reactors are used to compensate for the voltage rise effects of line capacitance with light load. The compensation of shunt reactors is useful to long transmission lines.

- *Shunt capacitors:* Shunt capacitors are widely used in transmission and distribution systems. At distribution level, the main purposes of shunt capacitors are power factor correction and voltage control. At transmission level, the purposes of shunt capacitors are to reduce the power losses and voltage drops and provide voltage support when the network is heavily loaded.

- *Series capacitors:* Series capacitors are used to compensate transmission line impedances such that transfer capability of transmission lines can be improved and voltage drops can be reduced. Although series capacitors can be used for voltage control, the control may not be efficient in comparison to the control by shunt capacitors. With use of series capacitors, one must be cautious about Subsynchronous Resonance (SSR) that may be caused by series capacitors.

- *Synchronous condensers:* Synchronous condensers are synchronous generators without active power generating capability. This means that they are pure reactive generating machines. Equipped with a voltage regulator, a synchronous condenser can automatically control its terminal voltage to the specified control target. In comparison to Static VAr Systems, synchronous condensers are expensive. Like synchronous generators, synchronous condensers have very good dynamic voltage control capabilities in power system transients.

- *Static VAr Systems* (SVS): A SVS system may consists of one more of the following components: saturated reactor, thyristor-controlled reactor, thyristor-switched capacitor, or thyristor-switched reactor. SVS are primarily used to control voltages at buses to which they are connected. SVS can also provide temporary overvoltage control, prevent the systems against voltage collapse, enhance transient stability, and increase the damping of power system oscillations. SVS can generate or absorb reactive power depending on the configurations of the systems.

- *Converter-based FACTS controllers:* Converter-based FACTS controllers such as the STATic Synchronous COMpensator (STATCOM), the Static Synchronous Series Compensator (SSSC), the Unified Power Flow Controller (UPFC), the Interline Power Flow Controller (IPFC), and the Generalized Unified Power Flow Controller (GUPFC) can be used to control of voltage, protect the systems against voltage collapse, enhance transient stability, and increase the damping of power system oscillations. In addition, those FACTS controllers that have series converters also have the ability to control active and reactive power flows. Other features of the converter-based FACTS controllers are their responses are very quick and they can provide continuous voltage control and reactive power support from inductive compensation to capacitive compensation. Converter-based shunt compensation, a STATCOM, has better control performance than a conventional Static Var System due to the excellent dynamic reactive power and voltage control capability of the converter-based FACTS controller.

- *HVDC light:* Back-to-back HVDC light based on converter technologies has very strong voltage control and power flow control capability. In addition, it also protects the system against voltage collapse, enhances transient stability, and increases the damping of power system oscillations. An HVDC light is usually used to link a wind farm to a distribution network.

- *Tap-changing transformers:* Tap-changing transformers are used for voltage control in transmission and distribution systems. The transformers are usually used to control the voltages at buses to which they are connected. The transformers are not reactive power generation devices, however, the tap-changing transformers can alternate the reactive power distribution of the network by changing their tap ratios such that active and reactive losses of the network may be minimized and the voltage profiles may be improved.

1.10.4 Optimal Voltage and Reactive Power Control

Optimal power flow (OPF) is security and economic control-based optimization, which selects actions to minimize an objective function subject to specific operating constraints. Most OPF programs can perform more than one specific function. One of the OPF applications in energy management systems is to minimize active power transmission losses while control of reactive power from generator and compensating devices and control of tap-changing transformers are scheduled and coordinated. The voltage control and reactive power management by OPF tends to reduce circulating reactive power flows, thereby promoting flatter voltage profiles.

1.10.5 Reactive Power Service Provisions in Electricity Markets

In the vertically integrated electricity company including generation, transmission, and distribution, voltage control and reactive power support service are provided together with active power to customers. In electricity market environments, voltage control and reactive power support is considered an ancillary service, which is unbundled from active power supply. Along with the worldwide development of energy markets, frequency control and active power reserve have already been developed in the framework of ancillary markets while the market mechanisms of reactive power service are still under development. In the US, reactive power provided by a generator is considered as an ancillary service and receives a fixed payment for its service. In the UK transmission system, according to the requirement of the grid code, all generators are required to provide frequency response and reactive power to specified capabilities. The mandatory reactive power service is referred to as the Obligatory Reactive Power Service in Schedule 3 of the Connection and Use of System Code (CUSC), where all providers of the Obligatory Reactive Power Service are paid utilization payments via a default mechanism in accordance with Schedule 3 of the CUSC [24].

Further discussions on the development of ancillary service markets for reactive power can be found in [24–27]. Previous research work has indicated that voltage control and reactive power support has significant impact on the outcomes of electricity markets, in particular, when there is congestion in the system or the system is heavily loaded [28, 29].

1.11   APPLICATIONS OF POWER ELECTRONICS TO POWER SYSTEM CONTROL

1.11.1   Flexible AC Transmission Systems (FACTS)

An electrical power network may consist of synchronous generators, transmission lines, transformers, and loads. As discussed in Section 1.1, transmission lines may be represented equivalently by series impedance and shunt capacitance. The series impedance of the transmission lines and the voltage angles and magnitudes at the ends of the transmission lines usually determine the maximum transfer powers on the transmission lines, while the capacitance effects the voltage profiles of the power system. For operating an AC power system, the minimal requirements that should normally be satisfied are a) power generated by the synchronous generators and the power consumed by the loads should be balanced at any instant; b) the synchronous generators should remain synchronously operated with the power system; and c) the voltages at the system buses should be kept within the operating limits.

The concepts of Flexible AC Transmission Systems (FACTS) were initiated in the US by the Electric Power Research Institute (EPRI) in the late 1980s, and FACTS are basically power electronic devices. The basic control principle of FACTS is that the impedances of a power system can be changed by suitable FACTS controls. Then the power flows and voltages of the power system can be controlled. In addition to the power flow and voltage control capabilities, FACTS can also be used to control voltage stability, dynamic stability or small signal stability, and transient stability or angular stability.

In the family of FACTS controllers, two categories can be further classified based the implementation principles. The first category of FACTS consists of static VAr compensator (SVC), thyristor-controlled series capacitor (TCSC), thyristor-switched series capacitor (TSSC), and phase-shifter (PS) based on thyristor technologies without gate-turn-off ability. These thyristor-based FACTS controllers can control voltage, impedance, and angle, respectively. The second category of FACTS controllers is based on self-commutated voltage sourced switching converters technologies to realize controllable, static, synchronous AC voltage or current sources. The FACTS controllers employing switching converter-based synchronous voltage sources, the STATCOM, the SSSC, the UPFC, the IPFC, and the GUPFC. The STATCOM is shunt connected with a bus via a transformer. The UPFC consists of two converters, which are coupled via a DC link. The shunt converter is shunt connected with a local bus while the series converter is series connected with a transmission line.

The next generation FACTS control equipment, named the convertible static compensator (CSC), for the transmission grid was recently installed at Marcy

Substation of New York Power Authority (NYPA) and can increase power transfer capability and maximize the use of the existing transmission network. The salient features of a CSC are its convertibility and expandability, which are becoming increasingly important as electric utilities are being transformed into highly competitive marketplaces. The functional convertibility enables the CSC to adapt to changing system operating requirements and changing power flow patterns. The expandability of the CSC is that a number of voltage-sourced converters coupled with a common DC bus can be operated. Additional compatible converters can be connected to the common DC bus to expand the functional capabilities of the CSC. The convertibility and expandability of the CSC enables it to be operated in various configurations. The CSC installed at NYPA consists of two converters, and it can operate as a STATCOM, a SSSC, a UPFC, or an IPFC, but not as a GUPFC, which requires at least three converters. The IPFC and GUPFC are significantly extended to control power flows of multi-lines or a subnetwork beyond that achievable by the UPFC or SSSC or STATCOM. With at least two converters, an IPFC can be configured. With at least three converters, a GUPFC can be configured.

1.11.2   Power System Control by FACTS

All shunt FACTS controllers such as the SVC and STATCOM can provide voltage and reactive power control, whereas series FACTS controllers can provide power flow control. In addition, FACTS controllers can be used to improve power system angle stability, enhance power system voltage stability, and provide damping of system oscillations. However, the FACTS controllers from the two categories have different control performance.

The SVC (consisting of Thyristor Switched Capacitor—TSC and Thyristor Controlled Reactor—TCR) and the STATCOM have very similar functional compensation capability. The difference of their operating principles accounts for the STATCOM's overall superior performance and greater application flexibility. Due to the fact that the STATCOM can maintain full capacitive output current at low system voltage, it is more effective to improve the power system angle stability than the SVC. The time response of the STATCOM is much quicker than that of the SVC. In contrast to the SVC, the STATCOM can be interfaced with an energy storage system via its DC link. This potential capability can provide a new control for enhancing dynamic compensation.

In comparison to the TCSC and TSSC, the SSSC has the possibility to interface with an energy storage system to increase the power damping. The multi-converter switching converter–based FACTS controllers such as the UPFC, IPFC, and GUPFC have voltage and power flow control capabilities. The IPFC and GUPFC can control power flows on multiple transmission lines.

In the past, there were numerous SVC installations. In recent years, there have also been installations of TCSC, STATCOM, SSSC, UPFC and IPFC.

Further discussions on the modeling of FACTS in power flow and optimal power flow can be found in [6]. Similar to other power system controls, FACTS control has significant impact on system power flows and voltage profiles and hence on electricity market outcomes. The FACTS and HVDC controllers, together with emerging WAMS (wide area measurement systems), will be cost-effective and innovative control devices to effectively manage network congestion while ensuring an electricity network flexible enough to meet new and less predictable supply and demand conditions in competitive electricity markets. In addition, FACTS and HVDC will be key technologies in shaping the way towards smart grids.

REFERENCES

1. Kunder P. *Power System Stability and Control*. New York: McGraw-Hill; 1994.
2. Sauer PW, Pai MA. *Power System Dynamics and Stability*. Englamovel Cliffs, NJ: Prentice-Hall; 1998.
3. Song Y-H, Johns AT (eds). *Flexible AC Transmission Systems* (FACTS). London: IEE; 1999.
4. Hingorani NG, Gyugyi L. *Understanding FACTS—Concepts and Technology of Flexible AC Transmission Systems*. New York: IEEE Press; 2000.
5. Mathur RM, Varma RK. *Thyristor-Based FACTS Controllers for Electrical Transmission Systems*. New York: IEEE Press; 2002.
6. Zhang X-P, Rehtanz C, Pal B. *Flexible AC Transmission Systems—Modelling and Control*. Berlin: Springer; 2006.
7. Lasseter RH, Paigi P. Microgrid: a conceptual solution. In Proceedings of 2004 IEEE 35th Annual Power Electronics Specialists Conference, Aachen, Germany, IEEE; 20–25 June 2004. pp. 4285–4290.
8. Hatziargyriou N, Asano H, Iravani R, Marnay C. Microgrids. *IEEE Power & Energy Magazine* 2007;5(4):78–94.
9. Pure Power: Wind energy targets for 2020 and 2030. The European Wind Energy Association, November 2009, http://www.ewea.org.
10. Kirby BJ. Frequency Regulation Basics and Trends. Report ORNL/TM-2004/291. Oak Ridge National Laboratory, USA, December 2004.
11. *NERC Operating Manual*, March 2008.
12. IEEE AGC Task Force. Understanding of automatic generation control. *IEEE Transactions on Power Systems* 1992;7(3):1106–1122.
13. Jaleeli N, VanSlyck LS. NERC's new control performance standards. *IEEE Transactions on Power Systems* 1999;14(3):1092–1099.
14. Yao M, Shoults RR, Kelm R. AGC logic based on NERC's new control performance standard and disturbance control standard. *IEEE Transactions on Power Systems* 2000;15(2):852–857.
15. Christie RD, Bose A. Load frequency control issues in power system operations after deregulation. *IEEE Transactions on Power Systems* 1996;11(3):1191–1200.
16. Chen CY, Strbac G, Zhang XP. Evaluating the impact of plant mix on frequency regulation requirements. In: Proceedings of the 2000 Universities Power Engineering Conference (UPEC), Queen's University Belfast, UK, 6–8 September 2000.
17. Zhang XP, Chen CY, Zheng Z, Strbac G, Kubokawa J. Allocation of frequency regulation services. In: Proceedings of the IEEE/IEE International Conference on Electric Utility Deregulation and Restructuring and Power Technologies 2000, City University, London, 4–7 April 2000, pp. 349–354.
18. Nobile E, Bose A, Tomsovic K. Feasibility of a bilateral market for load following. *IEEE Transactions on Power Systems* 2001;16(4):782–787.
19. Donde V, Pai MA, Hiskens IA. Simulation and optimization in an AGC system after deregulation. *IEEE Transactions on Power Systems* 2001;16(3):481–489.
20. Bhowmik S, Tomsovic K, Bose A. Communication models for third party load frequency control. *IEEE Transactions on Power Systems* 2004;19(1):543–548.
21. Verbic G, Gubina F. Cost-based models for the power reserve pricing of frequency control. *IEEE Transactions on Power Systems* 2004;19(4):1853–1858.
22. Kirby B, Hirst E. Allocating the costs of contingency reserves. *Electricity Journal* 2003; 16(10):39–47.
23. Kirsch LD, Morey MJ. Efficient allocation of reserve costs in RTO markets. *Electricity Journal* 2006;19(8):43–51.

24. Connection and Use of System Code, Schedule 3—Part I Balancing Services Market Mechanisms—Reactive Power. Version 1.4. National Grid, 26 Nov 2008.

25. Isemonger AG. Some guidelines for designing markets in reactive power. *Electricity Journal* 2007; 20(6):35–45.

26. Bhattacharya K, Zhong J. Reactive power as an ancillary service. *IEEE Transaction* 2001;16(2):294–300.

27. Ahmed S, Strbac G. A method for simulation and analysis of reactive power market. *IEEE Transaction* 2001;15(3):1047–1052.

28. Petoussis SG, Petoussis AG, Zhang X-P, Godfrey KR. Impact of the transformer tap-ratio control on the electricity market equilibrium. *IEEE Transactions on Power Systems* 2008;23(1):65–75.

29. Petoussis S, Zhang X-P, Petoussis A, Godfrey KR. The impact of reactive power on the electricity market equilibrium. In: Proceedings of the 2006 IEEE Power Engineering Society (PES) Power Systems Conference and Exposition, Atlanta, Georgia, USA, 29 Oct–1 Nov 2006, pp. 96–102.

30. Glimn AF, Kirschmayer LK, Stagg GW. Analysis of losses in interconnected systems. *AIEE Transactions* 1952;71(3):796–808.

31. Kirchmayer LK. *Economic Operation of Power Systems*. New York: John Wiley & Sons; 1958.

32. Kirchmayer LK. *Economic Control of Interconnected Systems*. New York: John Wiley & Sons; 1959.

33. Meyer WS. Efficient computer solution for Kron and Kron-early loss formulas. In: PICA Conference Proceedings, Minneapolis, MN June 3–6, 1973, pp. 428–432.

34. Alvarado AL. Penalty factors from Newton's method. *IEEE Transactions on PAS* 1978; 97(6):2031–2037.

35. Carpentier JL. Contribution a. 'l'etude du dispatching economique. *Bulletin de la Societe Francaise des Electriciens* 1962;3:431–447.

36. HH. Happ. Optimal power dispatch—a comprehensive survey. *IEEE Transactions on PAS* 1977;96(3):841–854.

37. IEEE Working Group. Description and bibliography of major economy—security functions. Part II—Bibliography (1959–1972). *IEEE Transactions on Power Apparatus and Systems* 1981; 100(1):215–223.

38. IEEE Working Group. Description and bibliography of major economy—security functions. Part III—Bibliography (1973–1979). *IEEE Transactions on Power Apparatus and Systems* 1981; 100(1):224–235.

39. Chowdhury BH, Rahman S. A review of recent advances in economic dispatch. *IEEE Transactions on Power Systems* 1990;5(4):1248–1257.

40. Dommel HW, Tinney WF. Optimal power flow solutions. *IEEE Transactions on PAS* 1968; 87(10):1866–1876.

41. Carpentier JL. Optimal power flows: uses, methods and developments. In: Proceedings of IFAC Conference on Planning and Operation of Electric Energy Systems, 1986.

42. Stott B, Alsc O, Monticelli A. Security and optimization. *Proceedings of the IEEE* 1987; 75(12):1623–1624.

43. Carpentier JL. Towards a secure and optimal automatic operation of power systems. In: Proceedings of Power Industry Computer Applications (PICA) conference, 1987, pp. 2–37.

44. Wu FF. Real-time network security monitoring, assessment and optimization. *International Journal of Electrical Power and Energy Systems* 1988;10(2):83–100.

45. Huneault M, Galiana FD. A survey of the optimal power flow literature. *IEEE Transactions on Power Systems* 1991;6(2):762–770.

46. IEEE Tutorial Course. Optimal power flow: solution techniques, requirements and challenges. IEEE Power Engineering Society, 1996.

47. Momoh JA, El-Haway ME, Adapa R. A review of selected optimal power flow literature to 1993 Part 1 and Part 2. *IEEE Transactions on Power Systems* 1999;14(1):96–111.

48. Carpentier JL. Differential injections method: A general method for secure and optimal load flows. In: Proceedings of PICA Conference, 1973.

49. Stott B, Marinho JL. Linear programming for power system network security applications. *IEEE Transactions on PAS* 1979;98(3):837–848.

50. Alsc O, Bright J, Praise M, Stott B. Further developments in LP-based optimal power flow. *IEEE Transactions on Power Systems* 1990;5(3):697–711.

51. Burchett RC, Happ HH, Wirgau KA. Large scale optimal power flow. *IEEE Transactions on PAS* 1982;101(10):3722–3732.

52. Burchett RC, Happ HH, Veirath DR. Quadratically convergent optimal power flow. *IEEE Transactions on PAS* 1984;103(11):3267–3275.

53 El-Kady MA, Bell BD, Carvalho VF, Burchett RC, Happ HH, Veirath DR. Quadratically convergent optimal power flow. *IEEE Transactions on Power Systems* 1986;1(2):98–105.

54. Glavitsch H, Spoerry M. Quadratic loss formula for reactive dispatch. *IEEE Transactions on PAS* 1983;102(12):3850–3858.

55. Sun DI, Ashley B, Brewer B, Hughes A, Tinney WF. Optimal power flow by Newton approach. *IEEE Transactions on PAS* 1984;103(10):2864–2880.

56. Maria GA, Findlay JA. A Newton optimal power flow program for Ontario Hydro EMS. *IEEE Transactions on Power Systems* 1987;2(3):576–584.

57. Tinny WF, Bright JM, Demaree KD, Hughes BA. Some deficiencies in optimal power flow. *IEEE Transactions on Power Systems* 1988;3(2):676–682.

58. Chang SK, Marks GE, Kato K. Optimal real-time voltage control. *IEEE Transactions on Power Systems* 1990;5(3):750–756.

59. Hollenstein W, Glavitch H. Linear programming as a tool for treating constraints in a Newton OPF. In: Proceedings of the 10th Power Systems Computation Conference (PSCC), Graz, Austria, August 19–24, 1990.

60. Karmarkar N. A new polynomial time algorithm for linear programming, *Combinatorica* 1984;4:373–395.

61. Vargas LS, Quintana VH, Vannelli A. A tutorial description of an interior point method and its applications to security-constrained economic dispatch. *IEEE Transactions on Power Systems* 1993;8(3):1315–1323.

62. Lu N, Unum MR. Network constrained security control using an interior point algorithm. *IEEE Transactions on Power Systems* 1993;8(3):1068–1076.

63. Zhang XP, Chen Z. Security-constrained economic dispatch through interior point methods. *Automation of Electric Power Systems* 1997;21(6):27–29.

64. Momoh JA, Guo SX, Ogbuobiri EC, Adapa R. The quadratic interior point method solving power system optimization problems. *IEEE Transactions on Power Systems* 1994;9(3);1327–1336.

65. Granville S. Optimal reactive power dispatch through interior point methods. *IEEE Transactions on Power Systems* 1994;9(1):136–146.

66. Wu Y-C, Debs A, Marsten RE. A direct nonlinear predictor-corrector primal-dual interior point algorithm for optimal power flows. *IEEE Transactions on Power Systems* 1994;9(2);876–883.

67. Irisarri GD, Wang X, Tong J, Mokhtari S. Maximum loadability of power systems using interior point nonlinear optimisation method. *IEEE Transactions on Power Systems* 1997;12(1):167–172.

68. Wei H, Sasaki H, Yokoyama R. An interior point nonlinear programming for optimal power flow problems within a novel data structure. *IEEE Transactions on Power Systems* 1998;13(3):870–877.

69. Torres GL, Quintana VH. An interior point method for non-linear optimal power flow using voltage rectangular coordinates. *IEEE Transactions on Power Systems* 1998;13(4):1211–1218.

70. Zhang XP, Petoussis SG, Godfrey KR. Novel nonlinear interior point optimal power flow (OPF) method based on current mismatch formulation. *IEEE Proceedings—Generation, Transmission & Distribution* 2005;152(6):795–805.

71. El-Bakry S, Tapia RA, Tsuchiya T, Zhang Y. On the formulation and theory of the Newton interior-point method for nonlinear programming. *Journal of Optimisation Theory and Applications* 1996;89(3):507–541.

72. La Scala M, Trovato M, Antonelli C. On-line dynamic preventive control: An algorithm for transient security constraints. *IEEE Transactions on Power Systems* 1998;13(2):601–610.

73. Gan D, Thomas RJ, Zimmermann RD. Stability constrained optimal power flow. *IEEE Transactions on Power Systems* 2000;15(2):535–540.

74. Chen L, Tada Y, Okamoto H, Tanabe R, Ono A. Optimal operation solutions of power systems with transient stability constraints. *IEEE Transactions on Circuit and Systems—I: Fundamental Theory and Applications* 2001;48(3):327–339.

RESTRUCTURED ELECTRIC POWER SYSTEMS AND ELECTRICITY MARKETS

Kwok W. Cheung,
Gary W. Rosenwald,
Xing Wang,
and David I. Sun

This chapter discusses the restructured electric power systems and electricity markets. Section 2.1 reviews the history of electric power systems restructuring. Section 2.2 discusses the structure and the evolution of electricity markets. Section 2.3 addresses the key market design objectives and fundamental market design principles. Section 2.4 discusses the operation of electricity markets and the criteria for its success. Typical business process timelines are presented with examples. Computational tools for electricity markets are presented in Section 2.5. Final remarks are given in Section 2.6.

2.1 HISTORY OF ELECTRIC POWER SYSTEMS RESTRUCTURING

Over the past two decades, the electric utility industry has been under dramatic restructuring throughout the world. The restructuring typically emphasizes functional unbundling of the vertically integrated system leading to the potential for competition. Different governance structures and market design models have been devised and implemented to reflect regional needs and preferences. For almost every competitive electricity market, it is an ongoing evolutionary process and the path of evolution was not very easy for some of them. Unlike other economic systems, electrical network interactions in power system complicated the design of the institutions and pricing arrangements of an electricity market. It took some time for the industry to recognize the market for transmission and the market for energy are inherently intertwined and the institutions for these markets cannot be constructed

Restructured Electric Power Systems: Analysis of Electricity Markets with Equilibrium Models,
Edited by Xiao-Ping Zhang
Copyright © 2010 Institute of Electrical and Electronics Engineers

independently. After years of trials and experimentation, some properly designed/ redesigned electricity markets have reached maturity and are able to successfully demonstrate the effectiveness of competition. General market design principles tend to converge as other electricity markets learn from the successful ones over the years. However, significant differences remain in detailed design and implementation among various markets as of this writing.

2.1.1   Vertically Integrated Utilities and Power Pools

Before the era of power system restructuring had begun in late 1980s, vertically integrated utilities managed the generation, transmission, and distribution of the overall power system. The utilities were the providers of the power system infrastructure and the electric services to electricity customers. Electric utilities operated as monopolies and exercised exclusive control over the sale of electricity with their service territories for many decades. Each vertically integrated utility managed the maintenance and operations of its generation assets and determined how much, when, and where generation assets would be developed in order to keep up with future demand. Ultimately, the vertically integrated utility needed to ensure that there was adequate power supply to meet the demand of the customer load on a day-to-day, hour-to-hour, and minute-to-minute basis. In addition, the vertically integrated utility also maintained the high voltage transmission system and lower voltage distribution system. As load demands increased or new generation assets came online in their systems, they upgraded existing facilities or constructed new transmission corridors to maintain the reliable delivery of energy.

The responsibility of ensuring the reliability of a control area within an interconnected power system is shared by all the operating entities inside the power system. In a traditional regulated power system, the responsibility for assuring reliability is shared by the vertically integrated utilities. The reliability criteria are established by a committee consisting of all the interconnected utilities in the control area, and in many cases a system operating organization responsible for the overall power dispatch in the control area, commonly known as a power pool, is in charge of enforcing the reliability rules. Most of the operating entities also belong to a larger regional reliability council, such as the Northeast Power Coordinating Council [1] in the United States, so that neighboring control areas can operate with similar reliability criteria. The regional reliability councils form the North American Electric Reliability Council [2], which establishes recommended standards on system reliability. Before power systems restructuring, maintaining system reliability was the most important aspect of the electric utility industry.

2.1.2   Worldwide Movement of Power Industry Restructuring

The earliest introduction of market concepts and privatization to electric power systems took place in Chile in the late 1970s. Argentina built on top of the Chilean model and tried to privatize existing generation assets (which had fallen into disrepair under the government-owned monopoly, resulting in frequent service interruptions) and to attract capital needed for rehabilitation of those assets and for system expansion. The World Bank was active in introducing a variety of hybrid markets in other Latin American nations, including Peru, Brazil, and Colombia, during the 1990s, with limited success.

Economic forces have been driving a dramatic restructuring of the power industry in recent years. In this "brave new world" of restructuring, energy price and ancillary service prices are no longer cost-driven, but energy and ancillary services themselves are treated as competitive commodities [3–6]. Pioneers of the power system restructuring in different continents of the world embrace the idea of introducing competition in the wholesale supply and purchase of electricity combined with an open access regime for the use of electricity networks. The diversity of the approaches can be seen in looking at some examples of the leading movements of restructuring of our industry in Europe including the Nordic countries [34] and Great Britain [33], New Zealand [35], Australia [18], and North America [12, 21, 28]. A brief summary of each of their restructuring processes is given below.

2.1.2.1 Nordic Countries The electricity industry of the Nordic countries includes Norway, Sweden, Finland, and Demark. The industry went through a major reform during the 1990s. Norway was the first Nordic country introducing market competition, institutionalized by the Energy Act of 1990. The Swedish restructuring that was decided in 1995 lead to the establishment of a common Norwegian-Swedish Exchange (Nord Pool). This first electricity market completely open to trade across national borders has been in operation since 1996. Finland joined the common market in 1998, West Demark in 1999, and East Demark in 2000. The Nordic electricity market is presently the only truly international electricity market. There is one market operator (Nord Pool), and there are presently five system operators: one in each country with the exception of Demark, which has two (East and West).

2.1.2.2 Great Britain Restructuring of the electricity industry in Great Britain consists of a series of reforms. The first organized market in England and Wales commenced in 1990 with the breakup of the Central Electricity Generation Board (CEGB) into three generation companies, which are National Power, PowerGen, Nuclear Electric, and one transmission company, which is National Grid Company (NGC). The 12 regional distribution boards were also privatized into 12 Regional Electricity Companies (RECs) with each being effectively a combination of a distribution company and retail or supply company. Competition was first introduced by setting up an electricity pool. The pool operated for over 10 years until it was subjected to a review process. Following this review, the pool was replaced by an entirely new trading structure known as the New Electricity Trading Arrangements (NETA) in 2001. NETA established a framework for bilateral trading and new power exchanges had been developed. The third wave of reform was called the British Electricity Trading and Transmission Arrangements (BETTA). BETTA commenced in 2005 was the extension of NETA to the regions of Scotland so that one single wholesale electricity market is for the whole of Great Britain.

2.1.2.3 Continental Europe The liberalization in the European Union (EU) has been a top-down process driven by the directives of the European Parliament and of the Council. The directives lay down the general principles and conditions to assure the creation of a single Internal Electricity Market (IEM) in Europe. The liberalization process first put in force in 1996 by Directive 96/92/EC led to the unbundling of activities. IEM is divided into submarkets according to the control zones of the various transmission system operators (TSOs). Most wholesale trade volume in the IEM is traded bilaterally in forward and over-the-counter (OTC) types of markets. Most consumption portfolio is covered by long-term and forward contracts. A small fraction of trade volume is traded in daily or even hourly contracts in the spot markets due to incomplete predictability of real-time consumption. Although member states of the EU have similar electricity market architectures, these markets are weakly integrated among national borders. The association of ETSO (European Transmission Operator) was founded in 1999 in response to the emergence of IEM. ETSO has been integrated to a larger association called European Network of Transmission System Operators for Electricity (ENTSO-E) in 2009. ENTSO-E pursue the co-operation of the European TSOs and have an active and important role in the European rule setting process in compliance with EU legislation. To improve cross-border exchanges, Regulation 1228/2003, issued together with Directive 2003/54/EC in 2003, established a compensation mechanism for cross-border flows of electricity, the setting of guidelines and principles on cross-border transmission charges, and the allocation of available transmission capacities between national transmission systems. However, more European-level regulation or coordinated regulatory actions will clearly be necessary for the goal of an efficient Internal Electricity Market in Europe.

2.1.2.4 New Zealand Up until 1994, New Zealand had a system of monopoly providers of generation, transmission, distribution, and retailing. Since then, a step-by-step process of industry reform has led to the separation of the monopoly elements from the contestable elements to create competition in energy generation and electricity retailing. The reformed wholesale electricity market (NZEM) introduced competition within the wholesale electricity sector through creation of a national electricity pool and a spot market of electricity. The wholesale market for electricity is administered by M-co (formerly called Electricity Market Company) on behalf of the market regulator, the Electricity Commission. The transmission system is owned and operated by a state-owned enterprise, Transpower, which performs the functions of grid owner, system operator, scheduler, and dispatcher for the wholesale market. The main participants are seven generator/retailers who trade at commercial trading nodes across the transmission grid. Prices and quantities are determined half-hourly at each node. NZEM began trading in 1996. NZEM was considered to be the first electricity market in the world based on the nodal pricing model, which is a predecessor of the locational marginal pricing (LMP) model later widely adopted in the US.

2.1.2.5 Australia The reform of the Australian electricity industry commenced in the early 1990s. Separate commercial structures have been developed for the monopoly transmission and distribution functions and the competitive generation and retailing functions of the industry. The major reform in the Australian electricity industry involved the establishment in southern and eastern Australia of the National Electricity Market (NEM). The NEM operates in the states of New South Wales,

Victoria, Queensland, South Australia, and Tasmania and in the Australian Capital Territory. The market operator for the NEM is the National Electricity Market Management Company (NEMMCO). NEMMCO was established in 1996 to fulfil the roles of both market operator of the NEM and operator of the power system that underpins NEM operation. The owners of the company are the five states and the territory within which the NEM operates. The NEM operates on one of the world's longest interconnected power systems. The NEM commenced operation on 13 December 1998 under a detailed set of rules called the National Electricity Rules. The NEM comprises a physical spot market with energy traded through a commodities-type pool and a spot price set every five minutes by the most expensive generator selected to run. All electricity sold at the wholesale level is accounted for through the pool. There are six geographical regions in the NEM and constraints on interconnectors can cause marginal spot prices to separate between the regions. NEMMCO is responsible for generator dispatch, reliability management and financial settlements in the NEM.

2.1.2.6 United States In the United States, the Energy Policy Acts of 1992 launched a national effort to restructure electricity institutions to allow greater reliance on markets. The Federal Energy Regulatory Commission (FERC) took the lead in Order 888 in 1996 by opening access to the electric transmission grid. This triggered the formation of independent system operators (ISO), including California ISO (CalISO), New York ISO (NYISO), ISO New England (ISO-NE), PJM (Pennsylvania-New Jersey-Maryland), Midwest ISO (MISO), Southwest Power Pool (SPP) and Electric Reliability Council of Texas (ERCOT). The ISOs have the responsibility of ensuring the reliability of one or more than one control areas. Order 888 also required each control area to post the available transfer capabilities (ATC) of major transmission paths of their system on an Open Access Same-time Information System (OASIS). Bilateral trades are then facilitated by submitting transmission requests to the control areas whose networks are used for the energy transactions. These transactions are evaluated and the transmission requests that can be accommodated within the ATC of the transmission network are accepted. Otherwise, the requests are rejected, regardless of individual transaction's valuation of the transmission capacity.

There was an intense debate between bilateral model and Poolco model of market design in the mid-1990s. Meanwhile, a few ISOs that had the background of operating as power pools started creating *voluntary* spot markets. The uniform or zonal pricing approach was initially adopted in these markets, such as 1997 PJM market, and the 1999 ISO-NE market [12, 27]. The creation of these spot markets marked the beginning of a diverging market evolution which will be discussed in detail in the next section. Later on FERC Order 2000 encouraged the formation of Regional Transmission Organizations (RTO), which had larger authorities and responsibilities than the ISOs to oversee a region to ensure proper market operations and system reliability. The same year of 2000 California had its electricity crisis. California's crisis has vividly demonstrated that power interruptions or blackouts can significantly impact the economy, and market design and reliability assurance are closely related issues and should not be addressed separately.

At the same time, an alternate market model based on locational marginal pricing and the concept of a multi-settlement system with financial transmission rights as a financial instrument to hedge against transmission congestion risk had emerged in North America. The market model was a generalization of state-of-the-art examples of excellence in market design being often time referred as the framework of standard market design (SMD) [17, 29, 32] in the beginning of this century. Since then, more RTOs/ISOs had started to follow a similar market model and later enhanced their wholesale energy-only markets with ancillary service markets [28]. The deregulated power industry had recognized that system reliability is an integral part of a properly designed electricity market and ancillary services should be simultaneously co-optimized with energy.

2.2 STRUCTURE OF ELECTRICITY MARKETS

The main purpose of restructuring is to let economic forces drive the price of electric supply and maximize social welfare via competition. Restructuring creates an open environment by allowing electric supply to compete and consumers to choose the supplier of electric energy. The power industry has been evolving toward a market-based approach in the United States and throughout the world for well over a decade. Restructuring is a very complex and tedious process and differs from place to place for various reasons. But basically, restructuring requires the decomposition of three components of electric industry, namely generation, transmission and distribution. It also requires the separation of transmission ownership from transmission control to ensure fair and nondiscriminatory access to the transmission services and ancillary services. The entity which controls the transmission system maintains real-time operation of the system and its grid reliability. The market structure should allow long-term wholesale bilateral trading and a voluntary short-term spot market with transparent and justifiable prices of energy and ancillary services. The spot market would include both a day-ahead function to coordinate resource commitment and a real-time balancing function.

2.2.1 Stakeholders

In a restructured or deregulated electricity market, a traditional utility is separated into three generic entities: generator companies (GENCOs), transmission companies (TRANSCOs), and the distribution companies (DISCOs). Along with other non-asset owners such as energy brokers and marketers, aggregators, GENCOs, DISCOs and retailers (RETAILCOs) are collectively known as the Market Participants (MPs). A competitive electricity market structure with energy flow and information flow among its entities is illustrated in Figure 2.1.

A GENCO is an entity that operates and maintains existing generating plants. GENCOs have the opportunity to sell electricity to entities with whom they have negotiated bilateral sales contracts or they may also opt to sell electricity to an organized market. GENCOs may also trade reactive power and other ancillary services. One key difference from the vertically integrated structure is that GENCOs

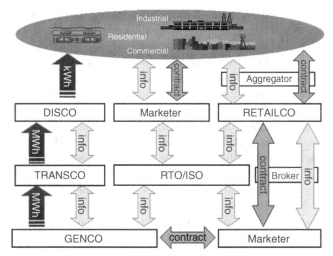

Figure 2.1   A competitive electricity market structure

will not be price-regulated in the market structure. GENCOs are also responsible to communicate generating unit outages for maintenance to the ISO/RTO.

A TRANSCO is an entity responsible for building, maintaining, and operating the transmission systems in a certain geographical region. TRANSCOs own the transmission assets but the control of the individual asset is up their regional ISO/ RTO to determine. TRANSCOs communicate with the ISO the list of equipment outages or any changes to the scheduled outages. Transmission maintenance and expansion is coordinated between TRANSCOs and the ISO/RTO.

A DISCO is an entity that receives bulk energy from the transmission grid and distributes the electricity through its facilities to customers in a certain geographical region. A DISCO, like a load serving entity (LSE), has the responsibility of responding to distribution network outages and power quality concerns and providing other services such as aggregating customers, purchasing power supply and transmission services for customers.

A RETAILCO is a newly created entity that has legal approval to sell retail electricity. A retailer may likely bundle electricity products and services in various packages for sale. A retailer may deal indirectly with customers through aggregators.

The responsibility of ensuring the reliability of a control area is delegated to an independent system operator (ISO) or a regional transmission organization (RTO). The ISO or RTO is a neutral, independent entity responsible for maintaining secure and economic operation of an open access transmission system on a regional basis. The ISO or RTO is responsible for maintaining the energy balance of the system by controlling the dispatch of flexible resources. An ISO or RTO provides transmission availability and pricing services to all users of the transmission grid. One of the main tasks for the ISO or RTO is the management of transmission congestion including the collection and distribution of congestion revenue. The ISO or RTO also coordinate maintenance scheduling and has a role in coordinating long-term planning.

Some ISOs and RTOs also act as a marketplace in wholesale power, especially since the late 1990s. Most ISOs and RTOs are set up as nonprofit corporations.

In the United States, FERC Orders 888 and 889 defined how independent power producers (IPPs) and power marketers would be allowed fair access to transmission systems, and mandated the implementation of the OASIS to facilitate the fair handling of transactions between electric power transmission suppliers and their customers.

In general, the MPs have responsibility for providing accurate data, certifying the performance of their equipment, and following the dispatch requested by the ISO/RTO. The ISO/RTO has the responsibility of ensuring that each MP meets its reliability rules and coordinating the dispatch of the electricity supply to meet the demand, such that the power system will also meet the operational reliability objectives at the lowest possible cost. NERC has already included ISOs/RTOs and MPs in its governance structure, with the traditional reliability criteria updated appropriately to meet the needs of restructured electricity markets.

2.2.2 Market Evolution

Electricity market design has focused on two aspects: reliability and economics (termed pricing-driven below). This evolution process is described in Figure 2.2, where the upper path represents the evolution of the reliability-driven design and the lower path the pricing-driven one. Both types of market design originated from the traditional integrated utility operating environment in which real-time operation is assisted with economic dispatch (ED) and security analysis (SA) for centralized control of the generation and transmission system via the Energy Management System (EMS).

Reliability-driven market design was stimulated with the issuance of FERC Orders 888 and 889. These FERC orders include requirements that each control area post the available transfer capabilities (ATC) of major transmission paths of their system on an OASIS node. They further required that transmission tariffs be charged based on FERC-approved revenue requirements related to the investment costs associated with transmission. Bilateral trades are facilitated by submitting

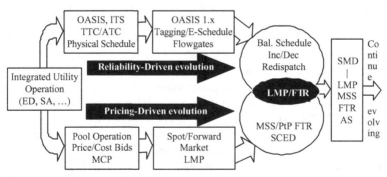

Figure 2.2 Market evolution

transmission requests to the control areas whose networks are used for the energy transactions. These transactions are evaluated and the transmission requests that can be accommodated within the ATC of the transmission network are accepted. Otherwise the requests are rejected, regardless of individual transaction's valuation of the transmission capacity. At the core of the reliability-driven market design is the rationing of limited transmission capacity.

However, the OASIS capacity reservation tariff approach is built around the unrealistic contract path model. When security violations occur in actual operation, transmission loading relief (TLR) procedures are activated and transactions are curtailed using pre-calculated power transfer distribution factors (PTDF) and the relative "firmness" (or schedule priority) of transactions. The TLR procedures do not take into account the economics of the congested transmission facility or individual transactions. The incorporation of a security-constrained economic dispatch (SCED) approach in the real-time operation reduced internal inconsistencies caused by the disconnect between PTDF-based physical model and cost/price-based economic model. This brought the reliability-driven market design closer to the pricing-driven market design, but it does not address many other issues required of an efficient and robust market [5, 24].

In the evolution of the pricing-driven market design, two main objectives, market liquidity which facilitates bilateral trading, and pricing efficiency which facilitates congestion management, have been pursued. The efforts to balance the seemingly conflicting objectives led to uniform or zonal marginal pricing in most market designs. Unlike locational marginal pricing, uniform or zonal pricing relies on being able to predefine regions within which congestions are insignificant and hence prices can be deemed uniform. The uniform (system-wide) marginal pricing design is biased toward achieving market liquidity but does not consider congestion. It is a matter of common practice that long or mid-term energy sale-purchase contracts are agreed upon between generators and loads in the form of contracts for difference (CfD). CfD can be a one-way or two-way hedge. In the one-way CfD, energy sellers provide purchaser with assurance of a certain strike price and when the market price is higher than the strike price, the sellers pay the purchasers the difference between the market and the strike price. In the two-way hedge, the purchasers also pay the sellers the price difference so that the strike price is guaranteed for both parties. The uniform pricing approach suits the common practice of CfD and it was adopted in several markets. The uniform pricing approach would work if there should exist ample transmission capacity so congestion was of no concern within the uniform pricing region. When this is not true, uniform pricing gives the wrong price signals and causes difficulties in physical system operation, which would often lead to command and control instructions, as experienced by the PJM, ISO-NE, and England and Wales operators. The frequent exercise of command and control goes against the goal of competitive markets and reduces market efficiency by requiring with significant uplift charges to fund payment of reliability instructions.

The development of the zonal pricing approach was an improvement over the uniform pricing model. In the zonal approach, it is assumed that transmission constraints are few, can be identified a priori, and can be used to delineate the network into several zones. A uniform energy price is then computed for each zone. However, practical experience has proved that the number of transmission constraints is not few, the congestion pattern is unpredictable, and that zonal price signals based on predelineated zones are ineffective in relieving congestion.

Unsatisfactory experiences with uniform marginal pricing and zonal pricing approaches led to the development of an alternative method: locational (or nodal) marginal pricing based on a full transmission network model with prices and quantities determined simultaneously via the SCED. The LMP at a specific location reflects the marginal cost of serving an increment of load considering the marginal production cost, the impact of locational injections on congestion constraints, and in some cases the marginal effect of transmission losses. The LMP method has proved its effectiveness in achieving congestion relief and market efficiency.

The unpredictability of transmission congestion implies greater uncertainty of LMPs, creating price volatility without appropriate financial hedging instruments. To address volatility and increase market liquidity, the point-to-point financial transmission rights (PtP FTRs) were developed. An FTR is a financial entitlement that can hedge its holder against congestion charges incurred on a specific transmission path defined by its source and sink location. The economic value of an FTR is determined by the difference in the LMP's congestion charge between its source and sink, independent of physical energy delivery. FTRs provide a mechanism overcome the congestion-incurred price uncertainty and market liquidity problems that are otherwise impediments to the LMP method.

In addition to energy and FTR markets, a robust, competitive market must also deal with ancillary services (AS) required to reliably operate the system. AS markets have been designed in different forms due to differences in operational practices and regional reliability standards between control areas. The broader differences in AS market design also manifest the fundamental differences between energy and AS products. Despite these design differences, it is recognized that AS and energy markets are closely coupled as the same resource and same capacity may be used to provide multiple products when economics justify. This capacity coupling between provision of energy and AS calls for joint optimization of AS and energy markets, though the exact form of joint optimization differs from market to market [28].

2.2.3 Market and Reliability Coordination

A key premise of a successful, competitive market is that the market works through the interaction of private, decentralized trading and investment decisions [5]. An effective electricity market should allow substantial commercial freedom to market buyers, sellers, and various types of traders. Trading rules would then allow the market participants the freedom to fashion and implement various trading and risk management arrangements with each other, at prices to which they mutually agree, in pursuit of their respective commercial interests. Market design should, on the other hand, recognize the fact that the laws of physics dictate certain essential characteristics of system operations and the complexities of electricity networks require a degree of centralized coordination over system operations to ensure system reliability.

In some electricity markets, the system operator, whose role is to maintain reliability, and the market operator, whose role is to settle supply and demand, are separate entities. This type of market structure requires a substantial amount of coordination between the two operators to be successful (Figure 2.3). This is sometimes referred to as the unbundled approach [6] was used in California (1998–2000). However, an electricity market in which the ISO or RTO functions both as the "system operator" for reliability coordination and the "market operator" for establishing market prices is more commonly used to provide commercial freedom and centralized economic and reliability coordination to co-exist harmoniously (Figure 2.4). Within this framework there are a number of variations which do not procure reliability ahead of time and only run real-time markets (e.g., Australia and New Zealand), and other practices like in the 2001 Texas market that clear energy and ancillary services sequentially. It is important to note that New Zealand does not have a single business entity of RTO/ISO although both its system operator and market operator use the same SCED model for dispatch and pricing, respectively. This next section focuses on the integrated (or SMD) approach, which has been adopted by most of the US electricity markets, including NYISO [7], PJM [8], ISO-NE [9], the new California market [10], and the new Texas Nodal market [13] in which reliability is addressed at many different levels.

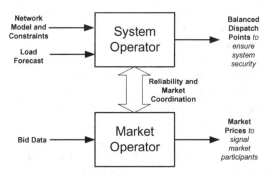

Figure 2.3 System operator and market operator as separate functional entities

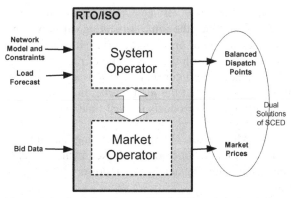

Figure 2.4 Dual functions of RTO/ISO and dual solutions of SCED

2.2.4   The SMD Framework

SMD is a generalization of state-of-the-art examples of excellence in market design. SMD is not a formal standard but rather a collection of effective market structure, proven strategies and best practices in market operations. At a high level, the SMD approach features the following essential functions for a robust and competitive market.

2.2.4.1   Transmission Service   The RTO/ISO must offer transmission customers nondiscriminatory, standard transmission services. In LMP-based market, this service is subject to congestion charges. To hedge against transmission congestion charges, financial rights are offered in the form of FTRs. FTRs may be offered as obligations or options [26]. Transmission congestion shall be managed using the LMP method. FTRs provide congestion price certainty, but they do not have any bearing on the operator's congestion relief redispatch decisions. Transactions may also be conducted without holding any FTRs. FTRs may be allocated as long-term, mid-term, or short-term rights. For interchange energy scheduling external to the RTO/ISO region, OASIS reservation is required.

2.2.4.2   Energy Market   The combination of a day-ahead energy market and a real-time energy market is a multi-settlement energy market design. The day-ahead market provides: (1) the opportunity for the system to commit sufficient generating units and transmission elements to meet the bid-in loads for the next day; (2) the opportunity for a generator to have an increased level of financial certainty with respect to the operational constraints and costs of generating units through the use of multi-part bids; and (3) better scheduling opportunities for the demand side to participate in the market. The day-ahead market shall be cleared to achieve the objective of maximizing the combined economic value of transmission service, energy, and ancillary services based on the submitted bids while ensuring reliability standard are met.

The real-time market shall be operated using the same LMP-based methodology. The real-time LMPs are applied for settling all deviations and imbalances from the schedules determined in the day-ahead market.

2.2.4.3   Ancillary Service Market   Ancillary services are necessary for real-time secure operation of a power system. Since the same generating capacity may be used to supply energy as well as regulation or operating reserves, it is beneficial to have the scheduling of energy coordinated with the scheduling of regulation and operating reserves. The AS markets could be co-optimized using bids and offers for energy and transmission services. The key is that the joint market must allow for flexible substitution of energy and different AS products for real-time operation.

2.2.4.4   Market Monitoring and Mitigation   Market monitoring and mitigation are essential to assure the quality of market operations. Market monitoring can be broadly divided into compliance monitoring and market power monitoring. Market monitoring should focus on two areas:

- Identifying any problems in market design resulting in inefficient outcomes and propose market rule changes;
- Detecting the exercises of market power in various forms such as physical or economic withholding.

Once identified, market power should be mitigated through structural solutions. There are also scenarios for which behavioral mitigation measures are necessary.

2.3   DESIGN OF ELECTRICITY MARKETS

Successful electricity market designs require combining fundamentals in market economics with physical properties of the underlying electricity products and systems [20]. At a high level, the designs for electricity markets are driven by a few key design objectives that, when achieved, would reflect successful market performance. Based on these market design objectives, a set of design principles can be established to guide more detailed design of specific markets.

2.3.1   Market Design Objectives

2.3.1.1   Secure and Reliable Operation of Power System   A major concern with electricity markets has been the potential adverse effects on secure and reliable operation of the physical power system. Societies have come to expect a certain quality of electricity service as part of people's normal lifestyle, and have little tolerance for noticeable service degradation. For competitive electricity markets to be sustainable, it is critical that they are designed to leverage competitive economic incentives to further promote secure and reliable electricity service.

Conversely, no competitive markets can operate effectively unless products in the market can be reliably delivered. Reliable physical operation must be naturally compatible with normal market forces that drive price quantity equilibrium. While administrative procedures must be established to handle emergency conditions, emergency suspension of normal market operation must be rare occurrences. No market can succeed if natural market activities are frequently suspended due to demand/supply imbalance or grid security emergencies.

2.3.1.2   Risk Management Facilities for Market Participants   Different market participants have different risk tolerance levels. As with most markets, normal market activities are direct results of market participants exercising different risk hedging instruments to manage overall business risks within their acceptable tolerance levels. While many risk hedging instruments already exist for managing electricity production risks, for example, fuel price contracts and options or weather derivatives, there are nevertheless additional risk management tools needed for electricity market participants. A particularly important area, which is unique to the electricity market, is management of risks associated with transmission grid congestion which can be addressed by FTRs.

In the design of early markets, there were extensive debates about whether bilateral contracts and/or centralized auction are required for electricity markets. As experience grew, it is now generally accepted that bilateral contracts and centralized auctions are both necessary and can be compatible. In fact, bilateral contracts for forward energy are a dominant hedging instrument against spot price volatility. In many major electricity markets, over 80% of physical energy deliveries are hedged with bilateral contracts.

2.3.1.3 Open and Transparent Market Performance   In addition to typical market monitoring practices at various nonelectricity commercial markets, there are several factors that contribute to additional monitoring requirements for electricity markets. They reflect the nature of the electricity products, the history of regulated noncompetitive environment, and the inherent nature of complex, highly specialized business processes for secure and reliable system operation.

Compliance monitoring covers performance by market participants and market operators with respect to complying with market protocols, including compliance by market participants to instructions issued by market operators. Market power monitoring checks for evidence of economic and/or physical withholding by pivotal suppliers. Since the quantitative analysis for pivot supplier can be strongly affected by actual system conditions, for example, a transmission constraint unexpectedly becomes binding due to unscheduled equipment outages, it is often necessary to exercise practical judgments and considerations when assessing market power scenarios. Besides, market power can result from overconcentration of generations, strategic locational power due to transmission bottlenecks, or the similar market power derived through other means, for example, owning FTRs. Many of the problems with energy markets identified in the last few years are caused by market design flaws. Market power monitoring is to identify these flaws, and recommend mitigation rules once detected.

Finally, it is imperative that the design for power monitoring takes into account current technical capabilities of the IT system for market operation; and that the requirements for market monitoring are explicitly designed as an integral part of the IT system.

2.3.1.4 Phased Implementation of Market Migration   Electric utility restructuring introduces broad scale social economical impacts far beyond engineering technology. Consequently, the pace of change is significantly constrained by institutional and human inertia. Rather than taking one giant step, market design should allow for a multi-step development of progressive, incremental enhancements. This gives market participants and market operators sufficient time to be trained and become proficient with the new systems; and the society at large to adapt itself to the broader impacts of restructuring.

2.3.2 Market Design Principles

To achieve the market design objectives as described above, it is useful to consider the following design principles for electricity markets:

2.3.2.1 Establish Trading Mechanisms for Energy Resources A fundamental characterization of any market is the products, or commodities, that are
traded in the market. For electricity market, electric energy is the dominant product.
Hence, the design of electricity market must establish effective trading mechanisms
for electric energy by both generation and demand resources [25]. This includes
trading in the spot market where physical delivery of the product(s) can be reliably
achieved. With a robust spot market, it is then possible to have comprehensive
forward bilateral contracts essential for overall market liquidity.

In ensuring physical deliverability of electric energy for robust spot energy
market, the market clearing function for energy spot market must be consistent with
real-time physical system operation practices. This means energy spot market clearing must incorporate demand/supply balance and transmission security constraints
within the paradigm of market economic equilibrium.

In addition to energy trading, electricity markets may also be developed for
several ancillary services that are suitable for competitive trading. They include
regulation services and various categories of MW reserves, for example, operating
and supplemental. Certain other ancillary services, such as, Volt/VAr support, and
black-start services are less amenable to competitive trading and may be arranged
more easily using cost or value-based service contracts between system/market
operators and service providers.

In trading energy and ancillary service products, market participants should
be allowed to set their offer/bid prices and quantities to reflect its particular business
practices, subject to market power and physical system operation constraints. For
example, a generator may place X% of capacity as self-scheduled energy, Y% as
one or more priced energy segments, and Z% for priced reserve services. If the
generator is deemed to have market power, then its offer can only be based on cost
and/or administered prices.

2.3.2.2 Establish Open Access for Transmission Services To support a
robust spot market for generation and demand resources, it is necessary to ensure
all market participants have open access to transmission services, and that the cleared
generation offers and demand bids do not exceed available transmission capacity.

Open access to the transmission system involves determining transmission
capacity (total and available) and methodology for designating to market participants
the rights to use the capacity. There are two basic approaches to designating transmission rights: Physical Transmission Rights (PTRs) and Financial Transmission
Rights. Holders of PTRs are assured access to transmission according to the terms
of the PTR (e.g., 100 MW firm services from A to B). Holders of FTRs, on the other
hand, are entitled to receive congestion credits according to terms of the FTR (e.g.,
100 MW off-peak from A to B). FTRs are relevant only in markets that support
market-based congestion management with congestion charges including in the
clearing prices.

Market mechanisms for acquiring and trading PTRs and FTRs are significantly
different and can strongly affect market liquidity. PTRs are typically acquired
and traded using OASIS-based facilities, where available transmission capacity
and prices are posted. Transmission customers (market participants) submit PTR

reservations requests on OASIS, and OASIS-operators approve/reject the requests based on available transmission capacity. FTRs are typically acquired and traded through centralized FTR auctions. FTR market participants submit bids/offers into periodic FTR auctions. The auction clearing process contains explicit and detailed models of security-constrained transmission network to ensure revenue adequacy in congestion credit payoff for FTR holders. By including transmission security constraints in the FTR market clearing process, it is possible to dynamically re-configure FTRs for efficient use of transmission capacity (e.g., sellers of a FTR from A to B may be paired with buyers of a FTR from C to D).

2.3.2.3 Harmonize System Operation with Market Operation

As previously stated, a robust spot market requires close coordination between market operation and system operation. This requires incorporating physical and operational characteristics of reliable system operation within the competitive market framework that maintains price-quantity equilibrium. More specifically, it means generation offers and demand bids cleared for real-time markets are consistent with physical system operation practices with respect to demand/supply balance, transmission security constraints, and economic efficiency.

Demand/supply balance means total generation is balanced against demand (including losses) for each energy-dispatch instruction cycle (e.g., 5 minutes). This dispatch instruction cycle should match the real-time market interval.

Consistency in managing transmission security constraints between system and market operation requires formal congestion management methodology with the following characteristics:

1. Consistent model of transmission constraints: constrained facility (e.g., transformer), grid topology, scheduled and unscheduled outages, contingency definition (for n-1 security constraints), limits, etc.
2. Consistent model of resources: physical resource limits, price/cost model, ramp rates, sensitivity factors of commercial locations to each relevant transmission constraints.
3. Robust analytical foundation, such as formal mathematical optimization, to ensure transparency and consistency of the cleared quantities and prices of the market clearing results.

2.3.3 Energy Market Design

Energy is the primary product for any electricity market. Strong energy market is the necessary foundation for other ancillary electricity product markets. The design of electric energy market starts with issues concerning the production, consumption, and transmission of energy in the real-time spot market. They need to closely align with practices in real-time system operations.

Real-time system operation involves balancing generation with demand and managing transmission security constraints. In the absence of transmission congestion, the real-time market operates very similarly to the classical economic dispatch problem except generator incremental costs are replaced by bid/offer prices. Outputs of online dispatchable resources are adjusted, in simple merit-order, to maintain system energy balance. When transmission congestions exist, SCED is performed to ensure transmission security is satisfied while maintaining economic efficiency and business process transparency. Instead of the traditional approach of manually redispatching units to alleviate limit violations, system operators perform congestion management for real-time market by using the SCED. SCED determines which resources to (re)dispatch so the operators can focus on managing the congested constraints and overall system reliability. As part of the real-time market clearing process, SCED also needs to determine LMPs that are mathematically consistent with the re-dispatch results. There is broad industry acceptance of congestion management based on LMP and dispatch results from SCED.

To participate in real-time operations, resources must be on-line and dispatchable. Since most resources require some finite time to come on-line and be dispatchable, practical system operation traditionally performs forward scheduling of resources, e.g. day-ahead unit commitment. In the market context, forward scheduling can be part of the financially-binding day-ahead market, as in many US RTOs, or be understood to provide purely indicative information, as in New Zealand and Australia. One benefit of the financially binding day-ahead scheme is that it provides an additional risk hedging mechanism for the market participants. In the two settlement scheme consisting of a day-ahead market (DAM) and a real-time market (RTM), a market participant can hedge real-time spot market risks by managing positions in the DAM and/or RTM.

To further assist participation by generators in the DAM, it is common practice for generator offers to be priced in three parts: startup, no-load, and incremental energy prices. Three-part offers including temporal operating constraints (e.g., minimum up and down time) are designed to improve cost discovery over single part offer, which would require blending of different cost components into a single component.

With three-part offers and the associated temporal operating constraints, the DAM is designed to support commitment decision as part of the day-ahead market clearing process. Furthermore, in order for the DAM to provide effective hedging for the RTM, the DAM must also emulate congestion management in the RTM. As a result, the DAM market clearing must be able to perform security-constrained unit commitment (SCUC) analysis and to determine day-ahead market clearing prices (e.g., DA LMP). The development of robust optimization application technology for DAM SCUC is a major technical achievement directly attributable to the development of competitive electricity markets.

While DAM can perform SCUC, market participants retain the option for self-commitment and/or self-dispatch. Based on preliminary observations, there appears to be a trend for increased preference by market participants to have the DAM make the commitment decision for them.

2.3.4   Financial Transmission Rights Market Design

FTRs are closely associated with LMP-based congestion management for the energy market. FTR holders hedge the risks of incurring congestion charges associated with energy schedules in the energy market (i.e., the DAM if present, otherwise, the RTM), by receiving congestion credits associated with the FTRs. Of course, market participants are not required to hold any FTR to participate in the energy market. For participants not holding any FTR, they can schedule energy but are exposed to the risk of energy-market congestion charges.

Just as it is not necessary to hold FTR to schedule energy in the energy market, it is also not necessary for FTR holders to schedule energy flow in order to receive FTR credits. All FTR holders will receive congestion credits regardless of their position in the energy market.

In order to maintain revenue adequacy between congestion charges collected in the energy market and congestion credits paid out to the FTR holders, it is necessary for RTOs, as FTR market operators, to have a consistent model of the transmission network, including all constraints, between the energy market and the FTR market.

The primary FTR markets are centralized auctions. In the FTR auction, available transmission capacities, which are consumed by FTR bids and offers, are subject to transmission capacity constraints that have the same representation as the transmission security constraints in the energy market. The market clearing process for FTR auction produces cleared MWh bids and offers, and market clearing prices, which are LMPs for the FTR auction. FTR prices for path AB is the FTR market LMP difference between A and B.

Since its first introduction in electricity markets in the eastern US, the design of the FTR as a hedging instrument had evolved significantly. For example, PJM's FTRs can span multi-time intervals (i.e., on-peak, off-peak, 24 hours, multi-months), can be forward obligations or options. The benefit of an Option FTR is that it removes downside (negative credit) risks to FTR holder when the direction of congestion in energy market is reversed from the direction of the FTR. The introduction of Option FTRs as an alternative that resembles standard financial options was designed to increase FTR market liquidity.

Net revenues received from FTR auctions are paid to holders of Auction Revenue Rights (ARR) as reimbursement to preexisting users/owners of the relinquished transmission capacity. The subject of converting and/or transferring transmission rights to provide open transmission access is a major political challenge in the path towards competitive electricity markets.

2.3.5   Ancillary Service Market Design

System reliability is an integral part of a properly designed electricity market. Issues of system reliability and market economics should be coherently treated so that dispatch instructions and price signals could be consistently provided to the marketplace [28]. The introduction of ancillary service markets into a wholesale electricity market is envisioned to correctly value and price the actions necessary to prepare for and maintain the reliability of power system dispatch in conjunction with operating an efficient energy market [6]. Reliable and secure system operation requires the following products and services:

1. Energy
2. Regulation
3. Synchronized reserve
4. Non-synchronized reserve
5. Operating reserve
6. Voltage support
7. Black start

Of the seven items above, we will address the first five, which directly affect market and system operation decisions in real-time. All commodities except the first one are considered as ancillary services.

Regulation under automatic generation control serves to continuously balance the balancing authority's supply resources with real-time demand variations in order to meet the national and/or regional control performance standards. A regulation bid will be in the form of dollars per hour to provide a specified MW amount of regulation service. Regulating capacity blocks are bounded by auto low and auto high limits.

Synchronized reserve is a resource capacity synchronized to the system, which is (a) able to immediately begin to supply energy or reduce demand, (b) fully available within 10 minutes, and (c) able to be sustained for a period of at least 30 minutes to provide first contingency protection. Synchronized reserve blocks may be positioned anywhere between low operating and high operating limits.

Nonsynchronized reserve is a resource capacity nonsynchronized to the system, which is (a) able to supply energy or reduce demand, (b) fully available within 10 minutes, and (c) able to be sustained for a period of at least 30 minutes to provide first contingency protection.

Operating reserve is a resource capacity nonsynchronized to the system, which is (a) able to supply energy or reduce demand, (b) fully available within 30 minutes, and (c) able to be sustained for a period of at least 60 minutes to provide second contingency protection.

The four types of ancillary services are suitable for and have been implemented in competitive markets. Among the key design features for competitive ancillary services markets are:

1. Reserve requirements are defined at system and zonal level for each type of AS product.
2. Prioritized order of AS service quality should be defined to allow surplus capacity for a higher quality AS offer to be eligible to fulfill requirements for lower quality AS service.
3. AS resources include both generation and demand resources that meet specific qualification criteria.
4. Qualified resources can offer a mix of energy and AS products. The total amount of cleared energy and AS for each resource must be within physical and operating limits of the resource.

5. Co-optimized market clearing of energy and AS should be performed to ensure proper allocation of resource capacity into different products, enforcements of all pertinent constraints, and that overall market merit is maximized.

6. Co-optimized market clearing results include energy and AS schedule for each unit and each time interval, and the associated market clearing prices.

2.4 OPERATION OF ELECTRICITY MARKETS

The implementation of the market design principles varies from region to region [12, 32–35, 37], however, successful operation of the electricity markets satisfy common criteria for success. This section investigates important criteria for success and provides examples of how the criteria are exemplified in the operation of selected market systems.

2.4.1 Criteria for Successful Market Operation

The operators of electricity markets provide services for a number of stakeholders including GENCOs, TRANSCOs, DISCOs, energy brokers, marketers, aggregators, retailers, and the end users of electricity. This provides market operators with multiple objectives to achieve success. Among the most important criteria are the interdependent objectives of

- Power system reliability
- Market transparency
- Financial certainty
- Operational efficiency

2.4.1.1 Power System Reliability Power system reliability [15, 16] is the dependable delivery of electrical energy to the demand. The market system operator supports system reliability through coordination with the system operator. The system operator should be involved in all phases of the market. During the design of the market, the system operator will contribute to the design of the market protocols to ensure that the commodities and rules defined will support reliable power system operations. During operations, the market operator exchanges information with the system operator about the state of the power system and reliability needs.

The operational communication between the power system and market operators is normally facilitated by the sharing of a common power system model that models the transmission grid, including connection points for supply and demand market resources. The use of a common model enables a precise description of the system operator's system reliability concerns, normally expressed in terms of transmission or other operational constraints which are included in clearing the market. Common examples of operation constraints raised by the system operator are branch (transformers or lines) flow limits, interface stability limits, and equipment outages. The market operator will be most successful in supporting system reliability when the communication from the system operator is reflected in the market system model with a high level of precision. When model or communication abstractions exist between the power system and market operator, system reliability and other market operational objects can be degraded. Precise communications between power system and market operations is required to enable market results to provide incentives that are consistent with system reliability.

Ensuring reliability of a power system is a multi-stage process. The part that has an immediate impact on a customer is the real-time dispatch. Less visible are the system planning and scheduling that prepare a power system for reliable operation. The envelope of power system reliability can encompass five major progressions starting from many years in the future, as shown in Figure 2.5 [30].

2.4.1.2 *Market Transparency*   For the electricity market participants to respond to market signals produced in market operations predictably (which enables the market to support system reliability), each individual participant must trust that following the market signals will result in the best results for that participant and the system as a whole. This is achieved through a market design in which actual power system operating conditions are reflected in the market results and the settlement of the market reinforces the market instructions. In this context, transparency is the ability of individual participants to understand that the market signals they receive are consistent with minimizing the system cost, with maximizing the individual participant's revenue, and with power system operating conditions. For example, transparent market signals will provide instructions in which market prices and resource output are consistent with a resource offer curve.

The market must provide information to participants that allows for understanding of the system operating conditions and the consistency of market pricing and instructions. For example, in an LMP market, transparency is supported by publishing the binding constraints and market prices. The prices are consistent with the constraints, providing incentives that reinforce participant behavior to maintain and/or increase reliability.

2.4.1.3 *Financial Certainty*   While maintaining system reliability and transparency, the market must also provide financial certainty to ensure that the market transactions meet participant expectations in settling according to the market

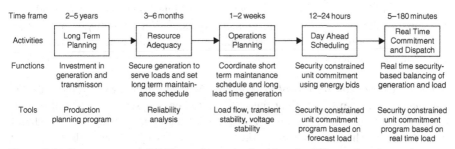

Time frame	2–5 years	3–6 months	1–2 weeks	12–24 hours	5–180 minutes
Activities	Long Term Planning	Resource Adequacy	Operations Planning	Day Ahead Scheduling	Real Time Commitment and Dispatch
Functions	Investment in generation and transmisson	Secure generation to serve loads and set long term maintain-ance schedule	Coordinate short term maintenance schedule and long lead time generation	Security constrained unit commitment using energy bids	Real time security-based balancing of generation and load
Tools	Production planning program	Reliability analysis	Load flow, transient stability, voltage stability	Security constrained unit commitment program based on forecast load	Security constrained unit commitment program based on real time load

Figure 2.5   Power system reliability and security functions in different time frame

protocols. The foundation for providing this financial certainty is an overall market design that is consistent with revenue adequacy principles, ensuring that the market collects enough revenue from purchasers to pay the suppliers. For electricity markets that simultaneously clear energy with transmission, the key to revenue adequacy is that the congestion charges collected to price the transmission congestion is sufficient to support the price incentives given to supply and demand to control the congested transmission and maintain system reliability.

In a simple settlement market based on LMP where market action is taken to enhance system reliability by maintaining transmission within operational limits, the market signals will indicate a preference (increased price) for increased supply and reduced consumption on the "to" side of the constraint and a preference (decreased price) for decreased supply or increased consumption on the "from" side of the constraint. With load generally on the "to" side of the constraint and the generation higher on the "from" side of the constraint, there is a net overcollection of revenue by the market, making the market revenue adequate. The overcollection is normally returned to the market participants, but in a manner that does not provide disincentive for following market signals, for example, through FTR credits.

Many markets allow participants to hedge their positions with forward price certainty in forward markets for transmission (such as FTRs) and/or combined energy and transmission markets (such as the DAM). In both cases, the forward market provides the participant with a financial right (or obligation) that carries forward and is settled against the next market for that time interval. Operating a multi-settlement market structure requires that markets are coordinated to ensure that the market collects enough revenue from those receiving services to pay those providing the service. Market rules are often designed such that the theoretical basis for revenue adequacy in multi-settlement markets requires that transmission constraint limits become no less restrictive as the forward market moves further from real time operations [40]. That is, when a time interval is included in a forward market, the transmission constraints should be at least as restrictive as the limits which will be used during that same time period in real time. When revenue adequacy is violated, market rules will often socialize the additional costs which will detract from the financial efficiency of the market.

Revenue adequacy concerns discourage overly optimistic assessments of transmission capabilities in future models of the power system. However, an effective market will not be overly conservative. To provide participants with the available hedging opportunities and to encourage liquidity through participation in the market, an effective market will reflect the available power system capabilities. Maximizing the availability of substantiated transmission capability in the market maximizes opportunity for market participation while maintaining the needed financial certainty for the participants.

2.4.1.4 *Operational Market Efficiency*   Market efficiency can refer to different measures. For example, financial market efficiency describes how well market prices reflect known information and provides information enabling effective direction of capital while operational market efficiency describes how well the market applies available resources to maximize the market objectives.

Operational efficiency can refer to the cost of delivering market services to operate the market, but in this discussion, the focus is on how efficiently the energy is produced and transmitted to the economically served demand. Efficient power system operation is included in the market clearing objective function which models the supply costs, demand willingness to pay, and the available transportation as part of the overall market security-constrained optimization. The resulting operational instructions (e.g., unit dispatch) and prices are consistent with the optimal dispatch, thus achieving transparency at the same time as operational efficiency.

The optimization of power system operation, meeting the needs of the economically served demand with the lowest feasible offer cost subject to transmission constraints, is applied at the various forward market time frames as well as in real-time operations. The uniform application of this model provides consistency between the various time horizons in market operations, supporting efficient forward market decisions (e.g., unit commitment) and consistent market results in multi-settlement markets.

2.4.2 Typical Business Processes Timeline

The structure of electricity markets commingles the operation of a power system to be commingled with the operation of the electricity market. This approach improves the ability of the market to successfully meet its objectives. For example, market signals are more apt to reflect real-time operating conditions when both the market and system operators are using the same or similar model and input data. However the organizational structures and business processes vary from market to market. In this section we will look at the business process of two LMP markets employing location-based marginal pricing and how they incorporate the operational objectives: the New Zealand Electricity Market and the PJM Market.

2.4.2.1 New Zealand Electricity Market   The New Zealand Electricity Market is a single settlement market which settles on ex-post prices [36]. The market provides forward looking results of the expected market conditions, real time market-based dispatch, and ex-post price calculation. A diagram of the timeline for the New Zealand Electricity Market (NZEM) is shown in Figure 2.6. Transpower and M-co combine to provide the services needed to run the market. Their roles will not be

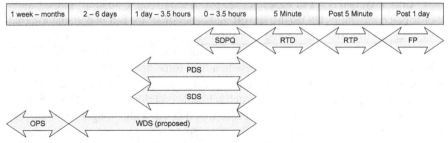

Figure 2.6   New Zealand Electricity Market timeline

distinguished here, but common market and power system models are used for the market functions. The major electricity market business processes are described in the following sections beginning with the earliest analysis through the real-time operations and ex-post pricing.

The New Zealand power system consists of two electrical islands (one for each physical island) with HVDC lines between the islands. The North Island has the majority of the New Zealand demand, while the South Island normally has surplus generation to send to send north over the HVDC lines. This creates power system operational challenges in moving the energy over long distances, but the market is designed to incorporate these conditions.

Also of note is that the New Zealand market is a single settlement market. This single settlement design has encouraged generation and load to join together to hedge the real time price risks. The participants' successes with the hedging supply against demand have marginalized the value FTRs to hedge congestion in this marketplace and no FTR market is currently in use.

2.4.2.1.1 Outage Planning Studies

Outage planning studies (OPS) are performed weeks or months in advance to confirm the power system reliability impact of outages submitted by transmission owners. The results of these studies help determine the scheduling of the outages but do not have a direct impact on market settlement. In addition to the considered outages, key inputs to these studies include load forecast and estimation of generation offers. The studies identify any reliability concerns that may arise from the outages and may result in operational settings suggested for use during the outage.

2.4.2.1.2 Market Schedules

In the single settlement New Zealand Electricity Market, forward schedules are indicative, nonbinding schedules that provide participants with information about the expected market clearing results. The schedules are published periodically to the market participants. The participants can use the results of the schedules to adjust their bids and offers (for energy, ancillary services, and transmission services) for future trading periods. The next time a schedule is published, the participants can confirm the impact of their changes on the forecasted market clearing results, including energy, 6 and 60 second reserves, and transmission congestion. This iterative process provides opportunity for participants to adjust their bids and offers and to plan their own operations for the real time market.

Currently, two types of schedules are published for up to 36 hours in the future. Pre-dispatch schedules (PDS) use the offered generation, the bid in demand, offered HVDC capability, and the offered transmission system configuration and limits to forecast market clearing results in future 30-minute trading periods. Security dispatch schedules (SDS) replace demand bids with a load forecast. The PDS and SDS alternate periodic publication throughout the day. As of the writing of this book, consideration is being given to a weekly dispatch schedule (WDS), which will be published daily. The WDS is expected to be similar to the SDS but with a longer time horizon.

Closer to real time, the next eight 30-minute trading periods are analyzed in the schedule of dispatch prices and quantities (SDPQ). The SDPQ provides a similar function of giving participations time to adjust their market submissions prior to real time. It uses offered generation, offered transmission, and the load forecast to provide more frequent updates of forecasted market clearing results for upcoming trading periods.

2.4.2.1.3 Real Time Dispatch Within a trading period, the market is cleared for five-minute trading intervals. At each five-minute interval and as needed for system reliability, the real-time dispatch (RTD) is run to optimally dispatch energy and reserves at generator and HVDC sites for the expected demand and offered transmission grid (offered status and limits). New dispatch instructions are sent to the generation and/or HVDC resources that have new dispatch instructions. The system operator monitors that the participants acknowledge the new dispatch instructions. The operators also update the reactive power dispatch of the resources as needed for system operating conditions.

The use of real-time measurement for the basis of system load does not anticipate the economic response of demand as prices change. This is a common approach taken in most electricity markets. Although there may be an impact on market efficiency with ex-ante clearing prices set by supply, the current operating conditions have been found to be the best predictor of the short-term demand behavior, even in New Zealand where there is relatively high penetration of demand-side management. The economic response will be observed and used as part of the next dispatch cycle, which will produce prices incorporating the response within five minutes. Use of demand bids in real time instead of real-time measurements would likely create a larger reliability issue.

The RTD is based on the transmission grid that is offered in by the transmission owner. This means that the dispatch for the next five minutes is not based on real time observations, but on intended operations. This could present challenges in meeting the market objectives. Given that this is the market rule, the operator must dispatch from the offered grid. However, to ensure reliability, the operator monitors the real time system conditions and can take action to maintain reliability, such as constraining the system to run within reliability margins or monitoring RTD dispatch instructions relative to a reliability concern.

2.4.2.1.4 Pricing Real-time pricing (RTP) and final pricing (FP) are ex-post pricing analyses that calculate market clearing prices by clearing the market with the observed demand rather than expected demand. By pricing after the fact based on observed demand, it ensures that prices will accurately represent actual operating conditions.

RTP employs the real-time telemetered measurement of load to calculate ex-post prices every five minutes. The key benefit of this study run at frequent intervals is that it can be used to identify deviations between the market model and system operating conditions. When a condition is identified it can be remedied quickly to minimize the impact on the market. This quick feedback contributes to the financial certainty and transparency of the market results and helps improve reliability through market participant trust in market signals.

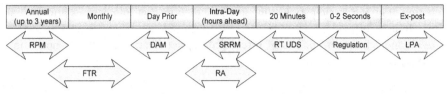

Figure 2.7 PJM Markets Timeline

FP calculates ex-post clearing prices for each 30-minute trading period using metered demand and market offers. Like RTP, it is otherwise similar to the RTD. The FP analysis is defined to use the operating conditions that existed at the start of the trading period. FP is the last step in the market clearing functions in the New Zealand Electricity Market. The pricing results of FP are posted for market participants and used for financial settlement of the real time market.

2.4.2.2 PJM Markets PJM electricity markets [8, 37] include real time (RT UDS) and day ahead (DAM) energy markets, financial transmission rights (FTR) auction, synchronized reserve and regulation markets (SRRM), and a reliability pricing model (RPM) used to clear capacity for up to three years in the future. A diagram of the timeline for the PJM Markets is shown in Figure 2.7. The PJM FTR auction, the day ahead and real time energy markets, and the synchronized reserve and regulation markets will be described.

In these PJM Markets, the FTR auction is run annually with residual auctions throughout the year. The owner of an FTR has a settlement in the FTR auction as well has a forward participation in the DAM transmission congestion pricing settlement. Similarly, an energy award in the DAM is settled in the DAM and is a determinant in the RTM settlement. This multi-settlement market arrangement provides the opportunity to participate in hedging opportunities to reduce the financial uncertainty associated with the real-time operations and prices.

2.4.2.2.1 FTR Market The FTR market is a forward financial market that provides an opportunity for participants to hedge the risk of congestion charges. The FTR awarded in the FTR Market provide a FTR credits that offset the uncertain future congestion prices in the DAM. PJM FTRs are point-to-point and net revenues from the FTR auctions are distributed to the participants using ARRs.

PJM conducts a number of different FTR auction time horizons, which include annual, quarterly, and monthly FTRs. The FTRs may be for on- or off-peak hours and may be obligations (positive or negative revenue) or option (only non-negative revenue). Regardless of the time horizon, the FTR provides a hedge against the DAM congestion prices [8].

The forward revenue (or payment) associated with an FTR is funded by the congestion revenue collected in the DAM. To ensure that the FTRs will be funded appropriately to provide financial certainty for participants, the available transmission system must support at least the level of transmission activity afforded by the FTRs; that is, the cleared FTRs must be simultaneously feasible from both financial and physical transmission capability point of view. The inputs to the simultaneous feasibility test (SFT) in the FTR auction are the candidate FTR awards and the electrical transmission system. Because the FTRs are cleared for a period of time, for example, monthly on- and off-peak, there is not one expected state of the transmission system, but a representative state is used.

The FTR Market must use one representative state of the transmission system for each FTR award period. The bid and offered FTR injections and withdrawals are modeled in the transmission network and the bids and offers cleared to maximize the overall value. The available capacity of the transmission system is dependent on the rules for that auction, for example, in PJM annual auctions, there are multiple rounds in which increasing amounts of transmission capacity are available. The FTR market operator will ensure that the cleared FTRs do not use more transmission that is expected to be available in the DAM, necessitating a tradeoff that must be made between conservative use of transmission capacity (and thus a conservative view of DAM congestion revenue) and increasing liquidity (thus enabling market participants to hedge to the full expected transmission capacity). For example, routine transmission outages are included (or excluded) in the monthly FTR analysis to match the expected transmission capacity. Including excessive transmission outages will reduce the quantity of cleared FTRs, but under-representing outages may cause a revenue shortfall in funding FTRs from congestion revenues. When FTRs are not fully funded, the value of the FTR revenue may be reduced to match the available funding which decreases the effectiveness of the hedge and reduces the financial certainty of the market participant.

2.4.2.2.2 Day Ahead Market   The DAM is cleared each day for the next operating day. The results of the DAM are energy schedules for supply and demand resources (both physical and virtual) and the associated LMPs. The cleared energy volumes are settled at the DAM LMPs. The FTR credits are also determined from the DAM congestion pricing between the FTR point of receipt and point of delivery.

The DAM provides a financially binding result for a 24-hour horizon that accommodates participants' operational constraints (e.g., unit minimum up and down times). Prior to the introduction of the DAM in PJM, there was only the single settlement of the Real Time Market (with FTRs settled against real time LMPs). In this market structure, the physical operating characteristics of generators and other equipment were more difficult to manage. In that design, the asset operator must anticipate the forward prices to determine when to be on- or off-line and estimate whether start up and minimum load costs would be recuperated. The introduction of the DAM allows the three parts of the generators cost (i.e., start up, minimum load, and incremental energy costs) and temporal constraints to be modeled in the market clearing process. The generators are scheduled in a way that minimizes the offered cost of the generation to reliably serve the bid-in economic demand. Generators with awarded DAM energy schedules are guaranteed to be paid at least their as offered operating costs when cleared in the DAM.

The cleared DAM award volumes are settled at DAM prices as are FTRs credits. The DAM awards provide day-ahead price certainty in that only the deviation from DAM quantities is settled at RTM prices.

Virtual bids and offers are supported in the DAM. Virtual bids and offers do not need to be backed by physical resources and allow participants to take a financial position in the DAM that can be closed at RTM prices as a real-time deviation when not backed by physical resources. Participation of virtual bids increases liquidity, allows transfer of FTR rights to real time for nonphysical participants, and creates arbitrage opportunities which tend to align DAM and RTM prices.

Similar to the FTR market, the DAM must be balanced in its hourly representation of the available transmission capability with its associated impact on the financial support of the RTM settlement and maximizing the use of the expected transmission capability by the participants in the DAM. The DAM clearing is also subjected to the SFT, ensuring physical and financial feasibility with the available transmission capability.

The results of the DAM, are used in a reliability assessment of tomorrow's operating day. In the day ahead reliability assessment (RA), the PJM load forecast and reserve requirements are used to determine if additional resources not cleared in the DAM need to be brought online to ensure sufficient capacity is available for real time operations. Additional resources are selected to minimize the cost of the additional commitments to meet the load forecast. Resources brought online to support reliability needs are guaranteed to recover at least their offered costs.

2.4.2.2.3 Synchronized Reserve and Regulation Market Hours before real time, PJM procures the required reserves and regulation services from the available resources. PJM uses a two-pass system to clear spinning reserve in which operating available capacity (Tier 1) is first identified and additional capacity (Tier 2) is cleared when needed. Regulation is also simultaneously cleared. Ancillary service requirements are specified by reserve zone to procure the services in the needed locations. The spinning reserve and regulation are co-optimized hourly with energy offers subject to transmission constraints to provide the best available selection of resources. The cleared reserves and regulation capacity are reserved in the RTM clearing.

2.4.2.2.4 Real-Time Market The PJM RTM optimizes the current system conditions, the expected change in load, and the available dispatchable resources subject to identified transmission constraints to provide near-term dispatch targets and ex-ante pricing signals that provide operating instructions and pricing signals to reinforce system reliability. The pricing for real-time settlement is calculated by an ex-post pricing process.

The key inputs to the RTM are the operating conditions of the power system, current resource outputs, short-term load changes, and expected interchange. The current operating conditions and resource outputs used in the RTM represent the state of the power system as identified by the state estimator based on telemetered power system data. The actively monitored transmission constraints may represent transmission constraints within the PJM reliability region or those influenced by PJM resources, but monitored by another reliability entity. The load is determined by a short-term load forecast function. The expected interchange is derived from interchange schedules. In real time, the operator controls the system power balance and transmission reliability by ensuring that there is adequate generation moving in response to load changes and that the constraints of concern for reliability are being respected and priced appropriately.

Transmission reliability constraints considered in the RTM may be constraints within PJM identified through real time security analysis which compares the current system conditions against current and N-1 contingencies and/or be constraints outside the PJM region which are of concern for a neighboring operator and influenced by PJM resources. These constraints from external areas may be the result of a TLR action or a coordinating action with a neighboring market operator to equalize control costs for increased economy and pricing consistency. When a constraint with external coordination opportunities is activated within PJM, the operator may request coordination with neighboring entities either through the TLR process or market to market coordination.

2.4.2.2.5 Ex-post Pricing The PJM RTM is settled using ex-post prices. The ex-post pricing process serves as a calculation of prices at observed (rather than forecast) real-time system conditions and as a performance monitor, encouraging the following of RTM dispatch suggestions. Because the prices are calculated after the fact, resources are encouraged to follow instructions so they participate in setting price and their LMP will not fall below their offer price. Nonperformance penalties to encourage units to follow dispatch are used in some markets, but not by PJM. Nonperformance penalties are prescribed and can be factored into resource decisions. The performance monitor function of PJM's ex-post LMP is more uncertain and thus more difficult to evaluate in a participant decision to deviate from the dispatch.

The ex-post pricing process (LPA) monitors observed resource performance relative to RTM dispatch results. When a resource over performs beyond a tolerance, the resource is not eligible to set price. When a resource under performs, it is eligible to set price only at the price corresponding to its achieved output. The market is cleared ex-post every five minutes with these resource attributes, with observed load, and with active constraints constrained to the level of their observed flows. In this way, the resulting ex-post prices reflect the price for the energy as needed for real-time operations. Hourly averages of the five minute ex-post prices are calculated for settlement.

2.5 COMPUTATION TOOLS FOR ELECTRICITY MARKETS

Electricity market operation requires a set of computation tools to facilitate various market clearing/analysis processes. Compared with the tools used in traditional grid control centers, market applications used by a RTO/ISO should be able to handle both price-based models and cost-based models to clear a market while fully respecting physical transmission network limitations. In addition, the market applications have higher standards in terms of solution accuracy, solution auditability, and computational performance.

The unified framework for competitive electricity market and grid reliability is founded on the duality theory that is concerned with the outcome in a competitive market and the mathematical solution to a constrained optimization problem. This duality result states that a competitive market solution consists of both the market clearing quantities and the associated market clearing prices that equate market demand to market supply in all interrelated parts of the market. Only when the market clearing quantities are priced at the logically associated prices is the overall solution incentive compatible. When prices are incentive compatible, participants will provide bids that reflect its actual cost or value, and produce or consume the efficient quantities. Participants' production or consumption of the efficient quantities leads to grid reliability. The duality theory clearly suggests that formal optimization techniques be utilized for electricity market clearing so that market based dispatches, when executed in real-time grid operation, may lead to harmonious unification of market efficiency and grid reliability. This fundamental principle of the duality theory has been reflected in most existing market designs. Most market rules require that RTO:

- Provide a reliable grid and coordinate grid operation to minimize costs consistent with secure operation of the grid,
- Schedule and dispatch generation to satisfy market demand for electricity at minimum price while taking into account the security of the grid,
- Purchase for the benefit of grid reliability the ancillary services and resources that are necessary for economic and secure delivery of electricity, and
- Provide information to participants in an open, nondiscriminatory manner to help them make consumption and production decisions.

These competitive market rules are also in line with the long-held industry principle about reliable and economic grid operation. Optimization-based economic dispatch and unit commitment has been practiced for decades to coordinate grid's economic and reliable operation. But traditional optimization methods for economic dispatch and unit commitment are less rigorous and the underlying grid models are oversimplified. The key to competitive market success lies in the dealings with modeling details. The constrained optimization problem behind electricity market dispatch and pricing must therefore model grid details in order to realize the market efficiency as promised in the theory of economics and mathematics. As a result, significant progress has been made in recent years in applications of formal optimization techniques for competitive market based resource commitment, scheduling, pricing and dispatch [31].

This section will start with introducing two important optimization based application components, that is, security constrained economic dispatch (SCED) and security constrained unit commitment (SCUC). All market functions described in the previous sections can be implemented with different forms and/or configurations of SCED and SCUC. Two examples, joint optimization of energy and ancillary services and reliability unit commitment problem, will be described to illustrate the application of SCED and SCUC in electricity market operation. At a high level, this section will also briefly illustrate some system integration aspects of market applications, such as modeling, data interface, audit support and market monitoring. At the end, future direction of market analysis tools will be discussed.

2.5.1 SCED and Associated Market Business Functions

2.5.1.1 Classic OPF SCED is evolved from the classic optimal power flow application (OPF) [38, 14, 19, 22, 23] in the Energy Management System (EMS). OPF is aimed to find a power flow solution that leads to both economic and reliable grid operation. It is common practice that the control variable values determined in the OPF solution are used as guidance to achieve optimal grid operation. The dual solution, that defines the marginal prices of relevant quantities, has been largely ignored until the advent of competitive markets. According to the duality theory, the dual solution in the form of prices is equivalent to the primal solution in the form of quantities in achieving the optimal objective.

The OPF problem is stated in mathematical terms as follows:

$$\min f(u_P, u_Q)$$

subject to

$$(\lambda) \quad g(u, x) = 0$$
$$(\mu) \quad h(u, x) \le 0$$
$$u^{\min} \le u \le u^{\max}$$

where

$u = [u_p, u_Q]^T$: Control variables, such as generator real and reactive power output and transformer tap positions

$f(u, x)$: Objective function typically represented with total production cost

$x = [x_\theta, x_V]^T$: State variables, such as bus voltage angles and magnitudes

$g(u, x) = [g_P(u, x), g_Q(u, x)]^T$: Bus real and reactive power balance equality constraints

$h(u, x) = [h_P(u, x), h_Q(u, x)]^T$: Inequality constraints, such as transmission line and transformer thermal limits and bus voltage magnitude limit constraints

$\lambda = [\lambda_P, \lambda_Q]^T$: Shadow prices of bus real and reactive power balance constraints

$\mu = [\mu_P, \mu_Q]^T$: Shadow prices of grid security constraints

$u^{\min} = [u_P^{\min}, u_Q^{\min}]^T$: Minimum limits for control variables

$u^{\max} = [u_P^{\max}, u_Q^{\max}]^T$: Maximum limits for control variables

In the OPF formulation, each of the equality or inequality constraints has a corresponding shadow price at its optima. The shadow price, μ, of a binding grid security constraint defines the marginal cost of maintaining the specific grid security constraint. The shadow price, λ_P of bus real power balance constraint can be used to define the short-run marginal prices of producing/consuming real power at a bus.

The shadow price, λ_Q, of bus reactive power balance constraint may be interpreted as the marginal cost of producing/consuming reactive power at a bus.

When generators or loads are allowed to respond to the spot price signals, the prices can induce generator responses to achieve economic and reliable grid operation as effective as the MW dispatches. In order for price signals to be able to convey effective control information for economic and reliable grid operation, price signals must be unambiguous, transparent and explainable.

While the full OPF formulation provides a global optimal solution for both real and reactive power, there is a need to solve decoupled OPF in electricity markets, where optimal real power flow solution is the results of market clearing process. In electricity markets reactive power control activities should be transparent to the participants and the impacts from reactive power is usually modeled as surrogate constraint limits.

The decoupled OPF formulation for real power can be described as follows:

$$\min f(u_P)$$

subject to

$$
\begin{array}{ll}
(\lambda_P) & g_P(u_P, x_\theta) = 0 \\
(\mu_P) & h_P(u_P, x_\theta) \leq 0 \\
& u_P^{\min} \leq u_P \leq u_P^{\max}
\end{array}
$$

This formulation is very similar to the previous full AC OPF formulation, except that reactive power related control variables, voltage magnitude variables and voltage security constraints are reduced to parameters with predetermined values. This is currently the most commonly used optimization framework in electricity market clearing and price calculation.

2.5.1.2 SCED for Market Clearing

The security constrained economic dispatch, which is essentially a reformulation of the decoupled OPF, is the commonly used algorithm for market clearing. The SCED algorithm has been successfully applied to solve market clearing problem with large scale grid models, such as 2000 bus ISO-NE market, 15,000 bus PJM market, and 30,000 bus MISO market in the US.

The SCED algorithm provides several advantages over the traditional optimal power flow:

- The SCED algorithm uses a detailed, linearized grid model, and is solved with state-of-the-art linear programming (LP) techniques.
- With the use of standard LP methods, the SCED algorithm provides almost unlimited capability to handle as many linear constraints as necessary.
- Both base and contingency grid security constraints can be effectively handled.
- Ancillary services can be incorporated in the LP-based optimization framework.
- Participant bid data models can be accurately reflected in the formulation. This modeling flexibility makes it possible for an electricity market to offer as many participant choices as needed for them to manage risks.

- The same framework can be applied for energy, ancillary service, and transmission right market clearing.

Different forms of SCED formulation can be applied to different market clearing processes, such as real-time energy market clearing, FTR market clearing, and so on. Below we discuss the applications of the SCED algorithm to solve the joint optimization of energy and ancillary services.

2.5.1.3 Joint Optimization of Energy and Ancillary Services

In many of the existing electricity markets, energy and ancillary services (AS) are both allowed for competitive bidding. The fact that the same resource and same capacity may be used to provide multiple products dictates that AS market operation should be closely coordinated with the energy market. This close coordination is best achieved through joint optimization of AS and energy markets, which minimizes the total market cost of meeting system demand and AS requirements while satisfying network security constraints.

In the development of AS markets, different market designs have been attempted. The designs can be categorized as three approaches: independent merit-order stack approach, sequential market clearing, and joint optimization.

1. *Independent merit order dispatch:* Independent merit order-based market clearing ignores the capacity coupling between energy production and supply of ancillary services. Each product is cleared separately from other products based on a separate merit order stack. This approach is simple, but it easily leads to solutions that are physically infeasible.

2. *Sequential market clearing:* The sequential approach recognizes that energy and reserves compete for the same generating capacity. In essence, a priority order is defined for each product. Available capacity of a resource (e.g., generating unit) is progressively reduced as higher priority products are dispatched from that resource. The degree of sophistication of recognizing the coupling varies from market to market. While the sequential dispatch is an improvement over purely independent merit-order dispatch, it needs further improvements for handling interdependencies among the coupled products. Explicit evaluation of costs associated with lost opportunity, production cost impacts, etc. are mechanisms that are used to provide quantitative indices for analyzing tradeoff decisions. They also help reduce the arbitrariness in the dispatch sequence.

3. *Joint optimization:* In the joint optimization approach, the objective is to minimize the total cost of providing ancillary services along with energy offers to meet forecast demands as well as AS requirements. The allocation of limited capacity among energy and ancillary services for a resource is determined in terms of its total cost of providing all the products relative to other resources. The effective cost for a resource to provide multiple products depends on its offer prices as well as the product substitution cost. Product substitution cost arises when a resource has to reduce its use of capacity for one product so that the capacity can be used for a different product (leading to an overall lower cost solution). The product substitution cost is determined internally as part

of the joint optimization. This product substitution cost plus its bid price reflects the marginal value of a specific product on the market. The marginal value, which is typically the market clearing price, represents the price for an extra unit of the product that is consistent with the marginal pricing principle for the energy product. Market clearing prices for ancillary services reflecting product substitution costs create price equity among the multiple products. Price equity refers to the fact that a market participant can expect to receive equivalent amount of profits no matter which kind of service the participant is assigned to provide. Price equity is an incentive for participants to follow dispatch instructions. Without the price equity, participants would tend to provide those services that produce the most profits, which could deviate significantly from dispatch instructions. Significant deviations from dispatch instructions could seriously degrade dispatcher's ability to maintain grid reliability. Therefore, joint optimization of energy and AS market has been accepted in many electricity markets, including PJM-RTO, New York-ISO, ISO-New England, New Zealand, and Australian markets.

2.5.1.4 SCED Formulation Example

Assume that a joint market allows bidding for multiple products, generation energy offer, price-responsive demand bid, and two types of ancillary service—10-minute reserve and 30-minute reserve. With unit commitment schedules given, an exemplar SCED formulation for this joint market may be described in mathematical terms as follows:

$$\min c(P) + cs(S) + co(O) - cd(D)$$

subject to

System power balance constraint:

$$(\lambda) \quad \sum_i (P_i - D_i) - FD - P_L = 0$$

Ten-minute reserve requirement constraint:

$$(\gamma_S) \quad \sum_i S_i \geq S^{\max}$$

Thirty-minute operating reserve requirement constraint:

$$(\gamma_O) \quad \sum_i (S_i + O_i) \geq O^{\max}$$

Grid base-case and contingency constraint:

$$(\mu_l) \quad \sum_i a_{l,i} (P_i - D_i - d_i \times FD) \leq L_l$$

Generator minimum generation limit constraint:

$$(\tau_i^{\min}) \quad P_i \geq P_i^{\min}$$

Generator joint maximum generation limit constraint:

$$(\tau_i^{\max}) \quad P_i + S_i + O_i \leq P_i^{\max}$$

Price-responsive load dispatch range constraint:

$$(\eta_i^{\max}) \quad 0 \le D_i \le D_i^{\max}$$

Generator ramp-rate limit constraint:

$$(\phi_i) \quad |P_i - P_i^0| \le RR_i^{\max}$$

where Greek labels are corresponding constraint shadow prices

c, cs, co, cd: Cost or benefit associated with unit production, 10-minute reserve, 30-minute reserve, and demand bids

P_i, D_i, S_i, O_i: Decision variables for generation dispatch, load dispatch, spinning reserve dispatch, and operating reserve dispatch of resource i

$P_i^{\min}, P_i^{\max}$: Min and max dispatch limit for generator i

P_i^0: Initial generation output for generator i

$D_i^{\max}$: Maximum dispatch limit for load i

FD, P_L: Fixed demand and transmission losses

d_i: Load distribution factor for fixed demand

$S^{\max}, O^{\max}$: Ten- and thirty-minute reserve requirement

$L_l^{\max}$: Grid security limit for constraint l

$a_{l,i}$: Sensitivity of constraint l with respect to bus i

The optimal solution to this SCED problem determines the market clearing quantities and market clearing prices by location. The market clearing prices by location, called locational marginal prices, and AS market clearing prices are by-products of the optimization solution. Under the joint optimization based market for energy and multiple ancillary services, the market clearing prices for the multiple products have the following important characteristics:

- Locational marginal prices for energy: $LMP_i = \lambda - \lambda \dfrac{\partial P_L}{\partial P_i} - \sum_l a_{l,i} \mu_l$ The three terms in the above LMP equation could be interpreted as the three components of LMP, namely energy, loss, and congestion, respectively. These locational price results give precise representation of the cause-effect relationship that is consistent with grid reliability management.

- Higher prices for higher quality of ancillary services
  - Market clearing pricing for 10-minute reserve: $\rho_S = \gamma_s + \gamma_o$
  - Market clearing pricing for 30-minute operating reserves: $\rho_O = \gamma_o \le \rho_S$

- Marginal equity between energy and reserve prices:
  - $LMP_i - \dfrac{\partial C(P)}{\partial P_i} = \rho_S$, if $S > 0$
  - $LMP_i - \dfrac{\partial C(P)}{\partial P_i} = \rho_O$, if $S = 0$ and $O > 0$

- LMPs for energy are always higher than market clearing prices for ancillary services except that there exists ancillary service capacity shortage. This can be seen from the above equations.

These price relationships are essential to encouraging rational participants' responses to market signals. But these price relationships are true only under rigorous mathematical terms. Therefore, adopting a formal optimization technology in market clearing is central to market efficiency and grid reliability.

2.5.2 Optimization-Based Unit Commitment

2.5.2.1 Market-Oriented Unit Commitment Problem
Unit commitment determines unit startup and shutdown schedules that meet forecast demand at minimum startup cost and energy production cost. Traditional unit commitment problems in general did not explicitly take into account grid security constraints in the commitment process. Even with the simplified modeling assumption, finding an optimal solution to unit commitment has been a challenging task due to its nature of dimensionality.

Many existing electricity markets, such as PJM, ISO-New England, New York ISO, and Midwest ISO, allow three-part bids including startup cost, no load cost, and energy cost components. These markets also adopt a two-settlement market structure that consists of a day-ahead market and a real-time market. The day-ahead market is a financial market that determines the minimum cost commitment schedules based on participants' bid-in fixed demand, price-sensitive demands, virtual transactions, external interchange schedules, and transmission security constraints. Because many trading activities of the day-ahead market are used as mechanism to hedge financial risks in the real-time market, participants' bids for demands and virtual transactions can deviate significantly from the physical grid operation. To guarantee grid reliability, a reliability commitment is required after day-ahead market is cleared. In the reliability commitment, the units committed in day-ahead market remain committed. Additional units required to meet forecasted demand and reserve capacity requirements are committed in the form of reserve capacity by minimizing the startup and no load cost components of available units that are not commitment in the day-ahead market.

The market-oriented unit commitment problem becomes more complex compared to the traditional unit commitment problem [11]. The unit commitment schedules are applied in the SCED-based market clearing for determination of day-ahead settlement quantities and prices. From a market efficiency perspective, those units committed in the day-ahead market process should be able to collect sufficient financial proceeds from day-ahead market settlement to cover their production cost based on the day-ahead market clearing quantities and prices. With this result, uplift charges resulting from cost compensation for market committed units are reduced, cost socialization is minimized, and market efficiency is improved. More importantly, any units that would be able to make a profit in terms of the day-ahead market clearing prices, if they were committed, should not be left uncommitted in the day-ahead market solution. When a generator that could be profitable is not committed, a rational explanation must be presented to the generator. This is important for creating a fair and transparent market that encourages rational bidding behavior complying with grid reliability requirements.

The key to fair and transparent market that involves day-ahead unit commitment is to include grid security constraints in the commitment solution. As such, SCUC has become an integral part of many day-ahead markets. Successful solution to the SCUC problem presents great technical challenges, due to its requirement on global optimality and the need to deal with large-scale commitment problems of 1000 to 3000 units over a two- to seven-day time horizon with grid models of over 20,000 buses.

In addition to the requirements of commitment optimality and the ability to deal with large-scale problems, requirements on new functional capabilities increase. Hydrothermal coordination needs to be handled more effectively and consistently. Combined-cycle units need to be modeled to reflect the coupling effect of gas turbines and steam components.

In meeting the challenges, great efforts have been made and significant progress has been achieved in advancing unit commitment techniques.

2.5.2.2   Advances in Unit Commitment Methods

The unit commitment problem is a large-scale combinatorial scheduling problem. For N units over M time periods there are a maximum of $(2^N - 1)^M$ combinations to evaluate. For a small problem where $N = 5$ and $M = 24$, there are $(s^5 - 1)^{24} = 6.2 \times 10^{35}$ possible combinations! Solving even this small problem by direct enumeration is clearly an intractable approach.

Historically, unit commitment algorithms have focused on heuristic methods that intelligently reduce the number of combinations that need to be evaluated. A popular method is to assume that units are restricted to some fixed startup/shutdown order. This assumption reduces the number of possible combinations to $2^N \times M$, or for our example above to $2^5 \times 24 = 768$ combinations. This new problem can be solved optimally using a dynamic programming algorithm.

Later on, Lagrangian relaxation (LR) algorithms were also utilized. The LR approach decomposes the problem by unit. Each single unit problem can be optimally solved using a small single-unit dynamic programming (SUDP) algorithm. A coordination module is used to adjust parameters (Lagrange multipliers) sent to the SUDP to achieve feasibility of system constraints. This approach implicitly evaluates many more combinations than is possible using priority-based methods. The LR decomposition provides a duality gap, which is a measure of how far the current solution is from the estimated lower bound (the theoretical minimum).

Early LR algorithms had numerous convergence problems. Later a "sequential bidding" method was developed, which utilized many of the LR components, but avoided convergence issues. This method creates a priority order that is determined dynamically by the algorithm, rather than a priori. This worked well for non-market-based systems, and provided additional capabilities for fuel and emissions constraints.

Early experiences with the market-oriented unit commitment made it clear that priority-based methods were not appropriate for market clearing purposes. Temporal changes in both unit offers and availability effectively invalidate the assumption of using a fixed priority over all study periods. The requirements for optimal hydro-thermal coordination and explicit inclusion of many grid security constraints created nearly insurmountable difficulties for the traditional unit commitment algorithms, such as priority-based method and classic dynamic programming method.

The LR method was a commonly used algorithm for market based unit commitment. In some practical market applications, LR algorithms provided significantly improved results that typically have duality gaps of less than 1%, indicating near-optimum schedules. It is interesting to note that the LR decomposition is analogous to an economic market simulation. The LR coordinator can be viewed as the market clearing function, and the SUDP as the participant bidding mechanism. Based on current estimates of system prices, the SUDP determines individual unit bids that maximize each unit's profit. The LR coordinator clears the market and commits additional units as necessary to meet system constraints. The solution iterates to achieve a minimum cost solution in much the same way as the market clearing process is repeated several times each day. As a result, the solutions from the LR tend to satisfy the participants, since their profits are maximized, while also producing secure and minimum cost system results.

The LR method was successfully applied in clearing large-scale market of over 1000 units. But the LR method was frequently challenged due to its inability to prove that it has found a global solution. In search of a global optimal solution for the unit commitment problem, attempts were made decades ago to use mixed integer programming (MIP) method without success.

In recent years there have been significant improvements in MIP algorithms. These MIPs utilize a "branch and bound" method that can implicitly evaluate all combinations, which enables MIP to ascertain global optimality. This claim of a globally optimal solution cannot be made for any of the other methods.

Since its development in the early 1970s, the branch-and-bound method had suffered from poor and unpredictable performance when applied to practical sized mixed integer problems. However, the situation changed significantly during the last few years. A major contributor to MIP performance improvement is the drastic performance improvement of the LP, which is at the core of any MIP algorithm. In addition to improvements due to LP, there are a large number of schemes that have been developed over time for improving the intelligence of the branch and bound search logic. Although there was not a single monumental breakthrough in branch-and-bound, these different schemes collectively provide a tool-kit approach for tailoring to specific MIP problems. The use of MIP solvers is appealing in that they eliminate numerous heuristics utilized in other approaches, and theoretically allow for inclusion of complex constraints that are difficult to deal with using other methods.

Commercial MIP solvers, such as CPLEX, have been consistently showing significant performance improvement in recent releases. It should also be noted that Branch and Bound algorithms can take advantages of parallel processing, which may be used in longer term planning areas.

MIP has been successfully applied to market based unit commitments with over 1000 units for both day-ahead market clearing and the reserve adequacy (RA) process [39].

2.5.2.3 SCUC Example Problem: Reliability Commitment While DA market produces financially binding schedules for the next operating day, these schedules may not provide an adequate resource plan for physical system operation. RA analysis must be performed to generate resource operating plans that ensure secure and reliable operation of the power system. Additional generation capacity may have to be committed for reliability purposes to bridge the gap between the capacities as seen from financially binding day-ahead schedules and the capacities needed to meet RTO's forecast of the physical demand.

Although the RA and DA market clearing processes share many common characteristics, such as the conventional UC problem definition, these are two fundamentally different business processes and have important differences in their respective problem definitions. The RA process not only needs to respect physical system requirements at least cost, but also must minimize any interference or distortion to the real-time economic price signals. The latter is particularly affected by the physical resource commitment schedules produced by the RA process. If the demand bids cleared in the day-ahead market should fall short of the forecast/actual demand, then real-time energy prices must be allowed to rise. The rise of real-time energy price would be the result of additional energy produced by units that were committed by RA. These are low-cost providers of operating reserve but not necessarily energy. The decision criteria in RA commitment is based on start-up and cost of minimum generation, but not on incremental energy cost.

The transition from three-part energy bids in the DA market to the two-part energy bid in the RA process is a relatively simple function from the application software and optimization method perspective. However, this fundamental transition in problem definition from DA to RA leads to additional important and related aspects of RA definition. The principal areas of continued RA problem definition centers on the issues of: (a) modeling and analysis of transmission security and incremental capacity commitment; and (b) RA input data preparation (interpretation and extrapolation). These are indicative of the challenge for the RA analysis; that is, the transition between market operation and physical operation, which is essential for reliable operations. Thus, it is naturally subjected to significant redefinition as the market continues to evolve.

The early and principal focus of the RA process was to ensure that the RTO system would have sufficient incremental capacities committed, in the form of operating reserves, to meet forecast demand. In that model, the energy balance in the SCUC was maintained at the level of cleared DA demand bids and thus the resulting power flows on the transmission system does not reflect the level commensurate with the energy demand forecast.

The growing need for further improvement over the earlier RA process clearly points in the direction of more accurate modeling and analysis of transmission security constraints. This direction implies potentially fundamental augmentation of the RA business requirements. More than system-wide incremental capacity commitment process, the RA commitment decision will need to better recognize the impacts of transmission security. This requires increased emphasis on energy scheduling, including possible redispatching MW from the reference DA schedules, which in turn requires modeling energy cost and energy demand. On the other hand, it is also fundamental that committing for reserves (based on two-part bids of startup and min-generation cost but not energy cost) remains the primary business objective for RA. How to navigate through seemingly conflicting requirements to produce useful and consistent definition of the RA problem is important and challenging. Close co-operation with RTO customers is the key to resolving the challenges confronting the market operators and security coordinators.

2.5.2.4   SCUC Performance Consideration

All markets have definitive time-lines for specific market events. To provide sufficient flexibility and margin, the market-clearing applications need to be completed in less than half of the total available time. This can be quite challenging given the size and complexity of the problem. The adopted system architecture for solving the SCUC problem shall be able to withstand the challenges. Such a system design is shown in Figure 2.8. Some of the key system features include the following:

- *Increase speed by plugging additional processors.* For the seven-day SCUC RA analysis, 168 SFT solutions must be conducted (one for each hour), and each SFT solution involves the analysis of thousands of power flows depending on the number of contingencies. Furthermore, several iterations between SFT and SCUC may be needed to reach a secure commitment solution. The amount of SFT computation is tremendous. Since the MW dispatches for each of the 168 hours are available simultaneously from the SCUC solution, the hundreds of thousands of SFT power flow analyses may be performed in parallel to reduce the SCUC-RA problem solution time. As system size increases and requirements change, this system architecture would provide the capability to scale the system by plugging in more processors to meet performance requirements.

- *Improve performance through efficient data management.* For the seven-day SCUC RA analysis, the volume of data flowing between SFT and SCUC is huge when hundreds of security constraints may be detected. Generating the vast volume of topology and security constraint sensitivity data not only takes

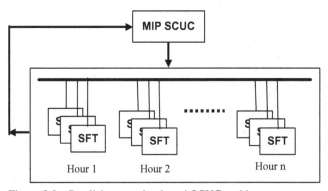

Figure 2.8   Parallel processing based SCUC architecture

up computer disk resources and system support personnel's burden to manage the data, it also decreases the performance of the analytical engines. To alleviate this problem, network topology for each of the hours is identified in advance so that hours with the same topology are grouped and one network topology is produced for each group of the same topology hours. This typically reduces the data volume dramatically.

As many RTO markets may grow in size and expand to provide more market products, performance, integration, and manageability are the critical system issues for the continual growth in size and complexity of the markets. This SCUC system architecture can better meet these challenges.

With the integration of the transmission network security analysis in the RA unit commitment, the SCUC commitment and dispatch MW solutions from one iteration are fed to the SFT application for security analysis. Network security violations for each hour detected by SFT are formulated as sensitivity-based constraints to be enforced in the next pass SCUC commitment solution. This iteration continues until no new violation is found in a commitment solution for any of the study hours. The process guarantees a reliable and secure commitment solution with minimum heuristics and operator's intervention.

This SCUC schema can handle the circumstances where detailed SFT analysis for all 168 hours in the RA study is required and parallel SFT processing may become necessary. However, flexibility will be built into the SCUC system that allows the operator to perform the RA commitment task as practically needed:

- Predefine security constraints for preemptive enforcement
- Perform SFT analysis for selected hours
- Control the iteration process between SFT and SCUC

2.5.3   System Implementation

In a practical market management system (MMS), there are not only the core optimization–based market clearing engines, but many other indispensible components. A good design for system integration should provide enough flexibility for future expansion while satisfying business functional requirements.

In recent years, the IEC common information model (CIM) and the service-oriented architecture (SOA) have become the trend for RTO/ISO system integration. Figure 2.9 shows a high-level example of SOA-based RTO system integration.

The key service components in Figure 2.9 are described below:

- Enterprise service bus (ESB) is the backbone of a SOA system. It provides high-performance data transfer between service components and facilitates comprehensive workflow management. Message definition for ESB data transfers should be CIM compliant in order to allow easy integration of a third party application service.
- Common source modeler (CSM) provides CIM-based network and commercial models for both EMS and MMS. It ensures model consistency between different services and significantly reduces RTO model maintenance efforts.

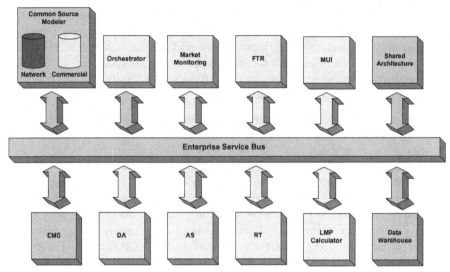

Figure 2.9   SOA-based system integration

- Shared architecture (SA) consists of a set of shared services that are available to all other business service components (BSC), for example, common logging service, etc.
- An orchestrator provides overall process control and coordination over the ESB.
- EMS. State estimation (SE) and real-time contingency analysis (RTCA) provide market systems with real-time network topology, generation and load MW values, and real-time transmission security constraints. On the other hand, automatic generation control (AGC) takes the unit base points from the real-time SCED solution for generation control.
- DA/AS/RT/FTR. Common market functional components for a RTO/ISO.
- MUI (market user interface) is for participants to submit bids and offers and for RTO to publish market clearing results.
- Data warehouse stores all market clearing related historical data for multiple years.
- Market monitoring is the post-market analysis to detect potential market power. It often requires the same core market clearing engines that are used to clear the markets and other statistical tools. Market monitoring system can access all historical market data stored in the data warehouse for play back, market simulation, and scenario analysis.

2.5.4   Future Direction

In order to meet the requirements from new electricity markets and new functional expansion in the existing markets, more innovation and development on the computation applications are expected in the many areas, including the following:

- *Modularization.* The optimization-based market computation applications should be modularized using the object-oriented design (OOD) approach, such that the common modules can be used to construct new market functions. For example, there are many pieces in common between SCED and SCUC. In particular, for intra-day resource scheduling processes (that fill the gap between DA and RT), the line between SCUC and SCED is becoming blurred. Therefore, it is important to modularize SCUC and SCED in order to build various RTO intra-day resource scheduling processes.

- *Performance.* With the increasing size of market footprints and more and more participation from the demand side, the optimization problems are becoming larger and more difficult to solve. The computational performance requirements for market applications can only be higher in the future, in order to obtain market clearing solutions in a timely manner. The latest enhancements on MIP and LP from the academic fields need to be applied to improve the performance of market applications.

- *New algorithms.* Besides the deterministic approaches that are used in the current market clearing functions, uncertainties and reachability have to be modeled effectively for long-term planning and look-ahead scheduling processes. It is expected to see more practical application of new algorithms, such as stochastic programming and robust programming, to be applied in market applications.

- *Coordination of multiple scheduling processes.* The market-based resource scheduling processes could have a wide time range, from multi-year planning to 5-minute real-time dispatch with various scheduling intervals. The system should be design in such a way that all the scheduling processes can seamlessly work together to help the operators to manage the markets with more complete consideration of grid security.

2.6   FINAL REMARKS

The movement of restructuring has transformed the power industry. Competitive electricity market is no longer a myth. We have seen many successful markets in operation and sharing of valuable experiences across the globe. While the pace of new market implementation might appear to have slowed down some, we expect power industry reform to continue unabated and become even more broad-scaled. Major drivers for future changes could include:

1. Financial drivers for liquidity, credits, and long-term investments
2. Environmental drivers for energy conservation, emission reduction, and inter-modal energy portfolio management
3. Business drivers: enterprise asset management, demand resource, long-term contracts
4. Technology drivers: smart devices, IT advances
5. Others: regulatory realignment

If the past decade of electric utility reform is any indication, we can all look forward to another decade of tremendous opportunities for the energy professionals.

REFERENCES

1. Northeast Power Coordinating Council. [Online] http://www.npcc.org.
2. Operating Policies and Planning Standards. North American Electric Reliability Council, 2004. [Online] http://www.nerc.com.
3. Schweppe FC, Caramanis MC, Tabors RD, Bohn RE. *Spot Pricing of Electricity*. Boston: Kluwer Academic Publishers; 1998.
4. Ilic M, Galiana F, Fink L, eds. *Power System Restructuring: Engineering and Economics*. Boston: Kluwer Academic Publishers; 1998.
5. Hogan W. *Competitive electricity market design: a wholesale primer*. J. F. Kennedy School of Government, Harvard University, 1998. [Online] http://ksghome.harvard.edu/~whogan/.
6. Wu T, Rothleder M, Alaywan Z, Papalexopoulos AD. Pricing energy and ancillary services in integrated market systems by optimal power flow. *IEEE Transactions on Power Systems* 2004;19(1):339–347.
7. Day ahead scheduling operations manual. New York Independent System Operator NYISO. Available online at http://www.nyiso.com/.
8. Scheduling operations manual. Pennsylvania-New Jersey-Maryland (PJM) Interconnection. Available online at http://www.pjm.com/.
9. Market operations. Independent System Operator (ISO) New England. [Online] http://www.iso-ne.com/.
10. CAISO 28-Jun MD02 Tariff Filing of Clean Tariff Sheets and Errata. [Online] http://www.caiso.com/.
11. Cohen AI, Brandwajn V, Chang SK. Security constrained unit commitment for open markets. In: *Proc. 21st Int. Conf. Power Ind. Comput. Applicat.*, May 16–21, 1999, pp. 39–44.
12. Cheung K, Shamsollahi P, Sun D, Milligan J, Potishnak M. Energy and ancillary service dispatch for the interim ISO New England electricity market. *IEEE Transactions on Power Systems* 2000;15(3):968–974.
13. Texas nodal market implementation. ERCOT. [Online] http://www.ercot.com/.
14. Wood AJ, Wollenberg BF. *Power Generation, Operation, and Control*, 2nd edition, New York: John Wiley & Sons; 1996.
15. DyLiaco TE. The adaptive reliability control system. *IEEE Transactions on Power Apparatus and Systems* 1967;86(5):517–528.
16. Schulz RP, Price WW. Classification and identification of power system Emergencies. *IEEE Transactions on Power Apparatus and Systems* 1984;103(12):3471–3479.
17. Impacts of the Federal Energy Regulatory Commission's proposal for standard market design. US Department of Energy Report to Congress, April 30, 2003.
18. An introduction to Australia's national electricity market. National Electricity Market Management Company, Ltd. [Online] http://www.nemmco.com.au.
19. Bergen AR, Vittal V. *Power System Analysis*, 2nd edition, Upper Saddle River, NJ: Prentice Hall; 1999.
20. Wilson RB. Architecture of power markets. *Econometrica* 2002;70(4):1299–1340.
21. Ma X, Sun DI, Cheung KW. Evolution toward standardized market design. invited paper. *IEEE Transactions on Power Systems* 2003;18(2):460–469.
22. Rao NS. Optimal dispatch of a system based on offers and bids—a mixed integer LP formulation. *IEEE Transactions on Power Systems* 1999;14(1):274–279.
23. Castronuovo ED, Campagnolo JM, Salgado R. New versions of interior point methods applied to the optimal power flow problem. In: Proceedings of IEEE/PES T&D, 2002. Also available online at http://www.optimization-online.org/DB_HTML/2001/11/405.html.
24. Cramton P. Electricity market design: the good, the bad, and the ugly. In: Proceedings of the Hawaii International Conference on System Sciences, January 2003.

25. Allocation of uplift costs to load and other entities associated with virtual trading. Technical Bulletin 82, New York ISO, 2001.

26. Ma X, Sun DI, Rosenwald GW, Ott AL. Advanced financial transmission rights in the PJM market. In: Proceedings of IEEE Power Engineering Society General Meeting, July 2003.

27. Shamsollahi P, Cheung KW, Chen Q, Germain EH. A neural network based very short term load forecaster for the interim ISO New England electricity market system. In: Proceedings of 2001 Power Industry Computer Applications Conference, May 2001, pp. 217–222.

28. Cheung KW. Ancillary service market design and implementation in North America: from theory to practice. In: Proceedings of the Third International Conference on Electric Utility Deregulation and Restructuring and Power Technologies, DRPT 2008, April 2008, pp. 66–73.

29. Cheung KW. Functional design of ancillary service markets under the framework of standard market design for ISO New England. In: Proceedings of the 2006 International Conference on Power System Technology (POWERCON 2006), Chongqing, China, 22–26 Oct 2006.

30. Chow JH, deMello R, Cheung KW. Electricity market design: an integrated approach to reliability assurance. Invited Paper for the Special Issue on Power Technology & Policy: Forty Years after the 1965 Blackout. *IEEE Proceeding* 2005;93(11):1956–1969.

31. Sun D, Ma X-W, Cheung KW. The application of optimization technology for electricity market operation. In: Proceedings of the 2nd IEEE/PES Transmission and Distribution Conference & Exhibition Asia Pacific, Dalian, China, 2005.

32. Cheung KW. Standard market design for ISO New England wholesale electricity market: an overview. In: Proceedings of the 2nd International Conference on Electric Utility Deregulation, Restructuring and Power Technologies (DRPT 2004), Hong Kong, 4–7 April 2004.

33. Cheng DTY. Economic analysis of the electricity market in England and Wales. *IEEE Power Engineering Review* 1999;19(4):57–59.

34. Flatabo N, Doorman G, Grande OS, Randen H, Wangensteen I. Experience with the Nord Pool design and implementation. *IEEE Transactions on Power Systems* 2003;18(2):541–547.

35. Alvey T, Goodwin D, Ma X, Streiffert D, Sun D. A security-constrained bid-clearing system for the New Zealand wholesale electricity market. *IEEE Transactions on Power Systems* 1998; 13(2):340–346.

36. Market design review. Electricity Commission of New Zealand. [Online] http://www.electricitycommission.govt.nz

37. Ott AL. Experience with PJM market operation, system design, and implementation. *IEEE Transactions on Power Systems* 2003;18(2):528–534.

38. Sun DI, Ashley B, Brewer B, Hughes BA, Tinney WF. Optimal power flow by Newton approach. *IEEE Transactions on Power Apparatus and Systems* 1984;103(10):2864–2880.

39. Streiffert D, Philbrick R, Ott A. Mixed integer programming solution for market clearing and reliability analysis. In: Proceedings of the IEEE Power Engineering Society General Meeting, vol. 3, pp 2724–2731.

40. Lesieutre BC, Hiskens I.A. Convexity of the set of feasible injections and revenue adequacy in FTR markets. *IEEE Transactions on Power Systems* 2005;20(4):1790–1798.

CHAPTER 3

OVERVIEW OF ELECTRICITY MARKET EQUILIBRIUM PROBLEMS AND MARKET POWER ANALYSIS

Xiao-Ping Zhang

3.1 GAME THEORY AND ITS APPLICATIONS

Game theory aims to help us understand situations in which decision makers interact [1]. Game theory is a branch of applied mathematics covering applications, for instance, in economics, social or behavioral science, biology, engineering, political science, computer science, and philosophy, and the scope of game theory is very broad. Game theory can help to mathematically capture behavior in strategic situations where an individual's success in making choices depends on the choices of others.

The major development of the ideas of game theory took place in the 1920s, particularly through the work by Émil Borel and John von Neumann. In 1944, John von Neumann and Osakar Morgentern published the famous book entitled *Theory of Games and Economic behaviour*, which laid the foundations of game theory. The next significant development of game theory was made in the 1950s by John F. Nash, who developed the key concept and mathematical framework of Nash equilibrium. He shared the 1994 Nobel Prize in Economic Sciences with John C. Harsanyi and Reinhard Selten. The development of game theory can be described by the diagram shown Figure 3.1.

Along with the development of the game theory and its applications, the application of game theory in electricity market modeling and analysis flourished and bore fruit. In 1999, IEEE Power Engineering Society published the *Tutorial on Game Theory Applications in Electric Power Systems* [2], which can be considered as a landmark for the application of electricity market modeling and analysis. During past 10 years or so, we have seen continuous applications of game theory in electricity market problems. The trend has also been reflected by a large number of journal and conference publications.

Restructured Electric Power Systems: Analysis of Electricity Markets with Equilibrium Models,
Edited by Xiao-Ping Zhang
Copyright © 2010 Institute of Electrical and Electronics Engineers

A. Augustin Cournot (1838), Joseph L. F. Bertrand (1883), Edgeworth (1925): oligopoly pricing and production

John von Neumann and Osakar Morgentern (1944): the publication of the book *Theory of Games and Economic Behaviour*, foundations of game theory

John F. Nash (1950): the concept of Nash equilibrium

John F. Nash (1951): Nash bargaining solution for coalitional games

Lloyd Shapley (1953): the concept of the "Shapley value" and the "core" as solutions to coalitional games

H.W. Kuhn (1953): extensive games with "imperfect information"

William Vickrey (1961): the first formalization of "auctions"

R. Duncan Luce and Howard Raiffa (1957): iterated elimination of dominated strategies method for strategic normal games and the concept of "repeated game"

Reinhard Selten (1965) developed the concept of a "subgame perfect equilibrium" as a refined solution for extensive form games

John C. Harsanyi (1967/68): the concept of a "Bayesian Nash equilibrium" for Bayesian games (games with incomplete information)

Robert J. Aumann and Martin Shubik (1960s): cooperative game theory applications; solution concepts for coalitional games and early statements of the "folk theorems"

David Schmeidler (1969): the "Nucleolus" solution for coalitional games

Robert J. Aumann (1974): correlated equilibrium" for Bayesian games

John C. Harsanyi (1973): the concept of a "mixed strategy"

Reinhard Selten (1975): "trembling hand equilibrium" for Bayesian games

Robert J. Aumann (1976): the concept of "common knowledge"

Herbert Simon (1978): Nobel Prize in Economic Sciences for his pioneering research into the decision-making process within economic organizations.

Thomas C. Schelling (1960, 1981): evolutionary games

Ariel Rubinstein (1982): transformed the co-operative Nash bargaining solution into a noncooperative strategic extensive game of sequential bargaining

David M. Kreps and Robert Wilson (1982): the concept of "sequential equilibrium" for extensive games with imperfect information

B.D. Bernheim and D.G. Pearce (1984): the concept of "rationalizability"

Elon Kohlberg and Jean-François Mertens (1986): the concept of "forward induction" for extensive games

D. Fudenberg and E.S. Maskin (1986): "perfect folk theorems" for infinitely repeated games

J.C. Harsanyi and R.Selten (1988): "equilibrium selection" for any type of game

D. Fudenberg and Jean Tirole (1991): the "Bayesian perfect equilibrium" for extensive Bayesian games

John Nash, J.C. Harsanyi and R. Selten (1994): Nobel Prize in Economic Sciences "for their pioneering analysis of equilibria in the theory of non-cooperative games"

William Vickrey and James Mirrlees (1996): Nobel Prize in Economic Sciences "for their fundamental contributions to the economic theory of incentives under asymmetric information"

Figure 3.1   The development of game theory

3.2   ELECTRICITY MARKETS AND MARKET POWER

3.2.1   Types of Electricity Markets

3.2.1.1   *Bid-Based Auction Pool / PoolCo / Spot Market*   The pool electricity market is based on the principle of centralized power dispatch. It can be either

mandatory, meaning that all power trade should be performed within the pool, or voluntary, where the main volume of power is traded with bilateral contracts and the pool transactions are restricted for the spot market for the last minute dealing, covering for imbalances or demand gaps arising from mechanical faults. In both cases the electricity market participants submit price-quantity bids to the pool, specifying the amount of power that they are willing to trade. If a seller bids too high it may not be able to sell power through the pool. On the other hand, if a buyer bids too low it may not be possible to acquire any amount of power [20]. If the market allows wholesale competition only, the sellers are generating firms and the buyers are distribution companies that resell power to the consumers. If the market allows retail competition, buyers could also be individual customers from the industry, suppliers, retailers, or other large consumers [3].

When the bidding period is over, the market operator governing the pool, based on the bids and considering demand, would implement the economic dispatch algorithm based usually on a social welfare maximizing software to produce a single spot market clearing price for electricity for that period [4]. The most popular market clearing mechanism is the simple matching algorithm. The market operator firstly chooses the generating quantity with the lowest price and then utilizes, continuing by choosing the generation quantity with the next lowest price and so on, until meeting the total demand. When demand bidding is allowed, a similar procedure is followed. The market operator ranks them in a demand curve with decreasing price. The winning sellers and buyers can be represented by the supply and demand curve. The system-wide market clearing price is usually the price of the last generator scheduled to supply power in that period interval. After the market clearing price has been established, the system operator reviews the scheduling concluded by the market operator and checks the power flows to ensure the system stability and the security of supply. In certain cases it is possible to modify the dispatch to provide some spinning reserve or relief transmission congestion in the network. In some countries the market operator and the system operator is under the same independent regulatory body [3].

The nature and complexity of the electricity power transfer does not offer the opportunity of a true spot market for immediate electricity delivery. Instead, the electricity spot markets are operating some time in advance, usually one day ahead or some hours ahead of actual dispatch. The actual imbalances resulting at the last minute prior to dispatch are handled using noncompetitive market rules set by the market operator.

3.2.1.2 Bilateral Agreements, Forward Contracts, and Contracts for Differences
Bilateral contracting proved to be one of the most efficient ways of trading, being a standard trading technique in all markets. In markets in which bilateral trading is allowed, the generating firms can individually negotiate quantities and prices with buyers or even directly with the consumers, purchase transmission services from the system operator, and then form contracts with each other at the agreed prices, terms, and conditions. All bilateral transactions have to be reported in advance to the system operator that analyzes the bilateral trading in each period and determines without discrimination under certain rules which

contracts are feasible to dispatch according to electricity network transmission constraints [3].

Bilateral trading can be facilitated either through short-term contracts or forward contracts or even future contracts. Forward contracts are very popular for securing a profitable price for both the seller and buyer. They can agree well in advance from the delivery period, maybe weeks, months, or even years, on the quantity and price of trading power, but also on the penalties if either of the parties fails to honor the commitment. The price and quantity are valid at the time of physical delivery independent of to the price of power at the spot market at that time. Forward contracts can be bought or sold among the marketers. Some market participants may secure forward contracts for a certain price and then, at a later stage, they will try to sell them to other market participants at a higher price. The trading of forward contracts (future contracts) is termed future market [7].

Bilateral trading is considered to be a more flexible trading approach because, unlike the centralized pool auction market, it can coexist with nonmandatory pool market. On the other hand, bilateral trading has been criticized for two reasons. Firstly, because bilateral trading is not compatible with centralized optimization system dispatch and secondly because prolonged long-term bilateral contracts can give incentives to strategic firms to exercise market power [5].

In electricity markets in which trading through the pool is mandatory, the price volatility in the spot market has given incentives to buyers and sellers of electricity to discover an alternative way of trading through the pool that guarantees lower market uncertainty. This new idea is termed *contracts for differences*. Since trading in the pool, in this case, is mandatory, the buyer and seller, prior to bidding, guarantee a fixed quantity at a fixed price of power to each other. The seller of power guarantees to provide to the buyer a certain amount of power at specified price (strike price) and the buyer is committed to buy this amount of power at that price, over the terms of a contract. Unlike in bilateral contracts, since the power has to be acquired through the pool, there is no physical interaction between the seller and the buyer [3]. The seller and the buyer take part in the centralized pool market like all the other market participants. Once the trading price is set in the pool for that period and market is cleared, if the strike price agreed to by the two parties in the contract is higher than the centralized market clearing price in the pool then the buyer pays the difference between the two prices to the seller. On the other hand, if the strike price is lower than the market clearing price then the seller pays the difference of the two prices to the buyer [7]. In this way both the sellers and the buyers can feel more secure against the price volatility of the spot market.

3.2.2 Competition Types

3.2.2.1 Perfect Competition
Perfect competition suggests that all market participants act as price takers and the market clearing price is set by the interaction of buyers and sellers. From the economic theory, the interaction among buyers and sellers results in a market price that is equal to the cost of producing the last unit sold. This is the most economically efficient solution [3]. Thus, for the case of the electricity market, each generating firm can increase production until the marginal

cost of generation is equal to the market clearing price [7]. This competition type models the ideal conditions for trade when there is fair trade between the producers and the consumers and no economic advantage is gained by any of the participants. The condition of competition in the electricity market is far from the competitive ideal case [8]. Jonathan Falk in [9] suggests: "Perfect competition represents an ideal state, like frictionless motion in physics. Real world markets in every commodity have frictions that cause prices to diverge from the ideal." Perfect competition assumes competition among a large number of competitive participants, whereas any firm can satisfy the whole demand or a large proportion in the market by simply offering lower prices than the rival firms. This cannot be applied in the electricity market, in which there are strong operational constraints.

3.2.2.2 Imperfect or Oligopolistic Competition Inevitably, in the electricity market there is a limited number of generating firms due to operational and capital constraints, which try to maximize their individual profit. Many of the generating firms control a large share in the electricity market, thus being able to control the market by acting in an uncompetitive manner. For example, they can withhold production to create a shortage of supply or take advantage of their strong presence in the market and raise the selling price. This implies that the electricity market is strictly based on imperfect competition. The strategic generating firms carefully consider the influence of their actions on the market in order to maximize the individual profits. For example, they should examine what the effect would be on the amount of power they sell if they increase the selling price or how the amount of power offered to the market would affect the market price. Provided that the exercise of the market power is successful it will result in a significantly higher market price than in perfect competition equilibrium, resulting in higher profits and less production [10]. In certain cases when a small number of large firms dominate in the market, they often agree among themselves as to the policies they follow for their mutual benefit [8]. This situation, where the firms do not act as price takers and try to increase the market price above competitive levels through their actions, is termed *imperfect competition* or *oligopolistic competition*.

3.3 MARKET POWER MONITORING, MODELING, AND ANALYSIS

3.3.1 The Concept of Market Power

The concept of perfect competition is assumed when the number of market participants is large enough to avoid having any strategic firm controlling a relatively large share in the market. Unfortunately, due to the complex nature of the electricity market, there are not enough market players in order to have a pure competitive market. Thus, unavoidably, the electricity market, after liberalization, is considered to be of an oligopolistic nature. Due to the fact that the primary scope of the strategic generating firms is to maximize their individual profits, each firm tries to influence the market clearing price, to rise above competitive levels, through individual strategic actions [6, 7]. The ability of a strategic firm to alter profitably the prices above competitive levels is defined as "market power."

Market power is usually exercised by strategic generating firms, either by asking for a higher price than the competitive levels or by withholding the generating capacity below the competitive demand levels. If such an action is successful, it would result in a higher market price, higher profits, and withheld generation output [10].

As a result, market power abuse by large strategic firms may be harmful for competition in the electricity market. It is imperative then, to develop the necessary tools to identify the levels of market power exercised by strategic firms.

3.3.2 Techniques for Measuring Market Power

Over the past 15 years several different approaches have been developed for assessing market power abuse by strategic firms and for measuring the competitiveness of the market. The following four most widely used techniques are discussed in [11]:

- Price-cost margin index
- Analysis of market share concentration, that is, the Herfindahl-Hirschan index (HHI)
- Estimation of pricing behavior through simulation analysis
- Oligopoly equilibrium analysis

3.3.2.1 The Price-Cost Margin Index The price-cost margin index is theoretically the most reasonable and straightforward technique, but practically is not efficient because the marginal costs of the strategic firms are classified as confidential information making the analysis very difficult.

3.3.2.2 The Herfindahl-Hirschan Index The HHI method involves the analysis of the structure of the electrical network in which the market share of each strategic firm should be analyzed along with the market concentration. This is usually performed with the aid of indices classifying the market concentration. The most widely known index is the Herfindahl-Hirschan Index (HHI), a measure of market concentration, which is defined as the sum of the squared market shares [12]. Unfortunately, this method has no supporting theory and is supported only by a rule of thumb [9]. In addition, as discussed by several researchers [12–14], the HHI method is considered inappropriate because it fails to take into account the elasticity of demand that directly affects the market outcome in the oligopolistic competitive environment. This is also accepted by the Federal Energy Regulation Committee (FERC) in the US, which has used this evaluation method for measuring the market shares in the past. The HHI is considered an indicator of potential market power. It is suggested that HHIs below 1000 are classified as unconcentrated markets while HHIs above 1800 are classified as highly concentrated markets. Basically, the HHI is a static index where dynamic market effects such as strategic bidding and capacity withholding cannot be captured. In addition, the HHI does not consider:

• market structure
• transmission constraints
• transmission costs
• the balance of supply and demand
• the pattern of ownership over the supply curve

3.3.2.3 Estimation of Pricing Behavior Through Simulation Analysis   The estimation of pricing behavior through simulation analysis method involves a series of successive studies that progressively assess the market power abuse by strategic firms. These are solely based on estimations. The probable clearing price based on bidding strategies of the strategic firms is compared with an estimation of the perfectly competitive price to estimate the level of market power exercised. Some of those methods rely on historical or statistical data, but this is unreliable because the commercial scope of the strategic behavior has changed after the liberalization of the market [11].

3.3.2.4 Oligopoly Equilibrium Analysis   The most popular method for assessing market power abuse in the electricity market is the oligopoly equilibrium analysis. Several studies have been performed in this area. Some of them were based on purely theoretical working from the economists' point of view, others were adjusted to meet the requirements of electricity networks, and the most recent studies involve equilibrium analysis using full AC electricity network representation and taking into consideration all operating aspects and constraints of the complex structure of the electricity network.

In the following sections, there are further discussions about the different oligopolistic equilibrium approaches, including a brief comparison about the advantages and drawbacks of each method with a comprehensive analysis based on the available literature.

3.3.3 Oligopolistic Equilibrium Models   For the past few years the majority of the market power evaluation studies and strategic interactions modeling studies in the electricity markets have been performed using oligopolistic equilibrium models. There are several different oligopolistic equilibrium models available. Among them the most popular models are:

• Bertrand equilibrium
• Cournot equilibrium
• Supply function equilibrium (SFE)
• Stackelberg equilibrium
• Conjecture variation (CV) and conjecture SFE (C.S.F.) equilibrium

Each of these methods is based upon different assumptions. However, in general, the market equilibrium in the electricity environment can be defined as the set of prices, generating power outputs, transmission line power flows, and load demand that can form an optimization problem. This problem simultaneously satisfies the first order KKT conditions for maximization of the profit of each generating firm while clearing the market [16], thus forming a Nash equilibrium [17], which occurs when no generating company has any incentives to unilaterally change its bids [18]. The basic principles of each model are discussed below.

3.3.3.1 *Bertrand Equilibrium*

In the Bertrand model, the firms compete by setting their prices to the market and then supplying whatever quantity is demanded by the consumers [12]. In principle, under this equilibrium no firm offers a price that is below its marginal cost as such an action will result in a loss [7]. If each firm offers a single price in the market for a product of the same quality and there are no capacity constraints, then this model will result in perfect competition [11]. However, such a method has been considered as a perfect competition model, which cannot be efficiently applied to the electricity market dominated by large strategic firms forcing an oligopolistic imperfect competition environment.

Furthermore, this model assumes that any firm can supply the whole demand in the market by simply offering lower prices than the rival firms. Such an assumption may not be true for the electricity market, where there are generation capacity constraints [11] and transmission line constraints. Consequently, this method may not be efficient when applied to the electricity market.

3.3.3.2 *Cournot Equilibrium*

The Cournot equilibrium model was considered as the most appropriate model to be applied to the electricity market at the early stage development of the electricity market equilibrium analysis. Even now, the Cournot model is still widely used. The Cournot competition is considered as supply quantity competition where each generating firm chooses for its output quantity to be available to the market and then accepts the resulting market price offered by the consumers [12]. In the Cournot model, the generating firms can maximize their profits under the assumption that rival firms will not change their outputs [10, 11]. In addition, this model assumes that all generating firms have the same information about the market demand function [10].

3.3.3.3 *Supply Function Equilibrium*

The Cournot competition leads to significantly higher market prices than the actual market prices. Many researchers have identified drawbacks that arise with the traditional oligopolistic imperfect competition equilibrium models when applied to the electricity market and tried to develop new alternatives [19]. In order to analyze the complexities of electricity markets, as an alternative approach for modeling the strategic interactions of the participants, the supply function equilibrium (SFE) model has been proposed.

In [20], the new concept of the SFE model, proposed in [21], has been applied to the model the England and Wales electricity market. In comparison to the earlier approaches concentrating either on price or quantity, the advantage of the SFE model is that it gives the generating firms the possibility of submitting bids to the pool in both price and quantity. Each generating firm has to define an entire supply curve with different prices for different quantities. In [21] it has been shown that for a given demand function, for any price near the competitive one, there would be a corresponding Nash equilibrium [17]. The interesting feature of the SFE approach is its capacity to accommodate random shocks in demand. In the past few years the SFE modeling of the electricity market has been recognized as a very promising alternative to the widely used Cournot model. It should be pointed out that the SFE is very difficult to compute for large power systems.

3.3.3.4 *Stackelberg Equilibrium*   This model is considered as a rather monopolistic approach, in which a large dominant firm acts in a privilege manner securing its profits maximization, and many other smaller firms act as followers to the dominant firm's strategic actions [22, 23]. Studies have been performed in applying this model into the electricity market analysis [22].

3.3.3.5 *Conjectured Supply Function Equilibrium*   The conjectural variation concept is also an alternative method that can be successfully applied to calculate the electricity market equilibrium. The basic idea of this concept is that each firm in an oligopolistic market maximizes its own profit while taking into account the expected reactions from its competitors. The conjecture of a firm is defined as its belief or expectation of how rival firms will react to the change of its output [24]. There conjecture models can be classified into two categories. The first one is referred to as the general conjectural variation model, which is based on the conjecture about the rival firm's price [25]. The second method is called the conjectured supply function equilibrium, which is the most popular among the two alternatives in the electricity market. In both cases, each firm is making an assumption about the influence of rival firm's actions on the market clearing price [26].

3.3.4   Market Power Modeling Using Equilibrium Models

Market power is a very important issue in the electricity market and its behavior analysis. Electricity markets are more vulnerable to the exercise of market power than other energy markets [27] since electrical power is a very special commodity to deliver due to the following reasons: First of all, the nature of the AC network requires all generators connected to the network to be synchronized and operated at the same speed and frequency. Secondly, the electrical power must be generated instantly to supply the demand required by the consumers because it cannot be stored economically and efficiently. In addition, the required reserve capacity should always be available to balance a possible sudden increase in demand or to cover a possible failure of generators. In terms of network operation and management, the situation is more complex because the power flow in the network cannot be directed easily as it will be distributed in the network following the physical laws of electrical circuits, while bus voltages of the network should be operated within given voltage limits and power flows of transmission lines should be run with the thermal limits. Hence, if the network is not strong enough to withstand a possible overload or a contingency, it may result in the violation of voltage and thermal limits of the network and even a chain of cascading failures of components in the network, leaving large number of customers without electricity within seconds. During the support of market operations, the system operator must ensure that the electricity network remains electrically secure and stable to guarantee the supply of electricity. The oligopolistic structure, limited number of participants, and complex network operating constraints may make monitoring and detection of market power very difficult.

In the past, a number of approaches for assessing the market power in the electricity market have been presented using the market equilibrium models. Early studies were conducted mainly by economists and the voltage and thermal constraints were not considered. Nevertheless, most of the studies were successfully applied at the time to provide a rough qualitative analysis of the electricity markets.

The England and Wales pool was investigated in [20, 30–35]. For the early bid-based pool in England and Wales market most of the studies were based on the SFE model proposed in [21]. Such a model was presented [20] in the analysis of the England and Wales pool electricity market, where each generating firm's strategic behavior is described by a supply function giving the relationship between its generating quantity and selling price. Although the prices using the SFE model were significantly higher than the actual prices, they were much less than prices of the Cournot model [12]. With the introduction of the linear SFE model [30], the model was improved in comparison with that in [20]. Further development has been seen with the incorporation of contracts into the SFE model [31, 32]. In [34], a two-level approach was presented to compute the SFE equilibrium.

The California market as well as other US markets were also studied [12, 13, 27–29]. In [13], it was demonstrated that transmission capacity plays an important role in the Californian market using the Cournot model. It has been pointed out that the limitation of transmission capacity could give incentives to strategic firms to restrict their generation output while increasing prices. Similar studies using the Cournot model for the California market are presented in [14]. It was found that there is a potential market power abuse during high demand periods, in particular when hydroelectric generation is at the lowest generation level. It is widely accepted that market power is mainly exercised by strategic firms, either by increasing prices or by withholding capacity. However, in [28] it was suggested that strategic firms can successfully exercise market power in a different manner than in other more familiar product markets. For instance, large firms could exercise horizontal market power by increasing their own production, lowering some prices, and exploiting the necessary feasibility constraints in the network to foreclose competition from others by considering the special properties of electric networks. In [29], a market power model based on the Cournot quantity approach has been presented to assess strategic interactions in electricity networks, where it has been shown that transmission constraints cause strong interactions between different parts of a system and horizontal market power can further be attained by increasing profit and production, utilizing feasibility constraints and reducing prices.

Similar studies were carried out for the Nordic [36], Australian [37], and Dutch markets [38]. In [36], the linear SFE model was employed to investigate the market power in the Nordic electricity market. It was suggested that large power producers have incentives to withdraw some generation capacity in order to raise the spot market prices. In [37], a variant of the Cournot model with statistical load data was used in modeling the bidding strategies of the generating firms for the Australia market. It has been found that there are noncompetitive behaviors of a small number of large strategic firms. In [39], the Australian market was also studied using a Nash equilibrium-based model. Based on the SFE equilibrium model in [35], in [38] a more general model was used to incorporate forward contracts where the prepositions in [31] were considered. In this model, the different options of forward contracts in the electricity market were addressed.

3.4 APPLICATION OF THE EQUILIBRIUM MODELS IN THE ELECTRICITY MARKETS

Based on the early techniques applied to the electricity market modeling and market power analysis, some promising studies have been conducted to adjust the equilibrium methods in order to fit to the electricity market rules and sensitivities. In this section a review of different equilibrium models is presented.

3.4.1 Bertrand Equilibrium Model

In [64], perfect competition behavior in the electricity market with the maximization of social welfare has been presented where a conventional nonlinear OPF algorithm with full network representation was modified to consider the new objective function that maximizes social welfare. In [65] the electricity market equilibrium under Bertrand competition has been studied where the Bertrand equilibrium based on price competition was evaluated using linear programming techniques. Both the methods in [64] and [65] are applicable to large-scale power systems.

3.4.2 Cournot Equilibrium Model

Several methods using the Cournot model for the electricity market equilibrium modeling and analysis have been presented [15, 16, 26, 43–55], where they are based on different assumptions and solution techniques. In [16], two Cournot equilibrium models have been formulated as a mixed linear complementarity problem to simulate the electricity market behavior. Between the two models, the first one is for the bilateral market and the second one is for the pool market auction where a DC network representation is used with the consideration of transmission line capacity constraints. In [40], the models in [16] were extended to consider proper arbitrage in the modeling. In [40], two models have been proposed; the first one is a Stackelberg equilibrium model and the second one is based on the Cournot equilibrium model. Similar electricity market models based on the mixed complementarity problems with equilibrium constraints formulations are discussed in [41], and the Cournot-based model, formulated as a mathematical program with equilibrium constraints, is presented in [42], which provides the ability of modeling both the economic cost functions and the electricity network using DC load flow.

In [43], a two-settlement equilibrium in competitive electricity markets has been formulated as a subgame-perfect Nash equilibrium problem where each generation firm solves a mathematical program with equilibrium constraints (MPEC), given other firms' forward and spot strategies and the DC approximation load flow for electricity network modeling has been used. In [43], two computational approaches, such as penalty interior point algorithm and steepest descent approach, have been presented and demonstrated on a six node system. The Nash equilibrium in two-settlement competitive electricity markets with horizontal market power, network congestion, demand uncertainties, and probabilistic system contingencies have been studied in [44], where the equilibrium is formulated as a stochastic equilibrium problem with equilibrium constraints (EPEC) and each firm solves a stochastic MPEC. It was assumed that there is no-arbitrage relationship between the forward prices and the spot prices. It has been found that with two settlements formulation, the generation firms have incentives to commit forward contracts with increased social surplus and decreased spot energy prices. In [45], the market equilibrium as a stochastic EPEC capturing congestion effects, probabilistic contingencies, and market power has been presented with the modeling of price caps. Numerical results on the 53-bus Belgian electricity network have indicated that, when comparing to two-settlement systems without price caps, the price capping either in the forward contract prices or in the spot market prices results in reduced forward contracting. Furthermore the reduction in spot prices due to forward contracting is smaller. Further studies on the solution techniques of MPEC and EPEC are described in [46].

The impact of the transmission line capacity constraint on the electricity Cournot equilibrium has been studied on a simple example network in [15], where there are three market players in a transmission-constrained system considering nonconstant marginal cost. Scenarios have been presented to show that a pure strategy equilibrium can break down even when a transmission constraint exceeds the value of the unconstrained Cournot equilibrium line flow.

When electricity market equilibrium studies were used, the Cournot market equilibrium model could not produce realistic results in regard to the market clearing prices. Typically, the Cournot model produces much higher prices than that of the actual market. In contrast, the Bertrand model produces significantly lower market prices than the actual prices because of the oligopolistic imperfect nature of the electricity market. Based on the above consideration, combining the Cournot and Bertrand models to develop a hybrid model could provide more realistic market prices. In [47], a hybrid Bertrand-Cournot approach was formed as a linear complementarity problem and it provides higher prices than the conventional Bertrand and lower than the unrealistic conventional Cournot outcome.

The proposed methodology [48] simulates the behavior of the energy market through an iterative method based on Wilson's rules proposed for the power exchange in the Californian market, which is based on a generalization of the Cournot and Bertrand equilibria where some firms have the possibility of modifying prices. The proposed iterative method performs in a dynamic way where, in each iteration, the market participants refine their beliefs about the other players' behavior through the information on the productions and prices, and players try to determine their best response to the market. In [49], a model for medium-term hydrothermal operation of electrical power systems has been proposed where the objective is to determine the oligopolistic market equilibrium point in which each firm maximizes its profit based on other firms' behavior. The strategic operational behavior of the hydro and thermal generation units in the electricity market environment is also analyzed based on a hybrid generalization of the Cournot model, including both the Cournot and Bertrand behaviors.

In [52], a three-level optimization problem for the strategic bidding in the electricity market under Cournot competition has been presented where a probabilistic bidding approach is applied. A statistical method has been applied to solve the problem formulated. In [53], an algorithm for the calculation of the Cournot equilibrium in the absence of transmission constraints has been presented. The algorithm is based on the framework of the Cournot game with a three level decision-making scheme with economic signal exchange, which is very efficient computationally and suitable for multi-period production scheduling models for the electricity market. In [54], a relaxation algorithm and the Nikaido-Isoda function have been presented for the calculation of Nash-Cournot equilibria where a DC power flow model is used.

In [26], a complementarity approach has been proposed to simulate the interaction between the electricity and pollution permit markets, considering contracts and the reserve market. In the studies, the electricity market has been formulated as the Cournot game while the pollution permits market has been modeled with the conjectured price response function. It has been shown that strategic behavior of strategic firms becomes more complex with the inclusion of the pollution permits market.

The incomplete information of the cost functions of the rival firms has been addressed in [55], where predicting the optimal generation cost function of the rival generating firms using Cournot noncooperative game theory has been proposed based on market equilibrium estimates and uncertainty of rivals information.

In [22], the Stackelberg equilibrium with the AC power flow representation has been formulated as a bilevel mathematical optimization problem where there is a dominant firm acting in a privileged position and there are a group of subordinated ones acting as followers. The Stackelberg equilibrium problem has been solved using a noninterior point algorithm with a relaxed-penalized Karush-Kuhn-Tucker (KKT) formulation.

In [72], a multimarket Cournot model, which deals with co-optimization of multiple commodities, namely energy and various classes of ancillary services, and with transmission constraints that separate these commodity markets spatially, is presented. The extension of the Cournot model to deal with bilateral contracts for energy is also discussed.

3.4.3   Supply Function Equilibrium Models in Electricity Markets

3.4.3.1   *Application of Supply Function Equilibrium Models*   The SFE has been developed specifically for calculating the maximum profit equilibrium under uncertain demand. Rather than competing only with fixed prices for Bertrand competition or fixed quantities for Cournot competition, the generating firms can compete in both price and quantity using supply functions. Thus, SFE can provide prices closer to the actual market prices than the Cournot model, which generally produces

significantly higher outcomes, and Bertrand, which, on the other hand, produces the lower, purely competitive prices [12]. Klemperer and Meyer [21] showed that all the SFE equilibria are bounded by the Cournot and the Bertrand outcomes [14]. The SFE outcome is close to Cournot equilibrium at peak time when generating capacities are almost near the limits and close to Bertrand equilibrium when there is significant excess capacity, that is, during off-peak demand [71].

In [20], the application of the SFE model in the England and Wales pool market is presented. In such a model, it was assumed that each firm submits a supply function schedule that relates the amount of power supplied with the marginal price. In this work, although the authors acknowledge the importance of the more realistic asymmetric firm duopoly model, they have decided to restrict their study to the simpler symmetric firm duopoly case. This is because, in the asymmetric case, the solution procedure involves higher order formulation, requires greater computational complexity, and hence is much more difficult to solve. This critical issue of numerical tractability was tackled in [30] using a linear supply function model where the more realistic asymmetric firm case that existed in the environment of the England and Wales market was formulated. The model in [30] has been further extended in [32] to incorporate forward contracts. The model proposed in [20] has also been extended to include contracts in [31].

The linear SFE model proposed in [30] has been further developed in [56] to the general SFE case. Between the linear SFE and the general SFE, the linear SFE has the ability to easily handle asymmetric firms in large-scale systems with the implementation of linear marginal costs. In contrast, the general SFE requires the solution of a set of differential equations, and such a model is considered to be difficult to solve, hence its application may be limited to the symmetric firm case. Furthermore, in terms of electricity market modeling, the asymmetric case is more realistic, and the application of the linear SFE in modeling the asymmetric case is more widely used.

In [35, 56], generalization of the linear SFE model proposed in [30] presented. The extended model can include heterogeneity, for example, to model the behavior of asymmetric firms with dissimilar cost functions, implement non-zero marginal cost intercepts into supply functions, and generation capacity constraints representation. The model has been applied to the England and Wales electricity market. This model has been further extended in [57] to study the interaction between generation capacity constraints, price-caps, and the time-period validity of the supply bids with the issue of multiple supply function equilibrium existence. In [20], it is suggested that the presence of generation capacity constraints limits the possible number of equilibrium solutions. In [57], the presence of price-caps further reduces the range of stable equilibrium points. The set of the possible equilibrium points can be further reduced by the presence of "pivotal suppliers" [58] where a generating firm, who at the peak or high demand withholds its production, is considered as a "pivotal supplier." In [59], the existence of multiple SFE equilibria has also been studied, and it has been found that under certain operating restrictions the resulting SFE equilibrium is guaranteed to be unique. It has also been proved that, in the case of symmetric producers, with the presence of inelastic demand, binding operating constraints, and imposed price-caps, the existence of an SFE equilibrium is unique.

In [60], it is the conditions under which an optimal SFE exists if demand is inelastic are discussed. In [61], a similar method has been applied to show the effect of limited generation production in the presence of price-caps, considering the impact of forward contracts on the behavior of the strategic firms and the existence of symmetric duopoly SFE equilibrium. The contracts modeling in the SFE equilibrium model is considered in [38] based on the Dutch electricity market. In [62], the influence of transmission line capacity constraints on the strategic behavior of the generating firms is examined using both the SFE model and the Bertrand model on a simple three-bus system.

3.4.3.2 *Electricity Network Modeling*   Most of the early studies did not consider the electricity network in their model. However, the presence of the generation capacity constraints can influence the set of SFE equilibrium points and hence the strategic behavior in general, even if there is significant excess in the generation capacity. A similar situation exists with the presence of the transmission line capacity constraints. For this reason, it would be useful to have an appropriate electricity network modeling in order to reflect the realistic strategic interactions in the electricity market. In [63], the linear SFE model has been developed to include a linearized DC load flow model where a transmission line capacity constraints can be considered. The SFE model presented in [75] has decomposed the optimization problem into equivalent subproblems, where in the base-case problem, the SFE equilibrium is calculated without taking into account the generation and network capacity constraints. In the subsequent optimization actions, the subproblems handle these constraints to minimize the network violations. In [18, 23, 83, 84] bi-level optimization models are used. The method proposed in [18] is based upon an iterative bi-level optimization problem to solve the linear SFE equilibrium. The two-level optimization problem has been formulated as an MPEC. The upper level chooses the parameters for the bid curves of the firm and the lower level maximizes the social welfare using a linearized DC power flow model where transmission line capacity limits are considered. In [68], a binary expansion (BE) solution approach to the problem of strategic bidding under uncertainty in short-term electricity markets is presented. The BE scheme transforms the products of variables in the nonlinear bidding problem into a mixed integer linear programming (MILP) formulation, and then the MILP is solved by commercially available algorithms. The proposed BE scheme is applicable to either pure price, or pure quantity, or joint price/quantity bidding models. Also, the model can represent transmission networks, uncertainties (scenarios for price, quantity, plant availability, and load), financial instruments, capacity reinforcement decisions, and unit commitment. In [77], a nonlinear complementarity approach based on the linear SFE equilibrium model using DC power flow representation has been proposed. In this approach a mixed nonlinear complementarity problem (NCP) was formulated as a single combined solution using the Levenberg-Marquardt algorithm. In [73, 74], an approach, which can be used for evaluating the performances of the electricity markets with network representation in presence of bidding behavior of the producers in a pool system, has been presented.

Although most of the methods that consider electricity network modeling implement the linearized DC power flow model are adequate for implementing the transmission line capacity limits for congestion modeling, the DC model has short-comings in terms of a number of basic electricity network properties [88]. A number of studies have been performed using full AC network representation [22, 69, 80, 104, 105, 135]. The model in [22] is based on the Stackelberg equilibrium model, which is quantity based and is solved using a relaxed-penalized, noninterior point algorithm. In [69], an algorithm for the maximization of the individual welfare of the electricity market participants with SFE bidding is presented. This method was further developed in [80] to search for multiple equilibrium points.

3.4.3.3 Modeling of Contracts
In SFE equilibrium model, bilateral forward contracts can be considered. In many electricity markets, the basic volume of electricity is traded with contracts while the rest is performed through the pool, which has been discussed in [31, 32] and followed in [70, 76, 82]. In [70], the interaction between the contract and the pool markets has been analyzed where only generating capacity constraints are considered. A more comprehensive method is reported in [76] using the linearized DC network model and incorporating forward contracts. In addition, the bidding strategies of the generating firms have been discussed with given forward contracts. The impact of long-term contracts using the SFE model is also addressed in [82].

3.4.3.4 Choosing the Appropriate Strategic Variable
In [86], the different approaches and characteristics of the strategic variables in the SFE bidding models are addressed. Four different situations are discussed. In the first situation, the strategic variable is the gradient parameter of the supply function where each generating firm can choose freely the value of the gradient of the supply function while the intercept is kept constant and the gradient was as strategic variables [18, 20, 30]. In the second situation, the strategic variable is considered to be the intercept parameter of the supply function while the gradient is kept constant where each generating firm can freely choose the value of the intercept [58]. In the third situation, both the gradient and the intercept can be changed but they should have a linear relationship [62, 69, 77]. In practical implementation, each generating firm multiplies its marginal cost with an arbitrary parameter, which is considered as the strategic variable. In the fourth situation, both the gradient and the intercept can be considered as strategic variables and can be freely chosen by the generating firms [86]. In [86], the fourth situation is suggested to be the most realistic model situation among the four, but the existence of multiple SFE equilibria is inevitable. Although the fourth provides the ability to the generating firms to consider all the possible supply functions for building their bidding strategies is subject to implementation problems, the flexibility of this approach would bring challenges to the solution. In [18], it is suggested that if both the gradient and the intercept are considered as strategic variables, then the supply bid functions can be of any form, and a unique equilibrium point rarely exists.

3.4.3.5 Conjecture Supply Function Equilibrium Model
The conjectured supply function equilibrium was introduced in [23], further developed in [78], and applied in large-scale electricity systems in [79]. The proposed model in [23] has been formulated as a mixed complementarity problem using the linearized DC network modeling. The interaction of competing generating firms is based on their

conjectures, where the conjecture is the belief of a generating firm with regard to how total supply from rival firms reacts to the price. In [78], this model has further been developed to accommodate mixed transmission pricing, including fixed transmission tariffs, congestion pricing, auction pricing, and export fees. In [79], this method has been applied to the Benelux electricity market. In [87], a multi-period market with the consideration of spinning reserve has been examined using a dynamic CSF approach. The equilibrium model has been developed considering linear complementarity conditions and a DC network representation. In [25], an empirical CSF model for the evaluation of optimal bidding strategies has been proposed using the available historical data on electricity prices.

3.4.4   Conjectural Variation and CSF Equilibrium Models

The conjectural variation equilibrium analysis of the electricity market is presented in [24, 66]. In [24], the conjectural variation equilibrium has been applied in the electricity spot market with imperfect information where the strategic bidding of the generating firm is based upon estimated conjectures of the rivals firm's bid and their reaction on generating firm's bidding. The classical game bidding strategies based on Cournot and Bertrand models are special cases of the conjectural variation Nash equilibrium models. In [66], the generating firm's bidding strategy was based on the dynamic learning behavior where each generating firm dynamically regulates its conjectures with respect to the reactions of the rival firms to its bidding strategies, drawn from the available information in the electricity market. In the above studies, the electricity network model was not taken into account. In [67], the impact of price caps on the electricity market under the Cournot conjectural variation model is presented, where a DC network model is used. In [81], the impact of learning behavior of electrical-power suppliers on electricity-spot-market equilibrium under repeated linear supply-function bidding is addressed. Basically, suppliers will conduct "learning" to improve their strategic bidding in order to obtain greater profit.

3.5   COMPUTATIONAL TOOLS FOR ELECTRICITY MARKET EQUILIBRIUM MODELING AND MARKET POWER ANALYSIS

Electricity market equilibrium analysis has been considered a very popular and interesting subject for research since the liberalization of the electricity sector. Moving from the monopolistic environment dominated by the large, primarily state-owned utility companies into a new liberalized electricity market environment has not been easy. Many complications arose due to the complexity of the nature of electricity, resulting in a small number of active participants in the market. The liberalization of the electricity market and the establishment of independent generating firms have brought forward the development of electricity market equilibrium analysis methods. The research area of electricity market equilibrium analysis, for the simulation of the behavior of the market participants, is very attractive, offering the opportunity to perform an extensive study in new challenging fields resulting in very interesting results.

Following up on the discussions in previous sections, this section introduces and discusses computational tools for electricity market modeling and market behavior analysis, such as mathematical programs with equilibrium constraints (MPEC), mathematical programs with complementarity constraints (MPCC), and equilibrium problems with equilibrium constraints (EPEC), where attention will be paid in particular to basic concepts, mathematical background, and mathematical formulations, while the strict mathematical proofs are left to references.

3.5.1 Mathematical Programs with Equilibrium Constraints (MPEC)

The term mathematical programs with equilibrium constraints (MPEC) was introduced in [89], and thereafter was widely used in the literature. Detailed discussions on MPEC are found in [90, 91, 96]. MPECs are optimization problems where some constraints are defined in terms of complementarity constraints. Many problems arising from engineering and economics can be treated by the MPEC methodology. Several iterative algorithms, such as a penalty-based interior point algorithm, an implicit programming algorithm, and a piecewise sequential quadratic programming algorithm, can be applied in solving MPEC problems. Mathematical programs with complementarity constraints (MPCCs) are considered a subclass of MPECs for which the lower level equilibrium system can be formulated as a complementarity problem. In principle, MPCCs can be easily formulated as nonlinear programming problems (NLPs), while the constraints of the complementary constraints lack some stability properties that are considered essential in analyzing and solving NLPs. Complementarities also arise in the solution of mathematical programs with MPECs and bi-level optimization problems. Both MPECs and MPCCs have the technical challenge of nonconvex characteristics, which is due to the consequence of the upper-level objective clashing with the lower-level objectives in the lower-level equilibrium program.

For the sake of illustration, an MPEC is a mathematical program constrained by a variational inequality, which takes the following general form:

$$\min f(x, y) \tag{3.1}$$

Subject to:

$$g(x, y) = 0, \quad (x, y) \in Z \tag{3.2}$$

$$(v - y)^T F(x, y) \geq 0, \, y \in C(x), \, \forall v \in C(x) \tag{3.3}$$

where Z denotes the feasible region for the upper level problem and $C(x)$ represents the feasible region for the lower level problem. The inequalities shown in (3.1)–(3.3) are referred to as variational inequalities, behind which a mathematical theory was proposed for the study of equilibrium problems that arise in many domains such as economics, finance, optimization, and game theory [92, 93]. The variational inequalities here are generally referred to as equilibrium constraints. The MPEC shown in (3.1)–(3.3) is also referred to by MPCC.

Due to the complementarity constraints or the variational inequalities, the MPEC is very difficult to solve. As pointed out in [94, 95], the Mangasarian-Fromovitz constraint qualification (MFCQ) does not hold at any feasible point for the MPEC, and some well-developed nonlinear programming theory and numerical methods may not be readily applicable for solving the MPEC. The failure of MFCQ implies:

- The set of Lagrange multipliers are unbounded
- Constraint gradients are linearly dependent
- A central path does not exist.

3.5.2 Bilevel Programming

A bilevel optimization problem may be given by:

$$\min f(x, y) \tag{3.4}$$

Subject to:

$$(x, y) \in Z \tag{3.5}$$

$$y = arg \min \theta(x, \hat{y}), \quad \hat{y} \in C(x) \tag{3.6}$$

Bilevel optimization problems shown in (3.4)–(3.6) have an upper-level and a lower-level problem. The lower-level problem is embedded into the upper-level problem and taken as constraints of the upper-level problem. The bilevel optimization problem here can be reformulated as a mathematical program with complementarity conditions (MPCCs) by writing the optimality conditions of the lower-level optimization problem as constraints on the upper-level optimization problem. If we assume that the function $F(x, y)$ is a gradient map with respect to the second argument $F(x, y) = \nabla_y \theta(x, y)$ and $C(x)$ has a finite representation, the MPCCs are equivalent to a lower-level minimization problem. This results in the MPEC problem shown in (3.1)–(3.3). Clearly, the bilevel programming (BLP) problem is a special case of the MPEC in terms of MPCC formulation. In principle, any solution technique for MPEC can be applicable for solving BLP. More about BLP is found in [96–100].

3.5.3 Equilibrium Problems with Equilibrium Constraints (EPEC)

An EPEC is a mathematical program to find an equilibrium point that simultaneously solves a set of MPECs where each MPEC is parameterized by decision variables of other MPECs. The EPEC problems often arise from noncooperative games, for instance, multi-leader-follower games [102], where each leader is solving a Stackelberg game formulated as an MPEC [90]. In the past, a number of EPEC models have been proposed to investigate the strategic behavior of generating firms in electricity markets [29, 85, 101–107] while the necessary optimality conditions of EPECs via multiobjective optimization has been studied in [108].

3.5.3.1 Formulation of Single-Leader-Follower Games as an MPEC
Before we introduce the multiple-leader-follower games such as EPEC, we will start with single-leader-follower games and show the properties of such games, which can be formulated in terms of an MPEC.

Suppose that a single-leader-follower game is played by a leader, who can make a decision, and a number of competitive followers, who react to the decision made by the leader. That is, given a strategy x for the leader, the followers choose their strategies such that [103]

$$\arg\min_{w_j \geq 0} b_j\left(x, \widehat{w}_j\right) \tag{3.7}$$

Subject to:

$$c_j\left(x, w_j\right) \geq 0 \tag{3.8}$$

where $j = 1, \ldots, \ell \cdot \widehat{w}_j = (w_1^*, \ldots, w_{j-1}^*, w_j, w_{j+1}^*, \ldots, w_\ell^*)$. Each player has its own objective and constraints, which are necessarily the same for all players. Such a problem, shown in (3.7) and (3.8), is a Nash game parametrized by the decision made by the leader. If the problem in (3.7) and (3.8) is convex and satisfies a constraint qualification for each follower, then the optimal condition of each follower's strategy is equivalent to the following parametric nonlinear complementarity problem (NCP):

$$0 \leq w_j \perp \nabla_{w_j} b_j(x, w) - \nabla_{w_j} c_j(x, w) z_j \geq 0, \quad \forall j = 1, \ldots, \ell \tag{3.9}$$

$$0 \leq z_j \perp c_j(x, w) \geq 0, \quad \forall j = 1, \ldots, \ell \tag{3.10}$$

where the parametric NCP is a set of KKT conditions for the optimization problems of the followers. Now we define variables

$$y = (w, z) \tag{3.11}$$

and functions

$$h_j(x, y_j) = \begin{pmatrix} \nabla_{w_j} b_j(x, w) - \nabla_{w_j} c_j(x, w) z_j \\ c_j(x, w) \end{pmatrix} \geq 0, \quad \forall j = 1, \ldots, \ell \tag{3.12}$$

We can re-write (3.12) in compact form:

$$h(x, y) \geq 0 \tag{3.13}$$

where $h(x, y) = [h_1(x, y_1), \ldots h_{j-1}(x, y_{j-1}), h_j(x, y_j), h_{j+1}(x, y_{j+1}), \ldots h_\ell(x, y_\ell)]^{\mathrm{T}}$.

Introducing slack variables s into (3.13), we have the following compact form of the parametric NCP shown in (3.9) and (3.10):

$$h(x, y) - s = 0 \tag{3.14}$$

$$0 \leq y \perp s \geq 0 \tag{3.15}$$

Now, given the parametric NCP shown in (3.14) and (3.15), the leader's optimization problem is to select a strategy by minimizing (or maximizing) its own objective function $f(x, y)$ subject to its own constraints $g(x, y) \geq 0$ while satisfying the parametrized NCP:

$$\min_{x \geq 0, y, s} f(x, y) \tag{3.16}$$

Subject to:

$$g(x, y) \geq 0 \tag{3.17}$$

$$h(x, y) - s = 0 \tag{3.18}$$

$$0 \le y \perp s \ge 0 \tag{3.19}$$

The optimization problem (3.16)–(3.19) is an MPEC. However, the limitations of the formulation in (3.16)–(3.19) should be pointed out:

- If the problems shown in (3.7) and (3.8) are nonconvex or they do not satisfy a constraint qualification, then the parametric NCP (3.14) and (3.15) are not equivalent to the original followers' problems as given by (3.7) and (3.8). In this situation, a solution to the leader's optimization problem (3.16)–(3.19) may not necessarily correspond to a solution to the single-leader-follower game.

- If a constraint qualification cannot be satisfied for each follower's optimization problem in (3.7) and (3.8), then the parametric NCP (3.14) and (3.15) may not have a solution because the multipliers z_j may not exist.

In (3.19), the complementarity condition can be replaced by a nonlinear inequality, the optimization problem becomes:

$$\min_{x \ge 0, y, s} f(x, y) \tag{3.20}$$

Subject to:

$$g(x, y) \ge 0 \tag{3.21}$$

$$h(x, y) - s = 0 \tag{3.22}$$

$$Ys \le 0, \quad y \ge 0, \quad s \ge 0 \tag{3.23}$$

where $Y = diag(y_j)$. This equivalent nonlinear program can then be solved by applying standard NLP solvers.

3.5.3.2 Formulation of Multi-Leader-Follower Games as an EPEC

For multi-leader-follower games, the leaders make decisions while the followers react to the decisions made. Such games can be modeled as equilibrium problems with equilibrium constraints (EPEC), which aim to find an equilibrium point where no leader can improve its objective given the strategies chosen by the other leaders and the reaction of the followers. The followers compute an equilibrium point where no follower can improve its objective given the strategies made by the leaders and those chosen by the other followers. For the case of electricity markets, generating firms can be considered as market leaders while ISO can be considered as the common follower.

Let K be the number of leaders, and $x_i(i = 1, \ldots, K)$ denotes the decision variables for leader i. The leaders' variables are given by $x = (x_1, \ldots, x_{i-1}, x_i, x_{i+1}, \ldots, x_K)$. In comparison to the MPEC problem for single-leader games shown (3.20)–(3.23), the optimization problem solved by leader i of the multi-leader-follower-games can be formulated as the following MPEC:
Minimize

$$\min_{\hat{x}_i \ge 0, y, s} f\left(\hat{x}_i, y\right) \tag{3.24}$$

Subject to:

$$g(\hat{x}_i, y) \geq 0 \tag{3.25}$$

$$h(\hat{x}_i, y) - s = 0 \tag{3.26}$$

$$Ys \leq 0, \quad y \geq 0, \quad s \geq 0 \tag{3.27}$$

where $\hat{x}_i = (x_1^*, \ldots, x_{i-1}^*, x_i, x_{i+1}^*, \ldots, x_K^*)$. In principle, each leader's optimization problem as an MPEC can be solved by applying standard NLP solvers. Basically in each leader's MPEC, it has its own objective $f(\hat{x}_i, y)$ and constraints $g(\hat{x}_i, y) \geq 0$ while the MPCCs $Ys \leq 0, y \geq 0, s \geq 0$ are common for all leaders' MPECs. In comparison to the single-leader games, the multi-leader-follower-games have k MPECs to solve. Such an optimization problem is also referred to EPEC.

3.5.4   NCP Functions for MPCCs

MPEC problems can be reformulated as nonlinear programming problems by replacing the complementarity constraint shown in (3.1)–(3.3) by a nonlinear complementarity problem (NCP) function. An NCP function is a function $\phi: R^2 \rightarrow R$. It is known that $\phi(a, b) = 0$ if and only if $a, b \geq 0$, and $a, b \leq 0$. There are several NCP functions that can be used in the reformulation and their practical performance was examined and compared on a range of test problems [110].

3.5.4.1   The Fischer-Burmeister Function   The NCP function shown in (3.28) is called the Fischer-Burmeister function [111].

$$\phi_{FB}(a, b) = a + b - \sqrt{a^2 + b^2} \tag{3.28}$$

It is nondifferentiable at the origin, and its Hessian is unbounded at the origin. It is a complementarity function for the NCP, which is called C-function or NCP function, that is,

$$\phi_{FB}(a, b) \Leftrightarrow a \geq 0, b \geq 0, ab = 0 \tag{3.29}$$

3.5.4.2   The Min-Function   The min-function [112] is a nonsmooth function, which is given by

$$\phi_{\min}(a, b) = \min(a, b) \tag{3.30}$$

Equation (3.30) can be written equivalently in terms of the natural residual function [112]:

$$\phi_{NR}(a, b) = \frac{1}{2}\left(a + b - \sqrt{(a-b)^2}\right) \tag{3.31}$$

The function $\phi_{NR}(a, b) = \frac{1}{2}\left(a + b - \sqrt{(a-b)^2}\right)$ in (3.31) is nondifferentiable at the origin and along the line $a = b$.

3.5.4.3   The   Chen-Chen-Kanzow   Function   The   Chen-Chen-Kanzow Function [113] is a convex combination of the Fischer-Burmeister function and the bilinear function, which is defined as

$$\phi_{CCK}(a, b) = \lambda\phi_{FB}(a, b) + (1 - \lambda)\phi_{BL}(a, b), \quad \lambda \in (0, 1) \tag{3.32}$$

where the bilinear function (BF) is defined as:

$$\phi_{BL}(a, b) = ab \tag{3.33}$$

Note that $\phi_{BL}(a, b)$ is an analytic function and its gradient vanishes at the origin; it is not an NCP function because $\phi_{BL}(a, b) = 0$ does not imply nonnegativity of a, b. It should also be mentioned that all the NCP functions are nondifferentiable at the origin where the Hessian of the Fischer-Burmeister function is unbounded at the origin. This property should be taken into account in the design of robust nonlinear optimization algorithms for MPCCs.

3.6 SOLUTION TECHNIQUES FOR MPECs

There are a number of techniques available for the numerical solution of MPECs, which are tailored to problems with a special structure [98–100, 109]. For solving a broader class of problems, at least three main approaches have been recognized:

1. *Nonlinear programming approach*: the equilibrium constraint is either converted to a smooth equation or augmented to the objective via a suitable error bound.

2. *Implicit programming approach (ImP)*: the equilibrium constraint is treated via a generalized implicit function theorem.

3. *Artificial intelligence approach.*

In the following sections, nonlinear programming approaches such as SQP, Interior Point algorithms, and artificial intelligence discussed, while the implicit programming approach is discussed in [90, 91].

3.6.1 SQP Methods

SQP methods have been applied in solving the MPEC problems as shown in (3.20)–(3.23). In [114], numerical experience of using SQP has been shown on a large collection of MPEC test systems. The SQP method used here is a trust region SQP algorithm, filter SQP. The description of the filter SQP is referred to in [139]. In [110], the SQP algorithm has been extended to solve MPECs using different NCP formulations as presented in Section 3.5.4. Numerical results have been presented to show the performance of the different NCP formulations. SQP methods exhibit fast local convergence near strongly stationary points under reasonable assumptions, and good convergence behavior has been observed on a large set of MPCC test problems. Local convergence of SQP methods for MPECs is examined in [115].

3.6.2 Interior Point Methods

3.6.2.1 *Interior Point Methods with Relaxed Complementarity Constraints* The interior point algorithms IPOPT (Interior Point OPTimizer) in [117, 118, 140] can be formulated as:

$$\min_{x \geq 0, y, s} f(x, y) \qquad (3.34)$$

Subject to:

$$g(x, y) \geq 0 \qquad (3.35)$$

$$h(x, y) - s = 0 \qquad (3.36)$$

$$y_j s_j \leq \delta\mu, \quad j = 1, \ldots, \ell \qquad (3.37)$$

$$y_j \geq 0, \quad s_j \geq 0, \quad j = 1, \ldots, \ell \qquad (3.38)$$

where δ is a fixed parameter while μ is the barrier parameter of the interior point formulations. The relaxed complementarity condition (3.37) has been proposed in [117–119, 140] while the theoretic aspects of the relaxed complementarity condition approach have been studied in [121, 122]. The problem in (3.34)–(3.38) can be solved by the interior point algorithms of IPOPT. The IPOPT algorithms proposed have the following features:

- The Newton-based interior point barrier algorithm has been applied.
- A filter line search has been used to improve the convergence from poor initial points.
- As an option, the nonlinear programming optimization problem can be solved by a full space or reduced space approach.
- The exact first and second derivatives can be obtained through the AMPL interface.

3.6.2.2 *Interior Point Methods with Two-Sided Relaxation*   A two-sided relaxation scheme for MPECs has been proposed [123]. In contrast to previous approaches [117, 118, 140], both the complementarity and the nonnegativity constraints have been relaxed. Such a two-sided relaxation scheme has been implemented in combination with a standard interior-point method to achieve superlinear convergence.

The nonlinear programming formulation with the two-sided relaxation scheme is presented as follows:

$$\min_{x \geq 0, y, s} f(x, y) \qquad (3.39)$$

Subject to:

$$g(x, y) \geq 0 \qquad (3.40)$$

$$h(x, y) - s = 0 \qquad (3.41)$$

$$Ys \leq \delta_c, \qquad (3.42)$$

$$y \geq -\theta, \quad s \geq -\theta, \quad j = 1, \ldots, \ell \qquad (3.43)$$

where δ and θ are vectors of strictly positive relaxation parameters. In comparison to (3.37), where the complementarity constraints are relaxed, the proposed new scheme in [123] relaxes both the complementarity constraints and the nonnegative constraints as shown in (3.42) and (3.43). It has been noted that if max $(\theta, \delta) > 0$, then the strictly feasible region of (3.42) and (3.43) is nonempty, and the active constraint gradients are linearly independent. In the algorithm it is assumed that either θ or δ, but not both, are driven to zero in order to recover a stationary point of the MPEC.

3.6.2.3 Interior Point Methods with Penalty

The interior-point algorithms LOQO (An Interior Point Code for Quadratic Programming) have been developed for solving the MPEC problems in [116]. In the formulations, the penalty functions such as the ℓ_1 penalty function and the ℓ_∞ penalty function are used.

In the interior-point algorithms with the ℓ_1 penalty function, the MPEC problem in (3.20)–(3.23) are reformulated as:

$$\min_{x \geq 0, y, s} f(x, y) + \rho \sum_j \xi_j \tag{3.44}$$

Subject to:

$$g(x, y) \geq 0 \tag{3.45}$$

$$h(x, y) - s = 0 \tag{3.46}$$

$$y_j s_j \leq \xi_j, \quad j = 1, \ldots, \ell \tag{3.47}$$

$$\xi_j \geq 0, \quad y_j \geq 0, \quad s_j \geq 0, \quad j = 1, \ldots, \ell \tag{3.48}$$

where $\rho > 0$ is a penalty parameter. In theory, if ρ is chosen large enough, the solution of the MPCC can be recast as the minimization of a single-penalty function. The appropriate value of the penalty parameter ρ is unknown in advance and must be estimated and updated during the course of the minimization.

In comparison to the MPEC problem with the relaxed complementarity condition in (3.20)–(3.23), the new formulation here in (3.44)–(3.48) includes the ℓ_1 penalty function in the objective while the relaxed complementarity condition is removed from the constraints.

In the interior-point algorithms with the ℓ_∞ penalty function, the MPEC problem in (3.20)–(3.23) is reformulated as:

$$\min_{x \geq 0, y, s} f(x, y) + \rho \max(\xi_j) \tag{3.49}$$

Subject to:

$$g(x, y) \geq 0 \tag{3.50}$$

$$h(x, y) - s = 0 \tag{3.51}$$

$$y_j s_j \leq \xi_j, \quad j = 1, \ldots, \ell \tag{3.52}$$

$$\xi_j \geq 0, \quad y_j \geq 0, \quad s_j \geq 0, \quad j = 1, \ldots, \ell \tag{3.53}$$

This desirable property of the ℓ_1 penalty function and the ℓ_∞ penalty function is that constraint qualification conditions hold at a solution to the true problem. The transformed problems in (3.44)–(3.48) and (3.49)–(3.53) can then be solved by the interior point algorithms LOQO.

In principle, for the penalty-based interior point algorithms, the appropriate value of the penalty parameter ρ is unknown in advance and must be estimated and updated during the course of the minimization. In [120], the theoretical and practical properties of interior-penalty methods for MPECs have been studied where the need for adaptive penalty update strategies was motivated with examples. Two strategies for updating the penalty parameter have been proposed while their efficiency and robustness have been demonstrated using numerical results.

3.6.3 Mixed-Integer Linear Program (MILP) Methods

In [68], a binary expansion (BE) solution approach to the problem of strategic bidding under uncertainty in short-term electricity markets is presented. The BE scheme transforms the products of variables in the nonlinear bidding problem into a mixed integer linear programming formulation, and then the MILP is solved by commercially available algorithms. The proposed BE scheme is applicable to either pure price, or pure quantity, or joint price/quantity bidding models. Also, the model can represent transmission networks, uncertainties (scenarios for price, quantity, plant availability, and load), financial instruments, capacity reinforcement decisions, and unit commitment.

In [136], for an electricity market cleared by a merit-order economic dispatch, the necessary conditions for the market outcomes supported by pure strategy Nash equilibria are found to exist when generating companies (Gencos) game through their incremental cost offers or supply functions. In the formulation, a Genco may consist of offers of multiple blocks and a mixed-integer linear programming (MILP) scheme was developed to find the NE without approximations or iterations. The MILP scheme was tested on several systems of up to 30 generating units, each with four incremental cost blocks.

In [137], a new approach for solving two-stage Stackelberg games with one leader based on disjunctive constraints and linearization is presented where the equilibrium constraints of the MPEC were replaced by integer restrictions in the form of disjunctive constraints. Finally the MPEC can be replaced by a MILP. The proposed discretely constrained MPEC problems can include discrete generation levels, fixed-cost problems involving binary variables, if-then logic relative to ramping constraints, and discrete investment levels.

3.6.4 Artificial Intelligence Approach

In [124], a genetic algorithm (GA) is used to model Bertrand and Cournot competition of an electricity pool. The advantages of using a GA over scenario analysis for applied market simulation are presented and the limitation of the proposed GA approach is discussed, with possible solutions suggested. In [51, 125], co-evolutionary methods have been studied in electricity market equilibrium analysis and it is recognized that the co-evolutionary programming models are effective and powerful for electricity market simulation and analysis. In [125], it is demonstrated that co-evolutionary programming, with the ability of parallel and global searching, can be applied to overcome the difficulty of the conventional optimization approach in identifying global optimization solution. In addition, a hybrid co-evolutionary programming was presented to deal with the poor convergence performance of the simple coevolutionary programming.

In [126], a co-evolutionary programming method has been applied to analyzing SFE models of an oligopolistic electricity market where the affine supply function model and the piecewise affine supply function model were considered. Simulation results have shown the rapid convergence characteristics of the SFE in the affine supply function model and hence the co-evolutionary programming method has the great potential to be used to solve the real equilibrium problems of electricity markets.

Application of genetic algorithms in bilevel programming solution has been reported [127, 128]. In [127], the bilevel programming problem consists of the objective of the leader at its first level and that of the follower at the second level. The KKT conditions for the second-level problem were derived first and then the bilevel programming problem was transformed into a single-level problem with complementary constraints. Then, the genetic algorithm has been used to solve the transformed single-level problem. Numerical results demonstrated that the proposed GA against hybrid Tabu-ascent algorithm is efficient in terms of computation while the quality of solutions was almost the same for the two methods compared.

In [129], a neural-network model for solving asymmetric linear variational inequalities has been proposed where linear variational inequality is a uniform approach for optimization and equilibrium problems. Computer simulations were carried out for linear programming and linear complementarity problems. The test results have demonstrated the good convergence characteristics of the neural-network model proposed.

In [130], a fuzzy goal programming approach for solving quadratic bilevel programming problems is presented. In the proposed approach, the membership functions for the defined fuzzy objective goals of the decision-makers at both the levels are discussed first. Then a quadratic programming model is formulated using the notion of distance function minimizing the degree of regret to the satisfaction of the decision-makers of both levels. The solution process consists of two phases. At the first phase of the solution, the quadratic programming model is transformed into an equivalent nonlinear goal programming model, maximizing the membership value of each of the fuzzy objective goals, while at the second phase, the concept of linear approximation technique in goal programming is introduced for measuring the degree of satisfaction of the decision-makers at both the levels.

In [131, 132], Tabu Search algorithms are presented in solving the bilevel programming problems. The Tabu Search algorithms have the potential to solve very large-scale bilevel programming problems with the incorporation of more rigorous mechanisms.

3.7 SOLUTION TECHNIQUES FOR EPECs

The solution techniques for EPECs can be classified into two categories. Diagonalization solution methods (DSM) belong to the first category while simultaneous solution methods (SSM) fall into the second category. The DSM have been frequently used in the literature.

3.7.1 Diagonalization Solution Methods

In the past, most algorithms have not been specifically designed for solving EPECs. The most widely used methods for solving EPECs are in the category of DSM [18, 29, 63] where the methods mainly rely on nonlinear programming approaches for solving MPECs. Typically, an EPEC problem, consists of a set of MPECs. The idea of the DSM is to use MPEC algorithms developed to solve one MPEC at a time and cyclically repeat the same procedure for every MPEC until an equilibrium point is found. In nature, the DSM are sequential iterating optimization algorithms. The DSM can be further classified into two types of methods, nonlinear Jacobi method and nonlinear Gauss-Seidel method.

3.7.1.1 Nonlinear Jacobi Method The nonlinear Jacobi method for the EPEC (3.24)–(3.27) can be presented as follows:

Step 0: Initialization

    (a) Set the starting point $x_i(0)$ $(i = 1, \ldots , K)$ and $y(0)$.

    (b) Set the maximum number of iterations N_t and the iteration count $t = 0$.

    (c) Set the accuracy tolerance ε_p.

Step 1: Loop over each MPEC

    (a) Suppose the current iteration point is $x_i(t)$, $y(t)(i = 1, \ldots , K)$.

    (b) For each $i = 1, \ldots , K$, the MPEC($\hat{x}_i(t)$) can be solved by either NLP solvers or MPEC solvers and $\hat{x}_i(t) = (x_1^*, \ldots , x_{i-1}^*, x_i, x_{i+1}^*, \ldots , x_K^*)$ while $x_1^*, \ldots , x_{i-1}^*, x_{i+1}^*, \ldots , x_K^*$ are fixed.

Step 2: Update solution

    (a) Set $(x_1, \ldots , x_{i-1}, x_i, x_{i+1}, \ldots , x_K) = x(t + 1) = x$.

    (b) Set $t = t + 1$.

Step 3: Check convergence

    (a) If the convergence tolerance is satisfied, go to *Step 4*, otherwise go to next.

    (b) If $t < N_t$, then go to *Step 1*, otherwise go to Step 5.

Step 4: Output the solution of equilibrium point.

Step 5: No equilibrium point found.

The nonlinear Jacobi method does not use the most recently available information when computing $x_i(t + 1)$.

3.7.1.2 Nonlinear Gauss-Seidel Method The difference between the nonlinear Jacobi method and the nonlinear Gauss-Seidel method is that, in the latter method, the most recently available information is used in computing $x_i(t + 1)$. The solution procedure of the nonlinear Gauss-Seidel method is very similar to that of the nonlinear Jacobi method above, except that Step 1 should be modified as follows:

For each $i = 1, \ldots, K$, the MPEC($\hat{x}_i(t)$) can be solved by either NLP solvers or MPEC solvers and $\hat{x}_i(t) = (x_1^*(t+1), \ldots, x_{i-1}^*(t+1), x_i(t), x_{i+1}^*(t), \ldots, x_K^*(t))$ while $x_1^*(t+1), \ldots, x_{i-1}^*(t+1)$, $x_{i+1}^*(t), \ldots, x_K^*(t)$ are fixed.

3.7.2 Simultaneous Solution Methods

A sequential nonlinear complementarity problem (SNCP) approach for solving EPECs is presented in [133], which is related to the relaxation technique for solving MPECs that relaxes the complementarity conditions and drives the relaxation parameter to zero [95]. In this proposed method, the complementarity constraints in each MPEC are simultaneously relaxed, and an EPEC is then solved by solving a sequence of nonlinear complementarity problems. In contrast, in Scholtes's regularized scheme, MPECs can be solved by solving a sequence of nonlinear programs (NLPs). The NCP of the EPEC in (3.24)–(3.27) can be relaxed and then a combined set of KKT conditions of MPECs can be solved by a single SNCP algorithm. The SNCP algorithm is considered to be in the category of simultaneous solution methods.

In [103], two approaches have been proposed to solve the multi-leader-common-follower games formulated as EPECs. The first approach is based on the strong stationarity conditions of each leader where NCP, NLP, and MPEC formulations have been derived. In the second approach, an additional restriction, called price consistency, is imposed and this results in a square nonlinear complementarity problem. In both approaches, standard nonlinear optimization software can be extended to solve EPECs. In comparison to the conventional approaches, that is, diagonalization solution methods, the EPEC is solved by a single optimization problem, which belong to the category of simultaneous solution methods.

In [77], the KKT conditions of the EPEC combine a set of KKT conditions of leaders' MPECs. Then the NCP function, namely, the Fischer-Burmeister function, is introduced to convert the complementarity constraints into the NCP functions. The inexact Levenberg-Marquardt algorithm for solving semi-smooth systems of equations by Newton-type method is applied to solve the aggregated KKT conditions of the EPEC involving nonlinear equations and NCP functions. The theoretic background of the algorithms is referred to in [134] while the application of such algorithm in SFE model is given by [77].

In [103], an interior point approach has been proposed to solve the EPEC problem where the KKT conditions of the EPEC combine a set of KKT conditions of leaders' MPECs and NCPs are transformed into NCP functions, that is, the Fischer-Burmeister functions. Similar to the approaches in [103, 133], the EPEC is solved by a single optimization problem.

In [105], an interior point approach has been proposed to solve the SFE problem as an EPEC where AC representation of the power system network is utilized. The benefit of the AC formulation is that there are no limitations in terms of network control modeling. In this approach, the NCP are transformed into NCP functions, that is, the Fischer-Burmeister functions. The combined set of KKT conditions of MPECs are then solved by Newton's method. Again, the EPEC problem here is formulated as a single optimization problem. The proposed interior point approach has also been applied in the analysis of the impact of transformer control on SFE in [135]. Case studies have been carried out on the IEEE test systems including the IEEE 118-bus system.

In [104], oligopolistic competition in a centralized power market, characterized by a multi-leader single-follower game, EPEC, has been formulated as a NLP problem where an AC network model is used in rectangular coordinates. The follower's problem has been transformed to mixed nonlinear complementarity constraints of the leaders' problems. The KKT conditions of each leader's MPEC can be combined. To overcome the violation of the constraint qualifications, all the nonlinear complementarity conditions have been formulated as NLP constraints, and all the terms of the nonlinear complementarity conditions are included in an objective function to be minimized. Then, the resulting problem can be solved by NLP solvers. In this work, the NLP solver SNOPT [146] has been used under GAMS, via NEOS. For the EPEC solution technique here, an objective value of zero means that a feasible solution has been found for the EPEC. Test results on 3-bus and 14-bus systems have been presented.

3.8   TECHNICAL CHALLENGES FOR SOLVING MPECs AND EPECs

In the past, various special algorithms have been developed for solving MPECs. The MPEC algorithms include branch-and-bound methods, implicit nonsmooth approaches, SQP methods, and perturbation and penalization approaches. Early attempts to solve MPECs using NLP approaches were not encouraging and solving MPECs using NLP solvers was not considered as numerically safe. The major technical challenge with NLP formulations is that they violate the Mangasarian-Fromovitz constraint qualification (MFCQ) at any feasible point. The consequence of this is that (a) the set of multipliers is unbounded; (b) the central path fails to exist; and (c) the active constraint normals are linearly dependent, and hence linearizations can become inconsistent arbitrarily close to a solution. In recent years, research work has demonstrated that MPECs can be solved reliably and efficiently by replacing the complementarity constraint $Ys = 0$, $y \geq 0$, $s \geq 0$ with $Ys \leq 0$, $y \geq 0$, $s \geq 0$, and then standard nonlinear optimization solvers can be applied.

EPECs have been considered a very useful tool in electricity market equilibrium modeling and analysis. For the case of electricity market modeling of bilevel game systems as EPECs, the technical difficulty is that, due to the strong dependence of the lower-level equilibrium solution on the upper-level parameters, nonconvexity in each player's objective function will be created. The consequence of this would be that there may be no equilibrium at all or the problem is very hard to solve. Although so far theoretic convergence analysis of EPECs algorithms has been relatively limited, EPECs algorithms have been very successful in practice, in particular, the algorithms of NLP solvers. In the past, EPECs have been solved using the so-called diagonalization solution methods, where each MPEC of the EPEC is solved one MPEC at a time and the same procedure for every MPEC is cyclically repeated until an equilibrium point is found. Recent success of simultaneous solution methods has encouraged researchers to solve more complicated formulations of EPECs in electricity market equilibrium modeling and market behavior analysis where AC network representation and various control devices can be incorporated, and hence the interactions between various controls can be modeled. In addition, both the diagonalization solution methods and simultaneous solution methods have the great potential to be applied in solving large-scale electricity market modeling and analysis. In light of this, new reliable and efficient NLP solvers are always welcome. In the meantime, convergence analysis of NLP algorithms for EPECs will be helpful to design more robust algorithms that can be used to deal with more realistic large-scale electricity market problems.

Most of the solution techniques for EPEC rely on efficient and robust MPEC solvers. However, there is not such an MPEC solver yet that can guarantee convergence to strongly stationary points under reasonable conditions. Another important question is how to formulate and presolve NCPs, MPECs, and EPECs so that the solvers can take advantage of special problem structures being solved. For instance, the electricity networks have very special structures where each node has only direct connection to a few adjacent nodes. The resulting MPECs or EPECs are also very sparse. Such property should be fully considered in the implementation of the computer code.

3.9   SOFTWARE RESOURCES FOR LARGE-SCALE NONLINEAR OPTIMIZATION

The MPEC, MPCC, and EPEC problems can be formulated as nonlinear programming problems, and they can then be solved by some large-scale nonlinear optimization software packages. There are a number of large-scale nonlinear optimization software packages available. Some of these include CONOPT [138], Filter [139], IPOPT [140], KNITRO [141], LANCELOT [142], LOQO [143], MINOS [144], PENNON [145], and SNOPT [146]. Further information about these software packages is given as follows:

CONOPT

Author: A. S. Drud

www: http://www.conopt.com/

Method: A feasible path solver based on the old proven GRG method with many newer extensions

Interface: AIMMS, AMPL, GAMS, LINGO, MPL, TOMLAB Optimization Environment for MATLAB, Fortran

Availability: Commercial

Remark: CONOPT is a solver for large-scale nonlinear optimization (NLP).

FILTER

Authors: R. Fletcher and S. Leyffer

www: http://neos.mcs.anl.gov/neos/solvers/nco:filter/AMPL.html

Method: SQP Filter

Interface: AMPL, CUTE, C/C++

Availability: NEOS

Remark: Suitable for solving large nonlinearly constrained problems.

IPOPT

Authors: A. Wächter, L. Biegler, A. Raghunathan, and Y.-D. Lang

www: http://www.coin-or.org/Ipopt/ipopt-fortran.html

Method: Interior point

Interface: Fortran/C, CUTE, AMPL

Availability: Commercial, test version free, NEOS

Remark: One of the most efficient codes for large-scale nonconvex problems.

KNITRO

Authors: R. Byrd, J. Nocedal

www: http://www.ziena.com/knitro.html

Method: Penalty method for constraints, subproblems solved by SQP-Trust Region

Interface: Fortran/C, AMPL, GAMS, Matlab

Availability: Commercial, test version free, NEOS

Remark: One of the most efficient codes for large-scale nonconvex problems.

LANCELOT

Authors: A. Conn, N. Gould, P. Toint

www: http://www.numerical.rl.ac.uk/lancelot/distribution.html

Method: Classic augmented Lagrangian method

Interface: Fortran/C, CUTE, AMPL

Availability: Commercial and academic version, NEOS

Remark: Older

LOQO

Authors: H. Benson, R. Vanderbei

www: http://www.princeton.edu/~rvdb/loqo/LOQO

Interface: Fortran/C, AMPL, Matlab

Availability: Commercial, test version free, NEOS

Remark: Very efficient for convex problems, not so robust for nonconvex ones.

MINOS

Authors: B. A. Murtagh and M. Saunders

www: http://www.sbsi-sol-optimize.com/asp/sol_product_minos.htm

Interface: Fortran, Matlab, CUTE, GAMS and AMPL

Availability: Commercial, test version free, NEOS

Remark: A software package for solving large-scale optimization problems (linear and nonlinear programs).

PENNON

Authors: M. Kočvara, M. Stingl

www: www.penopt.com

Method: Generalized Augmented Lagrangian method

Interface: Fortran/C, AMPL, matlab

Availability: Commercial, test version free, NEOS

SNOPT

Authors: P. Gill, W. Murray, M. Sounders

www: http://www.sbsi-sol-optimize.com/asp/sol_product_snopt.htm

Method: SQP-BFGS

Interface: Fortran/C, AMPL, GAMS, Matlab

Availability: Commercial and academic version, NEOS

Remark: One of the most efficient codes for highly nonlinear problems, first-order method

MPEC World can be found following the web link: http://www.gamsworld.org/mpec/

"MPEC World" is a forum for about all aspects of mathematical programs with equilibrium constraints, including practical software (MPEC Solvers), testing, comparison, and quality of solvers (MPECLib), research activities in both solution methods and model formulations, and in improving the communication between people interested in these topics.

A small collection of equilibrium problems with equilibrium constraints (EPECs) can be found at the website: http://www.mcs.anl.gov/~leyffer/macepec/

Complementarity Problem Net (CPNET) maintained by Professor Michael C. Ferris at the University of Wisconsin–Madison provides resources about mixed complementarity problem research. The web link for CPNET is as follows: http://www.cs.wisc.edu/cpnet/

Software packages for solving the mixed complementarity problem, which can be found at the above website, include PATH [147], MILES [148], SEMISMOOTH [149], NE/SQP [150], and SMOOTH [151].

REFERENCES

1. Osborne MJ. *An Introduction to Game Theory*. New York: Oxford University Press; 2004.
2. Singh H, editor. *IEEE Tutorial on Game Theory Applications in Electric Power Markets*. New York: IEEE Power Engineering Society; 1999.
3. Rothwell G, Gomez T. *Electricity Economics: Regulation and Deregulation*. Hoboken, NJ: IEEE Press, Wiley Interscience; 2003.
4. Shahidehpour M, Yamin H, Li Z. *Market Operations in Electric Power Systems: Forecasting, Scheduling and Risk Management*. Hoboken, NJ: IEEE Press, Wiley Interscience; 2002.
5. International Energy Agency. *Competition in Electricity Markets*. Paris: OECD/IEA; 2001.
6. Newbery DM. Power markets and market power. *Energy Journal* 1995;16(3):41–66.
7. Kirschen DS, Strbac G. *Fundamentals of Power System Economics*. Chichester: John Wiley & Sons Ltd; 2004.
8. Taylor EO, Boal GA. *Power System Economics*. London: Edward Arnold (Publishers) Ltd, London; 1969.
9. Falk J. Determining market power in deregulated generation markets by measuring price-cost margins. *The Electricity Journal* 1998;11(6):44–50.
10. Stoft S. *Power System Economics: Designing Markets for Electricity*. Hoboken, NJ: IEEE Press, Wiley Interscience; 2002.
11. David AK, Wen F. Market power in electricity supply. *IEEE Transactions on Power Systems*, 2001;16(4):352–360.
12. Borenstein S, Bushnell J, Kahn E, Stoft S. Market power in California electricity markets. *Utilities Policy* 1995;5(3/4):219–236.
13. Borenstein S, Bushnell J, Stoft S. The competitive effects of transmission capacity in a deregulated electricity industry. *RAND Journal of Economics* 2000;31(2):294–325.
14. Borenstein S, Bushnell J. An empirical analysis of potential for market power in California's electricity industry. *The Journal of Industrial Economics* 1999;47(3):285–323.
15. Cunningham LB, Baldick R, Baughman ML. An empirical study of applied game theory: transmission constrained Cournot behaviour. *IEEE Transactions on Power Systems* 2002;17(1): 166–172.
16. Hobbs BF. Linear complementarity models of Nash-Cournot competition in bilateral and POOLCO power markets. *IEEE Transactions on Power Systems* 2001;16(2):194–202.
17. Nash J. Non-cooperative games. *The Annals of Mathematics* 1951;54(2):286–295.
18. Hobbs BF, Metzler CB, Pang J-S. Strategic gaming analysis for electric power systems: an MPEC approach. *IEEE Transactions on Power Systems* 2000;15(2):638–645.
19. Kahn EP. Numerical techniques for analysing market power in electricity. *The Electricity Journal* 1998;11(6):34–43.
20. Green R, Newbery DM. Competition in the British electricity spot market. *Journal of Political Economy* 1992;100(5):929–953.
21. Klemperer PD, Meyer MA. Supply function equilibria in oligopoly under uncertainty. *Econometrica* 1989;157(6):1243–1277.
22. de Lujan Latorre M, Granville S. The Stackelberg equilibrium applied to AC power systems—A non-interior point algorithm. *IEEE Transactions on Power Systems* 2003;18(2):611–618.
23. Day J, Hobbs BF, Pang J-S. Oligopolistic competition in power networks: a conjecture supply function approach. *IEEE Transactions on Power Systems* 2002;17(3):597–607.
24. Song Y, Ni Y, Wen F, Hou Z, Wu FF. Conjectural variation based bidding strategy in spot markets: fundamentals and comparison with classical game theoretical bidding strategies. *Electric Power Systems Research* 2003;67(1):45–51.
25. Peng T, Tomsovic K. Optimal bidding strategies: an empirical conjectural approach. In: Lee KY, Shin M-C, eds. Proceedings of the IFAC Symposium on Power Plants and Power Systems Control, Seoul, Korea, September 2003.
26. Chen Y, Hobbs BF. An oligopolistic power market model with tradable NOx permits. *IEEE Transactions on Power Systems* 2005;20(1):119–129.
27. Borenstein S. Understanding competitive pricing and market power in wholesale electricity markets. *The Electricity Journal* 2000;13(6):49–57.

28. Hogan WW. A market power model with strategic interaction in electricity networks. *Energy Journal* 1997;18(4):107–141.

29. Cardell JB, Hitt CC, Hogan WW. Market power and strategic interaction in electricity networks. *Resource & Energy Economics* 1997;19(1):109–137.

30. Green R. Increasing competition in the British electricity spot market. *Journal of Industrial Economics* 1996;44(2):205–216.

31. Newbery DM. Competition, contracts, and entry in the electricity spot market. *RAND Journal of Economics* 1998;29(4):726–749.

32. Green R. The electricity contract market in England and Wales. *Journal of Industrial Economics* 1999;147(1):107–124.

33. Morch von der Fehr N-H, Harbord D. Spot market competition in the UK electricity industry. *The Economic Journal* 1993;103(418):531–546.

34. Day J, Bunn DW. Divestiture of generation assets in electricity pool of England and Wales: a computational approach to analysing market power. *Journal of Regulatory Economics* 2001;19(2): 123–141.

35. Baldick R, Grant R, Kahn E. Linear supply function equilibrium: generalisations, application, and limitations. Working Paper PWP-078, University of California, August 2000.

36. Halseth A. Market power in the Nordic electricity market. *Utilities Policy* 1998;7(4): 259–268.

37. Brennan D, Melanie J. Market power in the Australian power market. *Energy Economics* 1998;20(2):121–133.

38. Boisseleau FHAR, Kristiansen T, Petrov K, Van der Veen W. A supply function equilibrium model with forward contracts—An application to wholesale electricity markets. In: The Proceedings of the 6th IAEE European Conference: Modelling in Energy Economics and Policy, Zurich, Switzerland, 2–3 Sept. 2004, pp. 1–27.

39. Anderson J, Xu H. Nash equilibria in electricity markets with discrete prices. *Mathematical Methods of Operations Research* 2004;60(2):215–238.

40. Metzler C, Hobbs BF, Pang J-S. Nash-Cournot equilibria in power markets on a linearised DC network with arbitrage: formulations and properties. *Networks and Spatial Economics* 2003;3(2): 123–150.

41. Bunn DW, editor. *Modelling Process in Competitive Electricity Markets*. Chichester: John Wiley & Sons, Inc; 2004.

42. Ramos A, Ventosa M, Rivier M. Modelling competition in electric energy markets by equilibrium constraints. *Utilities Policy* 1999;7(4):233–242.

43. Yao J, Oren SS, Adler I. Computing Cournot equilibria in two settlement electricity markets with transmission constraints. In: Proceedings of the 37th Hawaii International Conference on System Sciences, HICSS37, Big Island, Hawaii, Jan. 5–8, 2004.

44. Yao J, Oren SS, Adler I. Cournot equilibria in two settlement electricity markets with system contingencies. *International Journal of Critical Infrastructure* 2007;3(1/2):142–160.

45. Yao J, Willems B, Oren SS, Adler I. Cournot equilibrium in price-capped two settlement electricity markets. In: Proceeding of the 38th Hawaii International Conference on Systems Sciences HICSS38. Big Island, Hawaii, January 3–6, 2005.

46. Yao J, Adler I, Oren SS. Modelling and computing two settlement oligopolistic equilibrium in a congested electricity network. University of California Energy Institute's Energy Policy and Economics Working Paper Series, University of California at Berkeley, June 2006.

47. Yao J, Oren SS, Hobbs BF. A hybrid-Bertrand-Cournot model of electricity markets with multiple sub-networks. Working Paper, Department of Industrial Engineering and Operations Research, University of California at Berkeley, March 2006.

48. Otero-Novas I, Meseguer C, Batlle C, Alba JJ. A simulation model for a competitive generation market. *IEEE Transactions on Power Systems* 2000;15(1):250–256.

49. Barquin J, Centeno E, Reneses J. Medium-term generation programming in competitive environments: a new optimisation approach for market equilibrium computing. *IEE Proceedings on Generation Transmission & Distribution* 2004;151(1):119–126.

50. Conteras J, Candiles O, de la Fuente JI, Gomez T. A Cobweb bidding model for competitive electricity markets. *IEEE Transactions on Power Systems* 2002;17(1):148–153.

51. Chen H, Wong KP, Nguyen DHM, Chung CY. Analysing oligopolistic electricity market using a co-evolutionary computation. *IEEE Transactions on Power Systems* 2006;21(1):143–152.

52. Peng T, Tomsovic K. Congestion influence on bidding strategies in an electricity market. *IEEE Transactions on Power Systems* 2003;18(3):1054–1061.
53. Gountis VP, Bakirtzis AG. Efficient determination of Cournot equilibria in electricity markets. *IEEE Transactions on Power Systems* 2004;19(4):1837–1844.
54. Conteras J, Klusch M, Krawczyk JB. Numerical solutions to Nash-Cournot equilibria in coupled constant electricity markets. *IEEE Transactions on Power Systems* 2004;19(1):195–206.
55. Wen FS, David AK. Oligopolistic electricity market production under incomplete information. *IEEE Power Engineering Review* 2001;21(4):58–61.
56. Baldick R, Grant R, Kahn E. Theory and application of linear supply function equilibrium in electricity markets. *Journal of Regulatory Economics* 2004;25(2): 143–167.
57. Baldick R, Hogan W. Capacity constrained supply function equilibrium models of electricity markets: stability, non-decreasing constraints, and function space iterations. Working Paper PWP-089, University of California, December 2001.
58. Genc T, Reynolds SS. Supply function equilibria with pivotal electricity suppliers. Working Paper, University of Arizona, March 2004.
59. Holmberg P. Unique Supply function equilibrium with capacity constraints. *Energy Economics* 2008;30(1):148–172.
60. Anderson EJ, Philpott AB. Using supply functions for offering generation into an electricity market. *Operational Research* 2002;50(3):477–489.
61. Anderson EJ, Xu H. Supply function equilibrium in electricity markets with contracts and price caps. *Journal of Optimization Theory & Applications* 2005;124(2):257–283.
62. Younes Z, Ilic M. Generation strategies for gaming transmission constraints: will the deregulated electric power market be an oligopoly? *Decision Support Systems* 1999;24(3):207–222.
63. Berry CA, Hobbs BF, Meroney WA, O'Neil RP, Stewart Jr WR. Understanding how market power can arise in network competition: a game theoretic approach. *Utilities Policy* 1999;8(3):139–158.
64. Weber JD, Overbye TJ, DeMarco CL. Modelling the consumer benefit in the optimal power flow. *Decision Support Systems* 1999;24(3/4):279–296.
65. Hobbs BF. Network models of spatial oligopoly with an application to deregulation of electricity generation. *Operations Research* 1986;34(3):395–409.
66. Song Y, Ni Y, Wen F, Wu FF. Conjectural variation based learning model of strategic bidding in spot market. *Electrical Power and Energy Systems* 2004;26(10):797–804.
67. Yu Z. A market power model with price-caps and compact DC power flow constraints. *Electrical Power & Energy Systems* 2003;25(4):301–307.
68. Pereira MV, Granville S, Fampa MHC, Dix R, Barroso LA. Strategic bidding under uncertainty: a binary expansion approach. *IEEE Transactions on Power Systems* 2005;20 (1):180–188.
69. Weber JD, Overbye TJ. An individual welfare maximisation algorithm for electricity markets. *IEEE Transactions on Power Systems* 2002;17(3):590–596.
70. Chung TS, Zhang SH, Wong KP, Yu CW, Chung CY. Strategic forward contracting in electricity markets: modelling and analysis by equilibrium method. *IEE Proceedings on Generation Transmission & Distribution* 2004;151(2):141–149.
71. Smeers Y. Computable equilibrium models and the restructuring of the European electricity and gas markets. *Energy Journal* 1997;18(4):1–31.
72. Chattopadhyay D. Multi-commodity spatial Cournot model for generator bidding analysis. *IEEE Transactions on Power Systems* 2004;19(1):267–275.
73. Bompard E, Lu W, Napoli R. Network constraint impacts on competitive electricity markets under supply-side strategic bidding. *IEEE Transactions on Power Systems* 2006;21(1):160–170.
74. Bompard E, Ma YC, Napoli R, Jiang CW. Assesing the market power due to the network constraints in competitive electricity markets. *Electric Power Systems Research* 2006;76(11):953–961.
75. Al-Agtash S, Yamin HY. Optimal supply curve bidding using Benders decomposition in competitive electricity markets. *Electric Power Systems Research* 2004;71(3):245–255.
76. Niu H, Baldick R, Zhu G. Supply function equilibrium bidding strategies with fix forward contracts. *IEEE Transactions on Power Systems* 2005;20(4);1859–1867.
77. Xian W, Yuzeng L, Shaohua Z. Oligopolistic equilibrium analysis for electricity markets: a nonlinear complementarity approach. *IEEE Transactions on Power Systems* 2004;19(3):1348–1355.

78. Hobbs BF, Rijkers FAM. Strategic generation with conjectured transmission price responses in a mixed transmission pricing system—Part I: Formulation. *IEEE Transactions on Power Systems* 2004;19(2):707–717.

79. Hobbs BF, Rijkers FAM, Wals AF. Strategic generation with conjectured transmission price responses in a mixed transmission pricing system—Part II: Application. *IEEE Transactions on Power Systems* 2004;19(2):872–879.

80. Correia PF, Overbye TJ, Hiskens IA. Searching for non-cooperative equilibria in centralised electricity markets. *IEEE Transactions on Power Systems* 2003;18(4):1417–1424.

81. Liu YF, Ni YX, Wu FF. Impacts of suppliers' learning behaviour on market equilibrium under repeated linear supply function bidding. *IEE Proceedings on Generation Transmission & Distribution* 2006;153(1):44–50.

82. Nam YW, Yoon YT, Hur D, Park J-K, Kim S-S. Effects of long-term contracts on firms exercising market power in transmission constrained electricity markets. *Electric Power Systems Research* 2006;76(6/7):435–444.

83. Li T, Shahidehpour M. Strategic bidding of transmission-constrained GENCOs with incomplete information. *IEEE Transactions on Power Systems* 2005;20(1):437–447.

84. Hu X, Ralph D, Bardsley P, Ferris MC. Electricity generation with looped transmission networks: bidding to an ISO. Paper no. 2004/16, Judge Institute of Management, University of Cambridge, 2004.

85. Hu X, Ralph D. Using EPECs to model bi-level games in restructured electricity markets with locational prices. *Operations Research* 2007;55(5):809–827.

86. Baldick R. Electricity market equilibrium models: the effect of parameterisation. *IEEE Transactions on Power Systems* 2002;17(4):1170–1176.

87. Bautista G, Quintana VH, Aguado JA. An oligopolistic model of an intergrated market for energy and spinning reserve. *IEEE Transactions on Power Systems* 2006;21(1):132–142.

88. Hogan WW. Markets in real electric networks require reactive prices. *Energy Journal* 1993;14(3):201–216.

89. Harker PT, Pang JS. On the existence of optimal solutions to mathematical program with equilibrium constraints. *Operations Research Letters* 1998;7(2):61–64.

90. Luo Z-Q, Pang J-S, Ralph D. *Mathematical Programs with Equilibrium Constraints*. Cambridge: Cambridge University Press; 1996.

91. Outrata JV, Kočvara M, Zowe J. *Non-smooth Approach to Optimization Problems with Equilibrium Constraints—Theory Applications and Numerical Results*. Dordrecht-Boston-London: Kluwer Academic Publisher; 1998.

92. Kinderlehrer D, Stampacchia G. *An Introduction to Variational Inequalities and Their Applications*. New York: Academic Press; 1980.

93. Facchinei F, Pang J-S. *Finite-Dimensional Variational Inequalities and Complementarity Problems*. Volumes I and II. New York: Springer; 2004.

94. Chen Y, Florian M. The nonlinear bilevel programming problem: formulations, regularity and optimality conditions. *Optimization* 1995;32(23):193–209.

95. Scheel H, Scholtes S. Mathematical program with complementarity constraints: stationarity, optimality and sensitivity. *Mathematics of Operations Research* 2000;25(1):1–22.

96. Dempe S. *Foundations of Bilevel Programming*. Dordrecht: Kluwer Academie Publishers; 2002.

97. Dempe S. Annotated bibliography on bilevel programming and mathematical programs with equilibrium constraints. *Optimization* 2003;52(3):333–359.

98. Bard JF. A grid search algorithm for the linear bilevel programming problem. In: Proceeding of the 14th Annual Meeting of the American Institute for Decision Science, pp. 256–258, 1982.

99. Dempe S. A simple algorithm for the linear bilevel programming problem. *Optimization* 1987;18:373–385.

100. Judice J, Faustino A. The linear-quadratic bilevel programming problem. *INFOR* 1994;32:87–98.

101. Su C-L. A sequential NCP algorithm for solving equilibrium problems with equilibrium constraints. Technical report, Department of Management Science and Engineering, Stanford University, 2004.

102. Pang J-S, Fukushima M. Quasi-variational inequalities, generalized nash equilibrium, and multi-leader-follower games. *Computational Management Science* 2005;2(1):21–56.

103. Leyffer S, Munson TS. Solving multi-leader-follower games. Preprint ANL/MCS-P1243–0405, Argonne National Laboratory, Mathematics and Computer Science Division, April 2005, Revised March 2007.

104. Bautista G, Anjos MF, Vannelli A. Formulation of oligopolistic competition in AC power networks: an NLP approach. *IEEE Transactions on Power Systems* 2007;22(1):105–115.

105. Petoussis S, Zhang X-P, Godfrey K. Electricity market equilibrium analysis based on nonlinear interior point algorithm with complementary constraints. *IET—Generation, Transmission and Distribution* 2007;1(4):603–612.

106. Yao J, Adler I, Oren SS. Modeling and computing two-settlement oligopolistic equilibrium in a congested electricity network. *Operations Research* 2008;56(1):34–47.

107. Zhang D, Xu H, Wu Y. A Stochastic EPEC Model for Electricity Markets with Two Way Contracts, www.optimization-online.org, Feb. 2008.

108. Mordukhovich BS. Equilibrium problems with equilibrium constraints via multiobjective optimization. *Optimization Methods and Software* 2004;19(5):479–492.

109. Liu GS, Zhang JZ. A new branch and bound algorithm for solving quadratic programs with linear complementarity cosntraints. *Journal of Computational and Applied Mathematics* 2002;146: 77–87.

110. Leyffer S. Complementarity constraints as nonlinear equations—theory and numerical experience. In S. Dempe and V. Kalashnikov, editors. *Optimization with Multivalued Mappings*. New York: Springer, pp. 169–208, 2006.

111. Fischer A. A Newton-type method for positive semi-definite linear complementarity problems. *Journal of Optimization Theory and Applications* 1995;86(3):585–608.

112. Chen B, Harker PT. Smooth approximations to nonlinear complementarity problems. *SIAM Journal of Optimization* 1997;7(2):403–420.

113. Chen B, Chen X, Kanzow C. A penalized Fischer-Burmeister NCP function. *Mathematical Programming* 2000;88(1):211–216.

114. Fletcher R, Leyffer S. Numerical experience with solving MPECs as NLPs. Numerical Analysis Report NA/210, Department of Mathematics and Computer Science, University of Dundee, Dundee, UK, 2002.

115. Fletcher R, Leyffer S, Ralph D, Scholtes S. Local convergence of SQP methods for mathematical programs with equilibrium constraints. *SIAM Journal on Optimization* 2006;17(1):259–286.

116. Benson HY, Sen A, Shanno DF, Vanderbei RJ. Interior-point algorithms, penalty methods and equilibrium problems. *Computational Optimization and Applications* 2006;34(2):155–182.

117. Raghunathan AU, Biegler LT. Mathematical programs with equilibrium constraints (MPECS) in process engineering. *Computers & Chemical Engineering* 2003;27(10):1381–1392.

118. Raghunathan AU, Biegler LT. An interior point method for mathematical programs with complementarity constraints (MPCCS). *SIAM Journal on Optimization* 2005;15(3):720–750.

119. Liu X, Sun J. Generalized stationary points and an interior-point method for mathematical programs with equilibrium constraints. *Mathematical Programming* 2004;101(1):231–261.

120. Leyffer S, Opez-Calva G, Nocedal J. Interior methods for mathematical programs with complementarity constraints. *SIAM Journal on Optimization* 2006;17(1):52–77.

121. Scholtes S. Convergence properties of a regularization scheme for mathematical programs with complementarity constraints. *SIAM Journal on Optimization* 2000;11(4):918–936.

122. Ralph D, Wright SJ. Some properties of regularization and penalization schemes for MPECs. *Optimization Methods and Software* 2004;19(5):527–556.

123. Demiguel V, Friedlander MP, Nogales FJ, Scholtes S. *SIAM Journal on Optimization* 2005;16(2):587–609.

124. Price TC. Using co-evolutionary programming to simulate strategic behaviour in markets. *Journal of Evolutionary Economics* 1997;7(3):219–313.

125. Son YS, Baldick R. Hybrid coevolutionary programming for nash equilibrium search in games with local optima. *IEEE Transactions on Evolutionary Computation* 2004;8(4):305–314.

126. Chen H, Wong KP, Chung CY, Nguyen DHM. A coevolutionary approach to analyzing supply function equilibrium model. *IEEE Transactions on Power Systems* 2006;21(3):1019–1028.

127. Hejazi SR, Memariani A, Jahanshahloo G, Sepehri MM. Linear bilevel programming solution by genetic algorithm. *Computers & Operations Research* 2002;29:1913–1925.

128. Calvete HI, Galé C, Mateo PM. A new approach for solving linear bilevel problems using genetic algorithms. *European Journal of Operational Research* 2008;188:14–28.

129. He B, Yang H. A neural-network model for monotone linear asymmetric variational inequalities. *IEEE Transactions on Neural Networks* 2000;11(1):3–16.

130. Pal BB, Moitra BN. A fuzzy goal programming procedure for solving quadratic bilevel programming problems. *International Journal of Intelligent Systems* 2003;18:529–540.

131. Wen UP, Huang AD. A simple Tabu search method to solve the mixed-integer linear bilevel programming problem. *European Journal of Operational Research* 1996;88:563–571.

132. Rajesh J, Gupta K, Kusumakar HS, Jayaraman VK, Kulkarni BD. A Tabu search based approach for solving a class of bilevel programming problems in chemical engineering. *Journal of Heuristics* 2003;9(4):307–319.

133. Su CL. A sequential NCP algorithm for solving equilibrium problems with equilibrium constraints. Technical Report, Department of Management Science and Engineering, Stanford University, 2004.

134. Facchinel F, Kanzow C. A non-smooth inexact Newton method for the solution of large scale nonlinear complementarity problems. *Mathematical Programming* 1997;76:493–512.

135. Petoussis SG, Petoussis AG, Zhang X-P, Godfrey KR. Impact of the transformer tap-ratio control on the electricity market equilibrium. *IEEE Transactions on Power Systems* 2008;23(1):65–75.

136. Hasan E, Galiana FD, Conejo AJ. Electricity markets cleared by merit order—Part I: Finding the market outcomes supported by pure strategy Nash Equilibria. *IEEE Transactions on Power Systems* 2008;23(2):361–371.

137. Gabriel SA, Leuthold FU. Solving discretely-constrained MPEC problems with applications in electric power markets. *Energy Economics* 2009, doi:10.1016/j.eneco.2009.03.008

138. Drud A. CONOPT—A large scale GRG code. *ORSA Journal on Computing* 1994;6(2):207–216.

139. Fletcher R, Leyffer S. Nonlinear programming without a penalty function. *Mathematical Programming* 2002;91(2):239–269.

140. Wächter A, Biegler LT. On the Implementation of an interior-point filter line-search algorithm for large-scale nonlinear programming. *Mathematical Programming* 2006;106(1):25–57.

141. Byrd RH, Nocedal J, Waltz RA. KNITRO: An integrated package for nonlinear optimization. In G. di Pillo and M. Roma, editors. *Large-Scale Nonlinear Optimization*, pp. 35–59, Heidelberg, Berlin, New York: Springer Verlag, 2006.

142. Conn AR, Gould NIM, Toint PL. *LANCELOT: A Fortran Package for Large-Scale Nonlinear Optimization, Release A.* Series in Computational Mathematics 17. Berlin: Springer-Verlag, 1992.

143. Benson HY, Sen A, Shanno DF, Vanderbei RJ. Interior-point algorithms, penalty methods, and equilibrium problems. *Computational Optimization and Applications* 2006;34(2):155–182.

144. Murtagh BA, Saunders MA. MINOS 5.5 User's Guide. Report SOL 83–20R, Dept. of Operations Research, Stanford University, 1998.

145. Kocvara M, Stingl M. PENNON—A code for convex nonlinear and semidefinite programming. *Optimization Methods and Software* 2003;18(3):17–333.

146. Gill PE, Murray W, Saunders MA. SNOPT: an SQP algorithm for large-scale constrained optimization. *SIAM Review* 2005;47(1):99–131.

147. Ferris MC, Munson TS. Complementarity problems in GAMS and the PATH solver. *Journal of Economic Dynamics and Control* 2000;24:165–188.

148. http://www.gams.com/dd/docs/solvers/miles.pdf

149. Munson TS, Facchinei F, Ferris MC, Fischer A, Kanzow C. The semismooth algorithm for large scale complementarity problems. *INFORMS Journal on Computing* 2001;13:294–311.

150. Billups SC, Ferris MC. QPCOMP: a quadratic programming based solver for mixed complementarity problems. *Mathematical Programming* 1996;76:533–562.

151. Chen C, Mangasarian OL. A class of smoothing functions for nonlinear and mixed complementarity problems. *Computational Optimization and Applications* 1996;5(2):97–138.

CHAPTER 4

COMPUTING THE ELECTRICITY MARKET EQUILIBRIUM: USES OF MARKET EQUILIBRIUM MODELS

Ross Baldick

In this chapter, we discuss the formulation of electricity market equilibrium models, distinguishing the physical, commercial, and economic models. We outline the uses of such models, qualified in the light of the many assumptions that must be made for them to be tractable. We also discuss the types of questions that can be sensibly answered by such models.

4.1 INTRODUCTION

Electricity market equilibrium modeling has progressed significantly in the last two decades, both in terms of formulation and computability. In this chapter, we discuss equilibrium formulations and offer an assessment of where these models are useful, where they are not, and the prospects for improving them. The focus will be on offer-based economic dispatch models, such as in typical restructured markets in the United States.

Modern equilibrium theory began, as is wellknown, with the notion of a Nash equilibrium [1]. This unifying principle for understanding the interaction of decision makers has been pervasive in economics. It is natural that it should be applied to analyze electricity markets, particularly given the relatively well-defined cost structure for electric power generation and the prevalence in restructured electricity markets of a relatively small number of large market participants, who might reasonably be expected to maximize their profits.

On the other hand, the details of generator operating cost components and the technical engineering constraints on power system operation greatly complicate the application of economic analysis to electricity markets. Operating cost of generation is not just characterized by a convex function of the quantity produced; however,

Restructured Electric Power Systems: Analysis of Electricity Markets with Equilibrium Models,
Edited by Xiao-Ping Zhang
Copyright © 2010 Institute of Electrical and Electronics Engineers

this assumption of neoclassical economics underlies many formulations of equilibrium models [2]. On the demand side, considerable demand is simply not exposed to wholesale price variation or is exposed to temporally and geographically averaged prices, which greatly complicates the specification of a demand model.

Furthermore, the specification of electricity rules can be prodigiously complex. For example, the "Protocols" describing the "Nodal market" to be introduced in the Electric Reliability Council of Texas (ERCOT) in late 2010 run to around 500 pages. Almost uniquely amongst markets, electricity markets are run by essentially automated exchanges, so that details of software implementations can be crucial in understanding outcomes.

Finally, regulators almost always reserve the right to intervene if market outcomes are not satisfactory. Consequently, participant behavior can also depend on anticipated responses by regulators to behavior that is otherwise permitted under market rules, including market power rules.

As in most fields, any attempt to develop a tractable model must abstract away from at least some of the detail. However, the choices in electricity markets are particularly difficult, in part because experience with electricity markets is still accumulating and in part because there are several features of electricity markets, as suggested above, that are not features of other markets.

This chapter investigates some formulations of electricity market models, relates the modeling assumptions to market rules, and discusses the uses of such equilibrium models. Section 4.2 discusses model formulation. Section 4.3 discusses market operation and price formation. Definition of, computation of, difficulties with, and uses of equilibrium models are discussed in Sections 4.4, 4.5, 4.6, and 4.7, respectively. Section 4.8 concludes the chapter.

4.2 MODEL FORMULATION

This section draws from various sources, including [3–17], concerning the formulation of equilibrium models to discuss the forms of:

- The transmission network model
- The generator cost function and operating characteristics
- The offer function in the market, including the representation of ancillary services in the market
- The demand model
- Uncertainty

In each topic, the "economic model" that is used in the equilibrium formulation is distinguished from:

- The underlying "physical model," that is, a (notionally) exact model of the physical characteristics.
- The "commercial model," that is, the model of the physical characteristics used in the actual market.

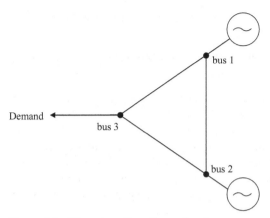

Figure 4.1   Three-bus, three-line network

For example, and as discussed further below, in the case of the offer function, the economic model involves the specification of the "strategic" variables for the market participants, which can differ significantly from the form of the offer required by the commercial model.

4.2.1   Transmission Network Model

4.2.1.1   Physical Model   Electric transmission systems consist of many buses and transmission lines, with power flow solutions specified by the nonlinear equality constraints arising from Kirchhoff's laws when power and reactive power are expressed in terms of voltage and current.

For example, consider the three-bus, three-line system illustrated in Figure 4.1. Each of the three buses is represented by a solid circle. To specify the state of the system, we need to specify the voltages at each of the buses. Under normal conditions, the voltages are well approximated by sinusoids, having a magnitude and phase-angle displacement from an arbitrary time reference. That is, the voltage at each bus is specified by two variables, corresponding to magnitude and angle.

Each pair of buses in Figure 4.1 is joined by a transmission line. At each bus, Kirchhoff's laws imply that the net real power leaving the bus is zero and that the net reactive power leaving the bus is zero [7]. The net real and reactive powers are both nonlinear functions of the voltage magnitudes and angles. Consequently, for each bus there are two nonlinear equations that must be satisfied.

In particular, at bus 1 in Figure 4.1, one of the nonlinear equations requires that the net real power injection by the generator at bus 1 equals the real power flows from the bus into the two lines connected to bus 1. The second equation requires that the net reactive power injection by the generator equals the reactive power flows into the lines. A similar pair of constraints applies at bus 2. At bus 3, one of the equations requires that the net real power withdrawn by the demand equals the real power flows into the bus out of the two lines connected to bus 3. The second equation requires that the net reactive power withdrawn by the demand equals the reactive power flows out of the lines. Satisfaction of these six equations for the system implies non-linear relationships between:

- The magnitudes and angles of the voltages at the buses in the system, and
- The injections by generators and withdrawals by demand.

The operation of electric transmissions systems is also subject to various linear and nonlinear inequality constraints such as line flows limited by thermal, voltage, and stability considerations [7]. For example, each line in Figure 4.1 will typically have a thermal limit, implying an inequality constraint corresponding to each line that can be expressed in terms of the voltage magnitudes and angles at the buses joined by the line, or approximated in terms of the power flow along the line.

In addition to the thermal limits on line flow, there may also be constraints on transmission that depend on particular generators being in-service. Furthermore, "nomogram" constraints that limit flows across corridors of lines are often represented as linear inequalities that are parameterized by the amount of generation in service as characterized in terms of its collective rotational inertia [9]. Again, these constraints are often approximated in terms of power flow along a line or lines.

When the inequality constraints on transmission bind, choices about dispatch of generation are more limited than they would otherwise be. In this case, the marginal value of generation, and by implication, the market clearing price, will vary with the location of the bus, or "nodally." This situation is sometimes referred to as "transmission congestion." Prices based on the value of generation as typically determined through a centralized optimization process are called "locational marginal prices" (LMPs).

4.2.1.2 Commercial Network Model Some of the detail of electricity networks is often hidden from market participants in that market prices may not be distinguished nodally, even when transmission constraints are binding. That is, the "commercial network model" may only represent some of the effects of the transmission network to the market.

At one extreme, there might be a single market price for most or all energy transacted. This was, roughly, the case in England and Wales in the 1990s [6] and was the case in the Electric Reliability Council of Texas (ERCOT) balancing market until 2002 [10].

At an intermediate level of representation, the commercial network model might include a representation of what are deemed to be major inter-regional transmission constraints. All the buses in each region are equivalenced into a single equivalent node, typically called a "zone." Consequently, "local" constraints within each zone are not apparent in market prices. For example, Figure 4.1 might be a three zone representation of a much more complicated network, with only three inter-regional transmission constraints represented by each "line" in Figure 4.1.

This approach has been used in ERCOT from 2002 [10] and was used in California in the late 1990s. In both cases, an approximate equivalent transmission system is defined to simplify Kirchhoff's laws and then proxy thermal limits on certain equivalent transmission lines or corridors are used to represent the limits on the original transmission system.

When proxy transmission constraints are represented in the commercial network model, there is a choice of whether to completely abstract from Kirchhoff's laws using a "transportation" model, as was used initially in the commercial network model for the California market, or use an equivalenced network that represents Kirchhoff's laws for the calculation of the flows on the lines that represent the proxy constraints.

Equivalenced networks could, in principle, preserve the nonlinear relationship between power and voltage; however, the nonlinear equations are typically linearized and the reactive power and voltage magnitude dependence eliminated to form the "DC power flow" [11].

As an example of DC power flow, consider again the system in Figure 4.1. For this system, the DC power flow would result in three equality constraints linking the real power injection and withdrawal at each bus and angles at each bus. An analogous linearization allows the real power flow on each line to be expressed in terms of the angles at each bus. Using the equality constraints to eliminate the angles then allows the real power flows on each line to be written in terms of the injections and withdrawals.

The approach of proxy constraints for an equivalenced network is taken in the ERCOT zonal market, in place from 2002 until late 2010, where a relatively small number of "congestion zones" are linked by "commercially significant constraints." In this commercial network model, all buses in each congestion zone are modeled essentially as being co-located in the commercial network model. Commercially significant constraints are proxy thermal limits assigned to certain groups of lines joining between the zones. The effect of dispatch in zones on flow on the commercially significant constraints is represented by average power transfer distribution factors for the buses in the zones.

At the other extreme, the commercial network model might represent considerable detail, as in the market models in the restructured markets in the Northeast United States, Midwest, and California and in the market model that will be introduced into ERCOT in late 2010. Large numbers of buses, lines, and line flow constraints, including contingency constraints, are explicitly represented in such commercial network models.

In all of these markets except that of the New York ISO, however, "DC power flow" is used to represent transmission in the day-ahead [11], with explicit consideration of reactive power and voltage magnitudes eliminated. That is, reactive power and voltage constraints are still, at best, represented as proxy thermal limits. Even in the New York ISO market, voltage constraints may not be enforced.

Even when the commercial network model is deliberately chosen to match the physical model as closely as possible, there will still typically be discrepancies. Consequently, in all markets there must also be some mechanism to deal with "out-of-market" issues. That is, there is some mechanism to deal with constraints that are not represented in the commercial network model or which are only approximately represented. For example, in a commercial network model that uses DC power flow, the proxy representation of voltage constraints in the commercial network model may not completely match the nature of the physical voltage constraints. When the outcome of the market would violate these physical constraints, some "out-of-market" measure must be taken, typically involving a "side payment" by the ISO to particular market participants in exchange for changing their generation in a way that aids satisfaction of the constraints.

Finally, in typical markets there are financial risk hedging instruments for energy and transmission prices. Energy price risk is primarily due to varying supply and demand conditions. Transmission price risk is due to the combination of regional variation of supply and demand together with limitations on transmission capacity.

Energy price risk hedging can be accomplished through bilateral contracts between generators and demand (or retailers). For example, a "contract for differences" arranges for a side-payment between demand and generator given by the product of:

- a contract quantity, multiplied by
- the difference between a strike price and the LMP at a particular bus.

If both generator and demand are interconnected at the particular bus and are both exposed to the LMP at that bus then the contract for differences costlessly hedges the variability of net payment for both of them for the contract quantity. There are many variations on this basic type of energy risk hedging contract.

If the generator and demand are not located at one bus or not exposed to the same price, however, the contract for differences cannot hedge all of their exposure to price risk: it can only, for example, hedge the variation in price at one or other of the buses. (Even if the generator and demand are located at one bus, typical commercial models expose demand to a "zonal" average of LMPs at several buses.) In this case, the generator or the demand is exposed to the difference in LMPs between the point of withdrawal minus the point of injection. The difference in LMPs is, by definition, the transmission price and so they are exposed to transmission price risk.

Financial transmission rights (FTRs) can be used to hedge this transmission price risk. FTRs are issued by the ISO to market participants either by allocation, or an auction process, or both [12]. FTRs are specified in terms of contract quantity, injection bus, and withdrawal bus, and pay based on the product of:

- a contract quantity, multiplied by
- the difference between an LMP at the bus of withdrawal (or zonal average of LMPs in the zone of withdrawal) minus the LMP at the bus of injection.

FTRs are funded out of the net revenue, or "congestion rent," that accrues to the ISO when transmission limits are binding and all energy is transacted at the LMPs of the associated buses.

To ensure that the congestion rent is adequate to fund the FTR obligations, it is sufficient that the "implied dispatch" of the FTRs is feasible for the system. For example, in Figure 4.1, there might be two types of FTRs issued:

1. injection at bus 1 and withdrawal at bus 3, and
2. injection at bus 2 and withdrawal at bus 3.

The implied dispatch for these FTRs would involve:

- Generation at bus 1 equal to the contract quantity of FTRs from bus 1 to bus 3
- Generation at bus 2 equal to the contract quantity of FTRs from bus 2 to bus 3, and
- Demand at bus 3 equal to the sum of the contract quantities of FTRs from bus 1 to bus 3 and from bus 2 to bus 3

If this implied dispatch satisfies the transmission constraints then the congestion rent is adequate to fund the FTRs [13].

4.2.1.3 Economic Model
Turning now to the implications for economic modeling of electricity markets, there are several issues involved. The first is that the economic model may only approximate the commercial network model. For example, in a commercial model with thermal transmission limits represented, the economic model might ignore these constraints or focus on time periods when the constraints are not binding [14].

The second issue is economic modeling of "out-of-market" actions. The most typical approach is to ignore out-of-market actions in the economic model, under the supposition that they make only a small qualitative difference to the overall outcome. However, in some cases, the effects of out-of-market actions may be part of the focus of economic modeling. An example of this would be an assessment of the economic significance of out-of-market actions.

As another example, the choice of the number and boundary of the zones in ERCOT affects the level of out-of-market actions. An analysis might seek to assess the effect of an increased number of zones on the cost of out-of-market actions.

The third issue is the modeling of FTRs, both in terms of the process of issuing them and how they affect market participant decisions.

A fourth consideration is the model of how participants represent the effect of their decisions on transmission constraints. For example, consider Figure 4.2, which shows, as a solid curve, the profit π_k of market participant k versus the value of its "strategic variables" s_k. As shown in Figure 4.2, the profit function has a kink at $s_k = 0$ because, as s_k varies from negative to positive, a particular transmission constraint changes from being binding to being nonbinding. Moreover, the profit is nonconcave, with two local maximizers, indicated by the solid circles.

The economic model of participant behavior may only partially reflect the effect of participant decisions on the transmission constraints, implicitly specifying that the participants are ignoring or cannot perceive some of the information available to them about transmission. For example, the economic model may approximate the profit function of market participant k by ignoring the change in functional form between positive and negative values of s_k. In Figure 4.2, the dashed curve extrapolates the profit function for positive values of s_k using the functional form that is applicable to negative values of s_k, simplifying the functional form of the profit function.

Such a simplification may be made in order to guarantee that the resulting profit function is concave, so that the first-order necessary conditions are sufficient for optimality. In the case of Figure 4.2, this simplification will result in a nonoptimal choice of strategic variable. This issue will be described further in Section 4.2.3.2

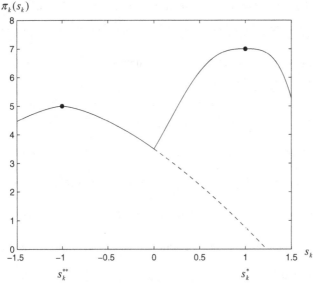

Figure 4.2   Profit function having multiple maximizers indicated by the solid circles

in the context of offer functions, since the offer function may implicitly determine this issue. The implications for equilibrium will be discussed in Section 4.4.

4.2.2   Generator Cost Function and Operating Characteristics

4.2.2.1 Physical Model Thermal generators have energy, start-up, and minimum load operating costs, together with ramp rate constraints and minimum up- and down-times. Ramp-rate constraints implicitly determine spinning reserve capability since spinning reserve is often defined as the sustained increased in generation available within 10 minutes, while generator start-up time sequences will implicitly determine the capacity to provide nonspinning reserves, since nonspinning reserves are typically required to be synchronized and injecting power into the system within, for example, 30 minutes.

Typically, energy cost functions for thermal generation are nonlinear functions of production, with marginal cost functions usually increasing with production, except at valving points and other changes in plant configuration. In contrast, typical hydro generators have low, roughly constant, marginal costs, but there is an opportunity cost associated with the use of the limited energy in a reservoir.

Nonzero minimum capacity limits for generators mean that the operating region for a typical generator is nonconvex, consisting of the operating point zero, together with an interval between minimum and maximum capacity. Even putting aside the nonconvexity of the operating region, start-up and minimum-load operating costs mean that the cost function is nonconvex, even if the energy costs are convex on the interval between minimum and maximum capacity.

4.2.2.2 Economic Model In typical restructured electricity markets, there are tens or hundreds of generators. However, historically, such generation assets have been owned by a relatively few market participants. In particular, since most restructured electricity markets were previously regulated monopolies or state-owned utilities, the ownership of generation will typically remain concentrated in each geographical area unless significant divestiture has occurred. Even when divestiture occurs, assets are often sold as a group, so that ownership may remain relatively concentrated.

Particularly when transmission limits are not modeled, it is common to model the cost function of participants by approximating the costs of a number of individual generators by a single equivalent portfolio-based cost function, typically corresponding to an affine or quadratic or a piecewise affine or piecewise quadratic marginal cost function. This abstracts various details such as unit commitment decisions and valving points, under the presumption that for any desired demand to be met by the participant, there is a well-defined commitment and dispatch configuration that would meet that demand. Naturally, this may be a poor approximation since the commitment configuration depends on more than just the demand level at a particular time.

4.2.3 Offer Function

4.2.3.1 Commercial Model Markets in the United States typically allow offers that represent a complex function that is aimed at capturing the details of the physical model of the generator cost function characteristics and parameters. Typical information in the offer would include, amongst other things:

- Minimum capacity
- Maximum capacity
- Start-up cost
- Minimum-load operating cost
- Offer for incremental energy above minimum capacity
- Ramp rate and reserve capability
- Start-up time

In contrast, the California Power Exchange market allowed for only energy offers.

In addition to continuous variables to represent the range of generation between minimum and maximum capacity, discrete variables are typically also used for thermal units in day-ahead, offer-based markets in the United States to model on-off status. There may be several continuous variables to represent not only energy but also "ancillary services" production. Ancillary services include spinning and non-spinning reserves and other services. Typical markets allow offers for energy that correspond to representing the marginal energy cost function as either a piecewise constant or piecewise linear function. Ancillary services are increasingly also being integrated into the same market offer structure as the energy offer [15].

Some markets, such as the day-ahead markets in England and Wales in the 1990s and in PJM, require that cost functions or parameters be held fixed over extended periods, such as a day, that consist of multiple pricing intervals having different demands. Other day-ahead markets allow for different offer costs every hour, although there may still be uncertainty at the time the offer is made about the demand in a particular hour. Typical "real-time" markets involve several pricing intervals over an hour, during which time offers are fixed. Some markets, such as New York, have restrictions on changes in some offer information if the change would lead to a substantive impact on market prices.

Bilateral energy contracts, such as contracts for differences, for hedging energy price variation are typically arranged between parties in addition to offers into the day-ahead and real-time markets [12]. Such contracts may not be publicly disclosed. Some markets also have "installed capacity markets," which provide payments for capacity with published clearing prices. Moreover, there are typically forward markets for energy and FTRs.

4.2.3.2 *Economic Model* Economic models typically abstract from the details of the commercial model, often paralleling, but simplifying, the form of the economic model of the cost function. For example, cost functions may be aggregated across portfolios in the economic model. If the economic model of the cost function is portfolio-based, then the economic model of the offer function may also be portfolio-based. For example, typical economic models of the England and Wales market in the 1990s use a portfolio offer representation, paralleling the economic model for the cost function, even though each individual generation unit was offered separately into the market [3–6]. In some cases, such as the ERCOT zonal balancing market, an economic model with portfolio offers may roughly match the commercial model. However, even in this case, there are likely to be differences between the economic and commercial model, particularly if transmission constraints are not fully represented in the economic model.

Although the commercial model typically requires an offer function, economic models may represent the effective outcome of the market through a hypothetical "strategic variable," such as quantity as in the Cournot model, that does not literally correspond to the commercial model.

For each participant that is modeled explicitly, the assumed form of its offer defines the strategic variable of the participant. For participant $k = 1, \ldots, n$, we will write s_k for the strategic variable of participant k. For example, in a single pricing interval Cournot model, s_k may specify the average power or the energy produced during the pricing interval. In a multiple pricing interval model, s_k may be a vector with elements corresponding to each pricing interval or to groups of pricing intervals. In these cases, the strategic variable s_k is finite dimensional.

The strategic variable may be more complex. For example, in a supply function model, the strategic variable s_k is a function or a set of parameters representing a function that is (almost everywhere) the inverse of the offer into the market. In the case that the strategic variable s_k is a function, it may be infinite dimensional. The relationship between the number of parameters specifying the strategic variable and the model of demand can have significant implications for the results of the model as will be discussed in Section 4.4 [16].

When transmission is represented in the economic model, the interaction between the values of strategic variables and the transmission constraints must be modeled. This may implicitly define the model of how participants believe that they affect transmission. For example, as discussed in Section 4.2.1.3, in models with transmission constraints, participants may be represented as not understanding that they can directly affect whether or not a particular transmission constraint is binding. That is, this effect is not accurately modeled in the representation of their profits. Whether this assumption is appropriate or not depends on the sophistication of market participants.

An analogous issue arises in the context of generator capacity constraints. Again, representation of this issue into participant profit functions may only be approximate. Finally, bilateral contracts for energy and financial transmission rights affect economic outcomes by changing the exposure of market participants to prices. Since quantities and prices in bilateral energy contracts may not be publicly disclosed, this poses problems for modeling outcomes.

4.2.4 Demand

4.2.4.1 Physical Model In many restructured electricity markets, demand is not exposed to wholesale electricity prices. Consequently, the price sensitivity of demand to wholesale price, at least in the short-term, is essentially zero. In some markets, particularly in day-ahead markets, large demands and retailers may have some facility to specify their willingness to pay for energy and consequently there may be some demand price elasticity at any given time.

Over time, demand varies, with typically a diurnal cycle that is modulated by weather, human cycles such as weekdays-weekends and public holidays, and season. The temporal variation of demand over a day is usually much larger than the amount of demand variation that can be effected by typical changes in price over a day. Moreover, uncertainty in issues such as weather that affect demand means that demand is itself uncertain. Uncertainty will be discussed in Section 4.2.5.

4.2.4.2 Commercial Model Demand is typically forecasted, both in the short and long-term, by various market participants. When there is price-responsive demand, a forecast of non-price-responsive demand can be added to the specification of price-responsive demand. Demand uncertainty in the forecast is often managed through the combination of day-ahead and real-time markets.

4.2.4.3 Economic Model

4.2.4.3.1 Price Elasticity As will be discussed below, economic equilibrium models, particularly Cournot models, are extremely sensitive to the specification of demand price elasticity. Because of this, demand is often specified in economic models as having far larger short-term price elasticity than is actually physically present in the market. There are at least two interpretations of this, both of which may apply simultaneously.

The first interpretation is that the assumed price elasticity is simply a calibration to the observed market behavior. While it is reasonable to calibrate model parameters to observed behavior, such calibration significantly undercuts the predictive value of such models, except possibly in the context of sensitivity analysis. As will be discussed in Section 4.3.3, specification of price elasticity may also attempt to mimic administrative price formation under conditions where supply and demand do not cross and it becomes necessary either to curtail demand or deploy reserves to meet demand.

The second interpretation is that the assumed price elasticity is not only due to actual demand response, but is also a proxy for unmodeled "competitive" market participants, such as owners of small shares of generation. That is, some of the price elasticity is a representation of participants whose market share is small enough that profit maximization for them involves offering at marginal cost. Since the anticipated behavior of such small participants can be determined in advance, it is therefore subsumed into a "residual demand" faced by the larger participants.

4.2.4.3.2 Temporal Variation Turning to temporal variation in demand, some models focus on a particular moment or pricing interval over which the demand forecast would be roughly constant. Such single pricing interval models do not capture interactions when market rules require that offers remain constant over multiple pricing intervals in a day, as in the day-ahead markets in England and Wales market in the 1990s, and in an hour, as in real-time markets. Single pricing interval models also do not capture demand uncertainty. In multiple pricing interval models that focus on day-ahead markets, the role of uncertainty in demand may be somewhat hidden or incorporated into the temporal variation of demand. Uncertainty will be discussed in more detail in Section 4.2.5.

4.2.5 Uncertainty

Many decisions in the operation of electric power systems must be taken in advance of full knowledge. Issues relating to decision making under uncertainty are therefore also relevant to electricity markets [17].

4.2.5.1 Physical Model Many electricity market parameters, such as demand, residual demand, fuel costs and availability, and equipment capacity, are stochastic.

4.2.5.2 Commercial Model Typical commercial models incorporate a recognition of the uncertainty of equipment availability through the incorporation of reserves. Stochastic demand is accommodated through the use of real-time markets with:

- net payments to generators based on the deviation of actual generation from day-ahead schedule or financial position, and
- net charges to demand based on the deviation of actual demand from day-ahead schedule, financial position, or forecast.

4.2.5.3 Economic Model While stochastic issues can be incorporated into the models, stochastic parameters other than demand are typically not explicitly modeled. In many cases, real-time markets may not be explicitly modeled either, under the assumption that the forecast of demand used in the day-ahead market is typically negligibly different from the real-time demand and that the addition of the real-time market to the economic model would not change the results significantly. Conversely, some models of real-time markets may not represent the day-ahead market or other forward markets.

4.3 MARKET OPERATION AND PRICE FORMATION

4.3.1 Physical Model

In most other markets besides electricity, there is at least some price elasticity of demand, the possibility of storage of product, and relatively lax transportation constraints, so that total supply and total demand can essentially always be assumed to intersect in a single market clearing price. However, in electricity markets, the lack of storability of electricity and the lack of price elasticity of demand implies that scarcity of electricity is not only possible, but would occur in the absence of active management of the market by an independent system operator. In fact, supply and demand is equated in electricity markets on a moment by moment basis by the operation of ancillary services. Moreover, because of limitations on transmission, market clearing prices will typically vary geographically.

4.3.2 Commercial Model

The commercial model typically abstracts from the active need for supply and demand to be balanced and models the crossing of supply and forecast demand. As mentioned above, ancillary services are used to deal with imbalances between supply and demand. When the imbalances persist for longer than a single pricing interval then energy prices may be set based on an administrative rule or may not be explicitly defined or may rely on post-market calculations.

As described in Section 4.2.1.2, most restructured markets in North America represent some of the effects of transmission constraints on market prices. LMPs are defined for each bus in the commercial network model; however, the commercial network model is typically simplified compared to the physical model.

Furthermore, electricity markets are typically organized as single clearing price markets. That is, at each bus, all accepted offers are paid and all demand pays the LMP, or some approximation to it. At each bus, the LMP is the market clearing price. However, there have been proposals for "pay-as-bid" markets where each accepted offer is paid its offer price instead of the LMP [18].

Moreover, for various other reasons, including market power concerns, prices may be set at a level other than the market clearing prices that would equate supply to demand. This is particularly prevalent under conditions of scarcity when ancillary services are used to match supply and demand. The presence of installed capacity markets and of unit commitment decisions in the market model implicitly couples decisions across multiple pricing intervals, even if energy offers are not required to be consistent across multiple pricing intervals.

Finally, there are typically both day-ahead and real-time markets. Settlement quantities in the real-time markets are based on deviations from forward positions in the day-ahead market.

4.3.3 Economic Model

Similarly, typical economic models abstract from the active need for supply and demand to be matched in electricity markets and simply model the crossing of supply and demand curves as specified by the economic model. In cases where supply and demand do not meet, there is a need to define the effect that would occur in the commercial model due to using ancillary services. As mentioned above, in some cases, such actions may be modeled in part by assumed price elasticity of demand, even if such price elasticity is actually absent from the market.

In a Cournot model, where the economic model involves participants setting quantities, it is essential to have demand price elasticity in order to have a well-defined price. In Cournot models with transmission constraints represented, in order to obtain well-defined nodal prices it is generally necessary to assume that demand at each bus is price responsive. This may be an extremely poor representation of actual demand elasticity.

Installed capacity markets and unit commitment issues may not be represented explicitly under the assumption that the energy market is not (directly) affected by installed capacity and unit commitment issues. As will be mentioned in Section 4.4, the incorporation of the discrete values required for a unit commitment model poses great difficulties for the calculation of equilibria because the nonconvexity of the feasible region of the underlying unit commitment problems can mean that there is no equilibrium or that there are multiple equlibria.

Moreover, many models consider only the day-ahead or only the real-time market, even though both markets operate jointly and an appropriate model would consider the joint equilibrium [19] and consider interaction with other forward markets for energy and FTRs. A related issue is that "virtual bids" allow arbitrage between the day-ahead and real-time markets, affecting the appropriate model of the linkage between these markets.

4.4 EQUILIBRIUM DEFINITION

The modeled Nash equilibrium of a market is a set of participant offers, as explicitly or implicitly represented by its strategic variables, such that no participant can improve its profit by unilaterally deviating from the offer. Nash equilibrium therefore ignores collusion. Formally, if there are modeled participants $k = 1, \dots, n$, with strategic variables $s_k: k = 1, \dots, n$, then we can implicitly calculate the profit in, for example, the day-ahead or real-time market, to any participant k due to the choice of strategic variables by all the participants. That is, the choice of the strategic variables leads to a market clearing price and quantities of production that, together with the cost function, allows specification of the profit.

The modeling choice of a particular strategic variable should be a reflection of the flexibility in the offer rules into the market and of the decision process undertaken by market participants. Some models, such as Cournot, only indirectly reflect the flexibility in offer rules.

Given the choice of strategic variables, we assume that the profit for participant k is specified by $\pi_k(s_k; s_{-k})$, where $s_{-k} = (s_\ell)_{\ell \neq k}$ is the collection of strategic variables of all the participants besides participant k. Then $(s_k^*)_{k=1,\ldots,n}$ is a "pure strategy" equilibrium if:

$$\forall k = 1, \ldots, n, s_k^* \in \arg\max_{s_k} \pi_k(s_k; s_{-k}^*), \tag{4.1}$$

where $s_{-k}^* = (s_\ell^*)_{\ell \neq k}$.

If there is no choice of $(s_k^*)_{k=1,\ldots,n}$ satisfying (4.1) then there is no pure strategy equilibrium. This can sometimes occur because of the nonconcavity of the profit function, as illustrated in Figure 4.2, or because of the nonconvexity of the feasible operating region, as discussed in Sections 4.2.2 and 4.3.3.

When there is no pure strategy equilibrium, there may be a "mixed strategy" equilibrium where the strategic variables are assumed to be chosen randomly and independently by each participant. Furthermore, relaxations of the equilibrium concept, such as allowing for local maximizers of the profit functions as in the discussion of Figure 4.2 in Section 4.2.1.3, can sometimes provide for a "local" pure strategy solution even when there are no choices satisfying (4.1). For example, if the profit function is as specified by the solid curve in Figure 4.2, it may be the case that there is no equilibrium. However, when the profit function for $s_k \geq 0$ is assumed incorrectly by participant k to be as in the dashed curve, there may be a choice of $(s_k^*)_{k=1,\ldots,n}$ satisfying (4.1).

In considering mixed strategy equilibria and relaxation of the equilibrium concept, it is necessary to consider whether the institutional framework is appropriate. This will be discussed further in Section 4.7 in the particular context of mixed-strategy equilibria. When there are choices satisfying (4.1) there may be multiple equilibria. To see how multiple equilibria can arise, consider the case where there is only a single pricing interval represented in the model and the supply function, S_k, of firm k is required to be affine for prices greater than or equal to a "price intercept". That is:

$$\forall k = 1, \ldots, n \forall p \geq \alpha_k, S_k(p) = \beta_k(p - \alpha_k),$$

where α_k is the price intercept and β_k is the slope of the supply function for prices greater than or equal to α_k.

If each market participant k is required to fix β_k at some particular value and is only allowed to choose α_k then the strategic variables are the price intercepts α_k, $k = 1, \ldots, n$. Under certain conditions on the form of the profit function for each participant k, including concavity of π_k as a function of α_k, there will be an equilibrium in price intercepts $(\alpha_k^*)_{k=1,\ldots,n}$ for each choice of $(\beta_k)_{k=1,\ldots,n}$. The equilibrium $(\alpha_k^*)_{k=1,\ldots,n}$ varies smoothly with $(\beta_k)_{k=1,\ldots,n}$. That is, for each choice of $(\beta_k)_{k=1,\ldots,n}$, there is a corresponding unique equilibrium $(\alpha_k^*(\beta))_{k=1,\ldots,n}$, where we have explicitly indicated the dependence of the equilibrium on the choice of $\beta = (\beta_k)_{k=1,\ldots,n}$ [16].

However, now consider the case where market participants are allowed to choose both the intercept and slope, so that the strategic variable of participant k is a vector $s_k = [\alpha_k, \beta_k]^t$. Under mild conditions it turns out that there is a continuum of equilibria. In particular, for any choice $(\beta_k^*)_{k=1,\ldots,n}$, there is an equilibrium speci-fied by $(s_k^*)_{k=1,\ldots,n}$, where $s_k^* = [\alpha_k^*(\beta), \beta_k^*]_{k=1,\ldots,n}^t$ with $\beta^* = (\beta_k^*)_{k=1,\ldots,n}$.

The multiplicity of equilibria occurs because the number of parameters in the strategic variable s_k for participant k exceeds the number of pricing intervals. In general, there can be multiple equilibria when there are more parameters than pricing intervals.

Conversely, when the number of pricing intervals is at least equal to the number of parameters then there will typically be a unique equilibrium. For example, if there are two pricing intervals, marginal costs are affine, and the strategic variable is the vector $s_k = [\alpha_k, \beta_k]^t$ consisting of the two parameters of supply function inter-cept and slope then, under mild conditions there is a unique equilibrium. Moreover, in this case, the equilibrium value of α_k is equal to the marginal cost intercept of participant k [6].

The above discussion suggests that the number of parameters in the strategic variable should be considered carefully in the model formulation. For example, when the number of parameters in the model is limited compared to the true flexibility in market rules, there may be only a single equilibrium, but this equilibrium may simply be an artifact of the particular limitation on the number of parameters. In the first part of the above example where only the intercept was the strategic variable, any particular choice of β would result in a particular equilibrium, but this choice may be inconsistent with the outcome of a realistic market.

Furthermore, as mentioned in Section 4.2.3.2, π_k may imperfectly represent the profit function, particularly in the context of transmission constraints. Another modeling issue is the representation of bilateral contracts, and forward markets for energy and FTRs [20].

The basic "single-shot" model of a day-ahead or real-time market in (4.1) can be extended to recognizing that the day-ahead market repeats on a daily basis and the real-time market repeats on an hourly basis. Such "repeated games" typically involve additional model assumptions, particularly when collusive "signaling" is possible [21].

4.5 COMPUTATION

Computation of a "single-shot" equilibrium can be easy or difficult depending on the specification of the form of the strategic variables, the market operation model, and the number of participants. There are a number of possible solution methods for equilibrium models as discussed in the following sections.

4.5.1 Analytical Models

In small models where dispatch and prices can be solved explicitly in terms of strategic variables, it may be possible to solve for equilibria analytically. For example, in a single pricing interval Cournot model without capacity constraints it is possible to solve analytically for the outcomes by simultaneously solving the first-order necessary conditions for each participant to maximize its profits. In other models, the analysis requires consideration of cases, but may still be susceptible to analytical techniques [22–24].

For example, consider an n firm electricity market with each firm represented by a portfolio cost function. We ignore transmission constraints and maximum capacity constraints and assume that the portfolio cost function $C_k: \mathbf{R}_+ \to \mathbf{R}$ of the k-th firm is quadratic and of the form:

$$\forall q_k \in \mathbf{R}_+, C_k(q_k) = \frac{1}{2}c_k q_k^2 + a_k q_k,$$

with $c_k \geq 0$ for convex costs. The marginal cost of firm k is C_k', with:

$$\forall q_k \in \mathbf{R}_+, C_k'(q_k) = c_k q_k + a_k \tag{4.2}$$

That is, the marginal costs are affine, with marginal cost intercept a_k and slope of marginal cost c_k. Each firm is assumed to be able to produce down to zero output, so that the minimum capacity constraints are all equal to zero. At minimum capacity, the marginal cost of firm k is equal to a_k, the marginal cost intercept.

We assume that the demand is of the form:

$$\forall p \in \mathbf{R}_+, D(p) = N - \gamma p,$$

with N characterizing the demand level and $\gamma \geq 0$ characterizing its response to price.

In the Cournot model, the strategic variables are quantities. Since total supply $\sum_\ell q_\ell$ must equal demand, we can invert the demand relationship to obtain the market clearing price $P : \mathbf{R} \to \mathbf{R}$ as a function of total supply:

$$\forall q_k, k = 1, \ldots, n, P\left(\sum_\ell q_\ell\right) = \left(N - \sum_\ell q_\ell\right) \Big/ \gamma$$

The operating profit for firm k is its revenue minus its operating costs:

$$\pi_k(q_k, q_{-k}) = q_k P\left(\sum_\ell q_\ell\right) - C_k(q_k)$$

which is concave and quadratic as a function of q_k for given q_{-k}, assuming that $c_k \geq 0$ and $\gamma \geq 0$. Necessary and sufficient conditions on q_k to maximize $\pi_k(q_k, q_{-k})$ are linear. In particular, the Cournot profit maximizing quantity $q_k^{Cournot}$ for firm k satisfies:

$$0 = \frac{\partial \pi_k(q_k^{Cournot}, q_{-k})}{\partial q_k}$$

$$= P\left(q_k^{Cournot} + \sum_{\ell \neq k} q_\ell\right) - q_k^{Cournot}(c_k + 1/\gamma) - a_k$$

Simultaneously satisfying this for all firms results in n equations. After some manipulation, the resulting "Cournot price" $p^{Cournot}$ is given by:

$$p^{Cournot} = \frac{N + \sum_{k=1}^{n} \dfrac{a_k}{(c_k + 1/\gamma)}}{\gamma + \sum_{k=1}^{n} \dfrac{1}{(c_k + 1/\gamma)}}$$

The corresponding "Cournot quantities" are calculated according to:

TABLE 4.1 Cost data based on five firm industry described in [25, 26]

Firm k	1	2	3	4	5
c_k (pounds per MWh per MWh)	2.687	4.615	1.789	1.93	4.615
a_k (pounds per MWh)	12	12	8	8	12

$$\forall k = 1, \ldots, n, \quad q_k^{Cournot} = \frac{1}{c_k + 1/\gamma} \left(p^{Cournot} - a_k \right)$$

As a specific numerical example, consider the $n = 5$ firm example described in sections 2.2, 6.3, 10.2 in [25] and based on the example in [26]. The cost data is shown in Table 4.1.

Assume that the demand level is specified by $N = 35$ and the demand slope is $\gamma = 0.1$ GW per (pound per MWh). We obtain:

$p^{Cournot} = 80$ pounds per MWh

$q_1^{Cournot} = 5.3911$ MW

$q_2^{Cournot} = 4.6799$ MW

$q_3^{Cournot} = 6.1410$ MW

$q_4^{Cournot} = 6.0684$ MW

$q_5^{Cournot} = 4.6799$ MW

4.5.2 Numerical Solution

In some models, the analytical approach results in first-order necessary conditions that are nonlinear or are differential equations. Numerical and differential equation solving methods may be used. Even if the first-order necessary conditions are linear, the form may not be as convenient to solve as in the Cournot model in Section 4.5.1 and may require use of a linear equation solver.

As an example of solving nonlinear equations, suppose that the firms have the same portfolio cost functions as in Section 4.5.1. However, instead of supposing that quantities are the strategic variable, consider the case where the strategic variable of firm k is the slope, β_k, of the offer function S_k. The intercept of the offer function α_k is assumed to be required to equal the intercept of the marginal cost function, a_k. That is, the offer function is required to be affine with intercept matching the marginal cost intercept. We consider equilibria in such supply functions. (As discussed in Section 4.4, this may be a restriction compared to the flexibility allowed in actual markets, with implications for the results. However, in the particular case that only affine offers are allowed, the equilibrium intercept of the affine offer will equal the marginal cost intercept [6].)

In [4, 25–27], linear and affine supply function equilibria (SFE) are exhibited for the case of affine marginal generation costs of the form (4.2), under the assumption that minimum capacity constraints are not binding. The affine SFE $S^{*affine} = (S_k^{*affine})_{k=1,\ldots,n}$ is of the form:

$$\forall k = 1, \ldots, n, \forall p \geq \max_{\ell} \{a_\ell\}, S_k^{*affine}(p) = \beta_k(p - a_k) \qquad (4.3)$$

where we note that for prices below $\max_{\ell}\{a_\ell\}$ the minimum capacity constraints are binding on one or more generators and (4.3) does not apply, and the slopes $\beta_k \in \mathbb{R}_{++}$, $k = 1, \ldots, n$ satisfy:

$$\forall k = 1, \ldots, n, \frac{\beta_k}{1 - c_k \beta_k} = \sum_{\ell \neq k} \beta_\ell + \gamma \qquad (4.4)$$

These are a system of non-linear simultaneous equations in β_k, $k = 1, \ldots, n$.

As a specific numerical example, for the data in Table 4.1 and a demand slope $\gamma = 0.1$ GW per (pound per MWh), the slopes of the affine solutions are:

$$\beta = \begin{bmatrix} 0.2840 \\ 0.1857 \\ 0.3718 \\ 0.3550 \\ 0.1857 \end{bmatrix}$$

and the affine SFE is given by:

$$\forall p \in [12, \infty), S^{*affine} \begin{bmatrix} 0.2840(p-12) \\ 0.1857(p-12) \\ 0.3718(p-8) \\ 0.3550(p-8) \\ 0.1857(p-12) \end{bmatrix}$$

For $N = 35$, the resulting price is between 32 and 33 pounds per MWh, which is considerably below the Cournot price $p^{Cournot}$.

4.5.3   Fictitious Play

When the offer-based dispatch and price calculations cannot be solved explicitly in terms of the strategic variables, a natural approach is to successively update a representation of the strategic variables and then numerically calculate the results of offer-based economic dispatch. For example, each participant may model the results of offer-based economic dispatch as a function of a representation of its strategic variables and then estimate its profit maximizing response to the other participants' strategic variables. The profit maximizing response can be used to update the participants' own strategic variables. A sequence of strategic variables is produced for each participant, starting from some initial guess of the value of the equilibrium, and the hope is that the sequences converge to the equilibrium. A number of variations on this basic idea are possible that use different approaches to finding the profit maximizing response [25, 28–30]. Furthermore, mixed strategy equilibria can sometimes be found using a fictitious play approach [31].

In [25], for example, supply function equilibria are estimated by approximating the supply functions as piecewise linear. The piecewise linear supply functions are updated at each iteration through the solution of a quadratic program for each

TABLE 4.2 Capacity data based on five firm industry described in [25, 26]

Firm k	1	2	3	4	5
$\bar{q}_k$(GW)	5.70945	3.35325	10.4482	9.70785	3.3609

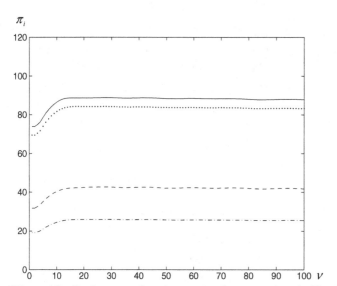

Figure 4.3   Profits versus iteration starting from capacitated affine SFE

participant to calculate a "step direction" for updating the supply functions. A fixed "step-size" is used to facilitate convergence to an equilibrium if it exists. Many issues, such as capacity constraints, can be incorporated into the model; however, the resulting profit maximization problems may not have concave profits, posing problems when standard local optimization software is used.

For example, consider again the $n = 5$ firm industry with costs as specified in Table 4.1, but suppose that the capacity $\bar{q}_k$ of each firm is as specified in Table 4.2. Instead of a fixed demand situation, we assume that firms must specify an offer that remains fixed over variation in demand, as specified by a uniform distribution of the parameter N from 35 to 10 GW. As example starting functions for the iteration, the affine SFE was used, except that the supply was capped at the capacity of the firm. This is referred to as a "capacitated affine SFE supply function" in [25].

Figure 4.3 shows the profits versus iteration ν starting from the capacitated affine SFE supply function. The leftmost points in Figure 4.3 show the profits if each firm were to offer the capacitated affine SFE supply function. Profits at iteration 100 are considerably higher than if all firms offer the capacitated affine starting function.

Figure 4.4 shows the supply functions at iteration 100. The price-duration curve for iteration 100 is shown in Figure 4.5.

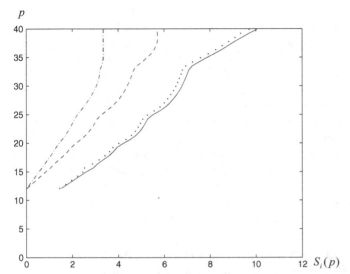

Figure 4.4   Supply functions at iteration 100 starting from capacitated affine SFE

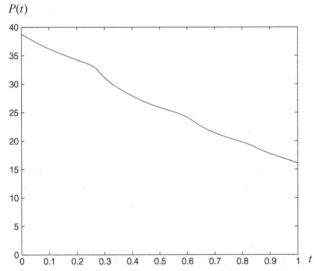

Figure 4.5   Price-duration curve at iteration 100 starting from capacitated affine SFE

It should be emphasized that fictitious play typically assumes myopia on the part of market participants. That is, the update does not anticipate changes in strategic variables by other participants. Consequently, there is no representation of the "repeated game" nature of electricity markets even though the fictitious play paradigm involves successive updates of the strategic variables. Myopic fictitious play can only reveal the "single-shot" equilibrium of the game as though it were actually only played once.

Under suitable conditions, the sequence of values of strategic variables may converge to a single-shot pure strategy equilibrium, if it exists. Figure 4.3 suggests that the iterations are approaching a limiting collection of supply functions. However, if the best response problems involve nonconcave objectives then it may be the case that there is no pure strategy equilibrium. Moreover, a local optimizer may result in iterations that are approaching a limit even if there is no pure strategy equilibrium. Finally, finite dimensional approximations of infinite dimensional supply functions involve approximations that can have subtle but significant effects on issues such as the multiplicity of equilibria [40].

"Agent-based" models fall into the fictitious play framework, although the "agent" may not be explicitly finding its profit maximizing response. On the other hand, agent-based models may have some model of anticipation of behavior of other participants, therefore changing the assumption of myopia. "Experimental economics" methods can also be set up in a fictitious play framework.

4.5.4 Mathematical Program with Equilibrium Constraints and Equilibrium Program with Equilibrium Constraints

A more systematic approach is to model the market clearing mechanism by its optimality conditions and then incorporate them into the optimization problems faced by each participant [32]. This results in a "mathematical program with equilibrium constraints." Modeling the equilibrium of multiple participants results in an "equilibrium program with equilibrium constraints" [33]. The application of such techniques to market models has grown with the increasing availability of software for solving such models; however, they have generally only been applied to single pricing interval models.

4.5.5 Specialized Solution Methods

In some cases, specialized algorithms may be applied to particular types of equilibria. For example, Anderson and Hu [34] describe a technique for finding supply function equilibria. In the case of two participants, the Lemke algorithm can find mixed strategy equilibria [35]. For multiple participants in the context of transmission constraints, heuristic solution methods are presented in [36].

4.6 DIFFICULTIES WITH EQUILIBRIUM MODELS

There are several difficulties in equilibrium models of electricity markets, including:

1. Modeling assumptions
2. Mixed strategy equilibria
3. Multiplicity of equilibria

A fundamental question in using equilibrium models is whether they are reasonable models of industry behavior. In some cases, the underlying economic modeling assumptions about knowledge, profit maximization, and rationality may not be consistent with observed behavior. For example, in the ERCOT balancing market, some smaller market participants' behavior is evidently not consistent with a model of profit maximization [14, 37], although some of the apparent deviations from profit maximization reported in [14, 37] may actually be due to discrepancies between the economic and commercial models.

Furthermore, in some cases, no "pure strategy" equilibrium exists and there are only "mixed strategy" equilibria [38, 39]. Unfortunately, in many cases there may be no straightforward interpretation of a mixed strategy equilibrium in the context of an electricity market where, empirically, there is very little evidence for randomized offers. In some cases, the existence of only mixed strategy equilibria may be evidence that assumptions in the model are inappropriate or that the model is ill-conditioned [25, Section 4.4].

Finally, in some models, particularly supply function equilibria, there may be multiple equilibria. Unfortunately, the presence of multiple equilibria is problematic since it significantly reduces the predictive value of the analysis. There have been several approaches to trying to "refine" multiple supply function equilibria by eliminating most of the range of equilibria using various criteria [25, 40–42].

In some cases, the choice of strategic variables in the economic model may limit the multiplicity of equilibria; however, if this choice artificially restricts the form of, for example, generator offers, then the results may simply be an artifact of the choice. As discussed in Section 4.4, this issue is particularly problematic in single pricing interval models with no uncertainty of demand [16].

4.7 USES OF EQUILIBRIUM MODELS

Even putting aside the three issues of the validity of equilibrium modeling, mixed strategy equilibria, and multiple equilibria, it remains that there are a large number of economic modeling assumptions made in equilibrium models as discussed in Sections 4.2 and 4.3. Given all of these assumptions, it is likely that the models are not capable of exact predictions of market prices and market outcomes.

Even in the absence of accurate predictive capability, however, an important role for such models is the principled analysis of the effect of changes in market rules or the effect of changes in market structure. Examples include analyses of the effect of:

- Detailed choices in the specification of alternative market rules, such as, allowing offers to change from pricing interval to interval versus requiring offers to remain fixed over multiple pricing intervals [25, 40], and single clearing price versus pay-as-bid prices [21, 23]
- Changes in market structure such as mandated divestitures [6, 29]
- Representation of transmission constraints [9, 32, 38, 43, 44]
- Level of contracts, such as:
  — bilateral energy contracts [20, 45] and forward energy markets, and
  — financial transmission rights [46, 47]
- Modeling assumptions, such as:

— the assumed form of cost functions or offer functions [16, 48],

— the use of portfolio-based versus unit-specific costs or offers, and

— the representation of unit commitment

In these cases, a general strategy for analysis is to hold most market rules and features constant and then vary one particular issue. In doing such a qualitative "sensitivity" analysis the hope is that, although the level of prices or the effect of other issues will not be calculated exactly, there will be a reasonable estimation of the change due to the modeled variation. This allows the potential for policy conclusions to be made from studies, even in the absence of perfect representation of market features or of fidelity to participant behavior. In the following sections, we summarize case studies of three such sensitivity analyses. Naturally, the results of any such modeling efforts must be evaluated with caution.

4.7.1 Market Rules Regarding the Changing of Offers

In an electricity market with multiple pricing intervals in a day, for example, it is possible to imagine that market rules allow for only a single set of energy offers that must apply across all pricing intervals in the day. Alternatively, market rules may allow for offers that can vary from hour to hour. Both alternatives have been adopted in practice, although there may be practical limits to varying offers hour by hour in a market that does not otherwise prohibit such changes, and there may be ways to effectively vary offers in a market that apparently prohibits such variation.

To consider the effect of a requirement that offers remain fixed across multiple pricing intervals compared to allowing more flexibility, it is possible to formulate a supply function equilibrium model representing both cases. All other issues are assumed the same for both models and, for the purposes of isolating this particular issue, many of the detailed features of electricity markets, including transmission constraints, might be ignored.

The result of such an analysis is presented in [25, 40]. A principal result of a rule requiring consistent offers is in the mitigation of market power. The requirement to offer consistently over a time horizon with multiple pricing intervals can help limit the exercise of market power, by depressing the prices that can be achieved in equilibrium compared to the equilibrium prices when offers can vary from pricing interval to interval. The basic insight is that market participants must compromise their offers between on- and off-peak. Naturally, the results of any such analysis must be tempered with the observation that much of the detail of the market was modeled in a simplified manner.

4.7.2 Single Clearing Price Versus Pay-as-Bid Prices

Single clearing price electricity markets are sometimes criticized as paying excessive prices to inframarginal generators compared to "pay-as-bid" markets. In a "pay-as-bid" market, each accepted offer is paid its offer price instead of the market clearing price. Therefore, even if the market clearing price is high due to market power, other offers will only receive their offer price.

As pointed out in [12], however, naive proposals for pay-as-bid markets neglect to realize that offers will change in response to changes in market rules. A result of economics, called the "revenue equivalence theorem" [49] suggests that the equilibrium prices in single clearing price and pay-as-bid markets should be the same and that market participants will be paid the same in equilibrium in both markets. However, not all of the assumptions required for the revenue equivalence theorem actually hold in electricity markets. Consequently, the conclusion of the revenue equivalence theorem might not hold, or might only hold approximately, in electricity markets.

As in the case of modeling the effect of consistency of offers, a simplified model of an electricity market can be used to obtain a sensitivity result for the change between single clearing price and pay-as-bid prices. In some models of electricity markets, pay-as-bid pricing can result in lower equilibrium prices than in single clearing price markets [21, 23]. The differences between single clearing price and pay-as-bid prices are typically much smaller than in the naive analyses that ignore the effect of market rules on offers. As previously discussed, the simplicity of the market model means that results should be interpreted with caution. Furthermore, there are serious drawbacks of pay-as-bid markets, including the likelihood of poor dispatch decisions when price predictions by market participants are imperfect.

4.7.3 Divestitures

In some markets, market structure has been changed by mandated divestitures. This occurred twice in the England and Wales market in the late 1990s. A model of this market, with demand elasticity calibrated to observed market prices, was used to verify the size of the change in market prices due to the divestitures. The results of such an analysis are reported in [6]. Generally speaking, the model was able to reproduce the change in prices from before to after the divestitures. However, the model was calibrated to observed demand prior to the divestiture, which somewhat weakens its predictive value.

4.8 CONCLUSION

In this chapter, equilibrium models, their solution, and uses have been discussed. There has been considerable effort in recent years in developing the theory and application of these models. There are strong prospects for improving such models, although their application should be tempered with the understanding that the actual market is likely to include a host of details that remain unmodeled. Several case study examples of qualitative sensitivity analysis were described.

ACKNOWLEDGMENT

This work was supported in part by the National Science Foundation under Grant Number ECS-0422914.

REFERENCES

1. Fudenberg D, Tirole J. *Game Theory*. Cambridge, MA, The MIT Press, 1991.
2. Varian HR. *Microeconomic Analysis*. Third Edition, New York: W.W. Norton and Co., 1992.
3. Green R, Newbery DM. Competition in the British electricity spot market. *Journal of Political Economy* 1992;100(5):929–953.
4. Green R. Increasing competition in the British electricity spot market. *The Journal of Industrial Economics* 1996;XLIV(2):205–216.
5. Singh H, editor, *Game Theory Applications in Electric Power Markets*. Piscataway, NJ: IEEE Power Engineering Society, 1999.
6. Baldick R, Grant R, Kahn E, Theory and application of linear supply function equilibrium in electricity markets. *Journal of Regulatory Economics* 2004:25(2);143–167.
7. Bergen AR. Vittal V. *Power Systems Analysis*. Second Edition, Upper Saddle River, NJ: Prentice-Hall, 2000.
8. Oren S, Spiller P, et al., Nodal prices and transmission rights: a critical appraisal. *The Electricity Journal* 1995;April:24–35.
9. Hogan WW. A market power model with strategic interaction in electricity markets. *The Energy Journal* 1997;18(4):107–141.
10. Baldick R, Niu H. Lessons learned: the Texas experience. In: *Electricity Deregulation: Choices and Challenges*, Griffin JM, Puller SL, editors. Chicago and London; University of Chicago Press, 2005.
11. Wood AJ. Wollenberg BF. *Power Generation, Operation, and Control*, Second Edition. New York; Wiley, 1996.
12. Stoft S, *Power System Economics: Designing Markets for Electricity*. Piscataway, NJ; IEEE Press and Wiley Interscience and John Wiley & Sons, Inc., 2002.
13. Hogan WW. Contract networks for electric power transmission. *Journal of Regulatory Economics* 1992;4(3):211–242.
14. Hortaçsu A, Puller SL. Understanding strategic bidding in multi-unit auctions: a case study of the Texas electricity spot market. *RAND Journal of Economics* 2008; 39(1):86–114.
15. Alsac O, Bright JM, Brignone S, Prais M, Silva C, Stott B, Vempati N. The rights to fight price volatility. *IEEE Power and Energy Magazine* 2004;2(4):47–57.
16. Baldick R. Electricity market equilibrium models: the effect of parametrization. *IEEE Transactions on Power Systems* 2002;17(4):1170–1176.
17. Barroso LA, Conejo AJ. Decision making under uncertainty in electricity markets. In: *Proceedings of the IEEE Power Engineering Society General Meeting*, Montreal, Canada, June 2006.
18. Kahn AE, Cramton PC, Porter RH, Tabors RD. Pricing in the California power exchange electricity market: should California switch from uniform pricing to pay-as-bid pricing. Study commissioned by the California Power Exchange, 2001.
19. Baldick R. Joint equilibrium of day-ahead and real-time markets. Unpublished manuscript, Department of Electrical and Computer Engineering, The University of Texas at Austin, August 2002.
20. Green R. The electricity contract market in England and Wales. *The Journal of Industrial Economics* 1999;XLVII(1):107–124.
21. Fabra N. Uniform pricing facilitates collusion: the case of electricity markets. Submitted to the Blue Ribbon Panel of the California Power Exchange, October 2000.
22. Song H, Liu C-C, Lawarrée J. Nash equilibrium bidding strategies in a bilateral electricity market. *IEEE Transactions on Power Systems* 2002;17(1):73–79.
23. Son YS, Baldick R, Lee K-H, Siddiqi S. Short-term electricity market auction game analysis: uniform and pay-as-bid pricing. *IEEE Transactions on Power Systems* 2004;19(4):1990–1998.
24. Lee K-H, Baldick R. Solving three-player games by the matrix approach with application to an electric power market, *IEEE Transactions on Power Systems* 2003;18(4):1573–1580.
25. Baldick R, Hogan W. Capacity constrained supply function equilibrium models of electricity markets: stability, non-decreasing constraints, and function space iterations. University of California Energy Institute POWER Paper PWP-089, www.ucei.berkeley.edu/ucei/PDF/pwp089.pdf, August 2002.

26. Baldick R, Grant R, Kahn EP. Linear supply function equilibrium: generalizations, application, and limitations, University of California Energy Institute POWER Paper PWP-078, www.ucei.berkeley.edu/ucei/PDF/pwp078.html, August 2000.

27. Rudkevich A. Supply function equilibrium in poolco type power markets: learning all the way. Tech. Rep. TCA Technical Paper Number 1299-1702, Tabors Caramanis and Associates, December 1999.

28. Stoft S. Using game theory to study market power in simple networks. In: *IEEE Tutorial on Game Theory Applications in Power Systems*. Singh H, editor. Piscataway, NJ: IEEE, 1999. pp. 33–40.

29. Day CJ, Bunn DW. Divestiture of generation assets in the electricity pool of England and Wales: a computational approach to analyzing market power. *Journal of Regulatory Economics* 2001;19(2):123–141.

30. Son YS, Baldick R. Hybrid coevolutionary programming for Nash equilibrium search in games with local optima. *IEEE Transactions on Evolutionary Computation* 2004;8(4):305–315.

31. Borenstein S, Bushnell J, Stoft S. Supplement to the competitive effects of transmission capacity in a deregulated electricity industry. Unpublished manuscript, available from http://www.ucei.berkeley.edu/ucei/PDF/bbsmixed.pdf, October 1998.

32. Hobbs B, Metzler C, Pang J-S. Strategic gaming analysis for electric power networks: an MPEC approach. *IEEE Transactions on Power Systems* 2000;15(2):638–645.

33. Hu X, Ralph D. Using EPECs to model bilevel games in restructured electricity markets with locational prices. *Operations Research* 2007;55(5):809–827.

34. Anderson EJ, Xu X. Finding supply function equilibria with asymmetric firms. Australian Graduate School of Management, The University of New South Wales, Sydney, NSW, 2007.

35. Lemke CE, Howson, Jr., T. Equilibrium points of bimatrix games. *The Journal of the Society for Industrial and Applied Mathematics* 1964;12(2):413–423.

36. Lee K-H, Baldick R. Nash equilibria of multi-player games in electricity markets with transmission congestion. Submitted to *IEEE Transactions on Power Systems*, 2007.

37. Sioshansi R, Oren S. How good are supply function equilibrium models: an empirical analysis of the ERCOT balancing market. Unpublished manuscript, April 2006.

38. Borenstein S, Bushnell J, Stoft S. The competitive effects of transmission capacity in a deregulated electricity industry. *RAND Journal of Economics* 2000;31(2):294–325.

39. Cunningham LB, Baldick R, Baughman ML. An empirical study of applied game theory: transmission constrained Cournot behavior. *IEEE Transactions on Power Systems* 2002;17(1):166–172.

40. Baldick R, Hogan W. Stability of supply function equilibrium: Implications for daily versus hourly bids in a poolco market. *Journal of Regulatory Economics* 2006;30(2):119–139.

41. Baldick R, Hogan W. Polynomial approximations and supply function equilibrium stability. In: *6th IAEE European Conference, Modeling in Energy Economics and Policy*, Zurich, Switzerland, September 2004, Swiss Association for Energy Economics, Centre for Energy Policy and Economics.

42. Genc T, Reynolds SS. Supply function equilibria with pivotal suppliers. Unpublished manuscript, March 2005.

43. Cardell JB, Hitt CC, Hogan WW. Market power and strategic interaction in electricity networks, *Resource and Energy Economics* 1997;19(1–2):109–137.

44. Hobbs BF. LCP models of Nash-Cournot competition in bilateral and POOLCO based power markets. In: *Proceedings of the IEEE Power Engineering Society Winter Meeting*, New York, NY, February 1999.

45. Anderson EJ, Xu H. Contracts and supply functions in electricity markets. Australian Graduate School of Management, The University of New South Wales, Sydney, NSW, 2001.

46. Stoft S. Financial transmission rights meet Cournot: how TCCs curb market power. *The Energy Journal* 1999;20:1–23.

47. Hogan W. Financial transmission right incentives: applications beyond hedging. John F. Kennedy School of Government, Harvard University, May 2002.

48. Morch von der Fehr NH, Harbord D. Spot market competition in the UK electricity industry. *The Economic Journal* 1993;103(418):531–546.

49. Klemperer PD. Auction theory: a guide to the literature. *Journal of Economic Surveys* 1999;13(3):227–286.

CHAPTER 5

HYBRID BERTRAND-COURNOT MODELS OF ELECTRICITY MARKETS WITH MULTIPLE STRATEGIC SUBNETWORKS AND COMMON KNOWLEDGE CONSTRAINTS

Jian Yao,

Shmuel S. Oren,

and Benjamin F. Hobbs

5.1 INTRODUCTION

Electricity market models are employed by market participants, policy markers, and stakeholders to characterize market agents' decisions and to predict market outcomes [21]. Many existing models follow game-theoretical approaches but are distinguished by the formulation of the interaction between generators and the independent system operator (ISO). Such distinctions are referred to as sequential vs. simultaneous clearing of energy and transmission markets [1], or as integrated vs. separated market designs [16]. From a perspective of generator-ISO interaction, these models can be grouped into four approaches: Stackelberg, Stackelberg approximations, pure Cournot, and pure Bertrand.

The Stackelberg approach assumes that the energy and transmission markets are sequentially cleared, with generators acting first and the ISO acting second. Thus, generators anticipate the impact of their strategies on transmission prices (equivalent to locational price differences when the ISO computes nodal prices). The resulting model is a multi-leader one-follower Stackelberg game [2, 3, 9, 16]. Mathematically, a producer's decision problem is a mathematical program with equilibrium constraints (MPEC) [14], and the market equilibrium is an equilibrium problem with equilibrium constraints (EPEC) [6]. However, this model introduces two difficulties

Restructured Electric Power Systems: Analysis of Electricity Markets with Equilibrium Models,
Edited by Xiao-Ping Zhang
Copyright © 2010 Institute of Electrical and Electronics Engineers

arising from the embedded optimality conditions for the ISO's problem in all the generation firms' problems. First, the game among generators is a generalized Nash equilibrium problem [7] because each firm's decision variables appear in the constraint sets of the other firms' problems. Second, the firms' sets of feasible decisions are non-convex. As a result, this model may lead to either zero or multiple pure-strategy equilibria (see, for example, [2]). Moreover, even if a solution is found, it may be degenerate; that is, firms will find it optimal to barely congest some transmission lines so as to avoid congestion rents (see [17]; counter examples are given in [20]). Finally, finding equilibria of this model, even if one exists, for a realistic size network is computationally challenging.

Some approximations of the Stackelberg game have also been proposed that are computationally more tractable. One proposal assumes that generators hold fixed a priori conjectures concerning the sensitivity of transmission costs with respect to changes in amounts of transmission services requested [11].[1] This model is formulated as a complementarity problem rather than a more difficult EPEC. However, exogenous response coefficients are unsatisfactory theoretically in that such responses should be the result of a game not an input, and problematic from an empirical viewpoint. Another approach [1] iterates between an ISO model and a generator model. The ISO model calculates sensitivities of transmission prices with respect to injections, and the generator models then calculate a Cournot equilibrium among generators, assuming that those sensitivities are constant. Given that equilibrium, the ISO model then checks if the same sensitivities indeed still hold; if not (which can happen if the set of binding transmission limits changes), new sensitivities are obtained, and passed back to the generators' models. However, this approach often fails to converge [1, 16].

The Nash approaches assume that the energy and transmission markets are cleared simultaneously [8, 9, 15, 18, 21, 23–25]. In these approaches, generation firms do not explicitly model transmission limits in their constraint sets, and the ISO becomes a Nash player acting simultaneously with the firms. The market equilibrium is determined by aggregating the optimality conditions for the firms' and the ISO's problems, which become a mixed complementarity problem or (quasi-) variational inequalities. This approach avoids the computational intractability of the Stackelberg approaches and, under a nondegeneracy assumption, can lead to a unique equilibrium.

One variant of the Nash approach is the pure Cournot representation of generator expectations of ISO actions. The Cambridge-I model in [16] and the spot market model in [23] and [24] fall in the above category. In these models, firms are assumed to behave *a la* Cournot with respect to the ISO's, or arbitrageurs', that is, they treat as given the ISO's imports/experts into a bus or region, and act monopolistically with respect to the residual demand they face which are the horizontally shifted local demand curves. Such a pure Cournot model may be suitable for networks with

---

[1]The hybrid model of this paper can be viewed as an extreme case of the conjectured price model in [11] in which the slope of the transmission price with respect to changes in flows is either zero (Bertrand) or infinite (Cournot).

relatively small interfaces between large markets that are frequently congested, especially for radial links such as the UK-France line. However, it is less realistic for general networks where generators may anticipate the impact of their outputs on interregional flows. Moreover, this model has the undesirable property that generators owned by one firm but located in different markets cannot coordinate decisions to their benefit; as a result, a company with plants in n different markets behaves the same as n separate firms (see Subsection 5.3.1.1 for more analysis).

Another variant of the Nash approach is the pure Bertrand model, in which generators take as given the nodal price markups due to congestion, or the locational price differences set by the ISO as congestion charges. In bilateral markets, this amounts to price taking behavior by generation firms with the respect to transmission services. In a POOLCO-like market, this model assumes that the residual demand function considered by generation firms take locational price differences as given by account for the fact that they can move all the prices up and down through their output decisions. As examples, Wei and Smeers [22] consider a Cournot game among generators with regulated transmission prices and solve a variational inequality problem to determine unique long-run equilibria. Smeers and Wei [18] consider a separated energy and transmission market, and show that such a market converges to the optimal dispatch with many marketers. Hobbs et al. [8, 15] present Cournot equilibria in both bilateral and POOLCO markets with affine demand and cost functions, with the models formulated as mixed linear complementarity problems. Hobbs and Pang [10] formulate a bilateral market with piece-wise linear demand as a linear complementarity problem with a co-positive matrix. The model of spot wholesale markets developed in [23] and [24] takes into consideration the financial settlements of forward contracts.

Price taking in transmission is a defensible assumption for highly meshed networks that have several players and variable patterns of congestion. However, it is also naive as swing generators would probably try to influence locational price differences in their favor. Although, unlike the Cournot model, the Bertrand model allows a firm with plants in several locations to profitably coordinate decisions, it too has an undesirable property often referred to as the "thin line phenomenon." Under the Bertrand approach, to transmission, establishment of a thin line (say 1 MW) between two previously unconnected markets causes the model to treat the two markets as strategically linked which resulting in more competitive outcomes in both markets. For instance, two symmetric monopoly markets connected by such a thin line will result in a duopoly solution while the line carries zero flow.

Real power markets often consist of multiple subnetworks. In these markets, subnetworks are connected with frequently saturated interfaces, and hence they are decoupled in terms of strategic interaction since the residual demand functions in each subnet is shifted horizontally but their slopes stay the same. On the other hand, congestion pattern within individual subnetworks is less predictable and hence generators within the subnetwork interact strategically. For instance, in Northwest Europe, the France-UK, France-Belgium, and Netherlands-Germany interconnections are usually congested, effectively decoupling the markets. A Cournot conjecture assuming that generators take interregional imports/exports as given regarding is reasonable in those cases. However, within the UK, German, and Benelux sub-markets, congestion occurs but is less easily predicted, and in that case the Bertrand conjecture where generation firms behave as price takers with regard to transmission prices is more defensible. Therefore, neither the pure Cournot nor Bertrand models would be appropriate for markets' multiple subnetworks. Even when the network cannot be divided into distict subnetworks, some transmission lines are systematically congested and such congestion is anticipated by all market participants who can predict how much power will flow across such interfaces and account for that in their strategic interaction. We refer to such transmission lines "common knowledge constraints," which can be accounted for within the Bertrand framework.

In this chapter, we first consider exogenous subnetwork structures and propose a hybrid Bertrand-Cournot model that represents generators' decision making in the presence of multiple subnetworks. This model assumes that firms behave *a la* Cournot with respect to the ISO's inter-subnetwork transmission quantities, but *a la* Bertrand with respect to intra-subnetwork transmission prices. We then formulate a Bertrand type equilibrium model with certain links designated exogenously as common knowledge constraints.

The remainder of this chapter is organized as follows. In the next section, we introduce the ISO's problem. Section 5.3 analyzes the shortcomings of the pure Cournot and pure models of generator-ISO interactions, and proposes a new hybrid model with multiple subnetworks; this section concludes with some results concerning solution uniqueness and computability. Section 5.4. reports numerical results and economic insights for the hybrid Bertrand-Cournot model applied to a stylized six node network. In Section 5.5 we introduce the formulation of the Bertrand model with public knowledge constraints and in Section 5.6 we apply this approach to a variant of the six node example introduced earlier. Concluding remarks are provided in Section 5.7.

5.2   ROLE OF THE ISO

Electricity restructuring in different markets has followed several different blueprints [19]. In the US, one prevailing design is for the ISO to maintain a pool as a broker or auctioneer for wholesale spot transactions. In addition, the ISO controls the grid and transmits power from generators to consumers while meeting network and security constraints. The ISO also sets locational energy prices and transmission charges for bilateral energy transactions.

We consider an electricity network that is composed of nodes $1...N$ and transmission lines $1...L$. This market consists of G competing firms, each firm $g = 1...G$ operating the units at $N_g \subseteq \{1...N\}$. We assume, without loss of generality, that there is one generation unit at each node: a demand-only node is denoted by a node with a zero-capacity generation unit, and a node with multiple units is split into multiple nodes.

Following the firms' decisions $\{q_i\}_{i=1}^{N}$, the ISO decides on nodal imports/exports $\{r_i\}_{i=1}^{N}$ that must obey the following constraints. Firstly, power flows should not exceed thermal or other limits $\{\bar{k}_i\}_{l=1}^{L}$ of transmission lines in both directions.

We use a lossless DC approximation of Kirchhoff's laws (see [4]) and define power flows in terms of the so-called power transfer distribution factors (PTDFs). Each PTDF D_{li} represents a proportion of the flow occurring on the line $l = 1...L$ resulting from an one-unit injection of electricity at the node $i = 1...N$ and a corresponding one-unit withdrawal at the reference bus. These network feasibility constraints are

$$-\bar{k}_l \le \sum_{i=1}^{N} D_{li} r_i \le \bar{k}_l, \quad l = 1...L$$

Secondly, because electricity is not economically storable, load and generation must be balanced at all times. This establishes an energy balancing constraint, which sets the total import/export in a lossless grid to zero:

$$\sum_{i=1}^{N} r_i = 0$$

The nonstorability of electricity also implies that the load at all nodes must be non-negative. Hence, the following constraints should also be considered in the ISO's decisions:

$$0 \le r_i + q_i, \quad i = 1...N \tag{5.1}$$

The objective of the ISO's transmission has been phrased as profit maximization in [8] and [10], cost minimization in [9] and social surplus maximization in [23]–[25]. In this chapter, we assume that the ISO aims to maximize social welfare, which denotes the total consumer willingness-to-pay, that is, the area under the nodal inverse demand functions, less the total generation cost. The ISO is assumed to act *a la* Cournot with respect to generation injections, so the generation quantities are treated as given in its objective. Mathematically, the ISO's decision problem is

$$\max_{\{r_i\}_{i=1}^{N}} \sum_{i=1}^{N} \left(\left(\int_{0}^{r_i+q_i} P_i(\tau_i)\,d\tau_i \right) - C_i(q_i) \right)$$

subject to:

$$\sum_{i=1}^{N} r_i = 0$$

$$\sum_{i=1}^{N} D_{li} r_i \ge -\bar{k}_l, \quad l = 1...L$$

$$\sum_{i=1}^{N} D_{li} r_i \le \bar{k}_l, \quad l = 1...L$$

$$r_i + q_i \ge 0, \quad i = 1...N$$

Let p, λ_i^-, λ_i^+, and η_i be the Lagrange multipliers corresponding to the constraints, then the first-order necessary optimality conditions (the Karush-Kuhn-Tucker, KKT) conditions) for the ISO's problem are

- with respect to r_i

$$P_i(r_i + q_i) - p + \sum_{l=1}^{L} (\lambda_i^- - \lambda_i^+) D_{li} + \eta_i = 0, \quad i = 1...N$$

- with respect to p

$$\sum_{i=1}^{N} r_i = 0$$

- with respect to λ_l^-

$$0 \le \lambda_l^- \perp \bar{k}_l + \sum_{i=1}^{N} D_{li} r_i \ge 0, \quad l = 1 \dots L$$

- with respect to λ_l^+

$$0 \le \lambda_l^+ \perp \bar{k}_l - \sum_{i=1}^{N} D_{li} r_i \ge 0, \quad l = 1 \dots L$$

- with respect to η_i

$$0 \le \eta_i \perp r_i + q_i \ge 0, \quad i = 1 \dots N$$

Here, the first KKT condition identifies two parts of nodal prices:

$$P_i(r_i + q_i) = p + \varphi_i, \quad i = 1 \dots N \tag{5.2}$$

where $\varphi_i = -\sum_{l=1}^{L} (\lambda_l^- - \lambda_l^+) D_{li} - \eta_i$.

We can interpret p as the reference energy price (when the nonnegative load constraint is not violated at the reference bus, this is just the price at the reference bus) and $\{\varphi_i\}_{i=1}^{N}$ as node-specific premiums. Consequently, the congestion charge for the bilateral transmission from node i to node j is $\varphi_j - \varphi_i$, and the total congestion charge in the system is $\sum_{i=1}^{N} \varphi_i r_i$.

The ISO's transmission flows may lead to total payment from load that differs from the total payment to generation. Hogan showed in [12] that the difference is nonnegative. In the following, we quantify this difference.

Proposition 1: The difference between the total payment from load and the total charge from generation is equal to the total congestion charge in the network.

Proof: The total payment from load is the consumptions charged at nodal prices:

$$\sum_{i=1}^{N} P_i(r_i + q_i) \cdot (r_i + q_i),$$

and the total charge from generation is the production compensated at nodal prices:

$$\sum_{i=1}^{N} P_i(r_i + q_i) \cdot q_i.$$

Their difference, denoted by Δ, is

$$\Delta = \sum_{i=1}^{N} P_i(r_i + q_i) \cdot r_i$$

By condition (5.2), we have

$$\Delta = \sum_{i=1}^{N} \varphi_i \cdot r_i$$

and this difference is equivalent to the total congestion charge. Furthermore, Δ can be written as

$$\Delta = \sum_{l=1}^{L} (\lambda_l^- + \lambda_l^+) \overline{k}_l + \sum_{i=1}^{N} q_i \eta_i = 0$$

Because $\{\lambda_l^-\}_{l=1}^{L}$, $\{\lambda_l^+\}_{l=1}^{L}$, $\{q_i\}_{i=1}^{N}$, and $\{\eta_i\}_{i=1}^{N}$ are nonnegative, this difference is also nonnegative.

5.3 THE HYBRID SUBNETWORK MODEL

5.3.1 Two Existing Models

Before introducing our model of multiple subnetworks, we review two existing Nash models and analyze their limitations. Both models assume the ISO to be Nash player, and present market equilibrium conditions by aggregating the KKT conditions for the firms' and the ISO's problems.

5.3.1.1 The Pure Cournot Model   The first model assumes that the firms behave purely *a la* Cournot with respect to the ISO's transmitted quantities (see, for example, [14], [23], [24]), that is, they take as given the ISO's import/exports at each node. Hence, a firm $g = 1 \dots G$ solves the following profit-maximization problem which is parameterized by $\{r_i\}_{i=1}^{N}$:

$$\max_{\{q_i\}_{i \in N_g}} \sum_{i \in N_g} P_i(r_i + q_i) \cdot q_i - \sum_{i \in N_g} C_i(q_i)$$

subject to:

$$q_i \geq 0, \quad i \in N_g$$
$$q_i \leq \overline{q}_i, \quad i \in N_g$$
$$r_i + q_i \geq 0, \quad i = N_g$$

Here, $\overline{q}_i$ and $C_i(q_i)$ are the capacity and cost function of the unit at node i, respectively.

Notice that firm g's problem can be decomposed into N_g subproblems, each corresponding to the firm's production decision at one node with a nodal demand function that has been shifted by the ISO chosen imports. As we noted earlier, this model will predict a market equilibrium that is not affected by whether or not generators in different locations are owned by the same firm. Moreover, under this formulation, the equilibrium solution for a network in which no transmission constraints are binding predicts average nodal prices that are higher than the Cournot equilibrium price corresponding to a single market with the aggregated system demand function. For example, in a two-node system with unlimited transmission capacity and with symmetric supply and demand (so that each node is self sufficient), this model yields the monopoly price at each node since it does not account for the reduced demand elasticity resulting from the merging of the two local monopoly markets into a duopoly. For the special case of zero generation cost, the monopoly price resulting from the model is twice the duopoly price corresponding to the merged markets.

5.3.1.2   *The Pure Bertrand Model*   The second model assumes that the firms behave purely *a la* Bertrand with regard to the ISO's transmission prices [25]. This is achieved by rewriting (5.2) as

$$r_i + q_i = P_i^{-1}(p + \varphi_i), \quad i = 1 \ldots N$$

and summing it up for all nodes:

$$\sum_{i=1}^{N} (r_i + q_i) = \sum_{i=1}^{N} P_i^{-1}(p + \varphi_i)$$

Due to the energy balance constraint on the ISO redispatch we get:

$$\sum_{i=1}^{N} q_i = \sum_{i=1}^{N} P_i^{-1}(p + \varphi_i) \tag{5.3}$$

which, implicitly, characterizes the residual demand function faced by each generation firm given the locational markups set by the ISO.

Now, the firms' competition can be modeled as a Nash-Cournot game among generators where each firm takes as given its competitors' production as well as the nodal price premiums $\{\varphi_i\}_{i=1}^{N}$, and acts as a monopolist with respect to the residual demand implied by (3). Mathematically, a firm g solves the following problem:

$$\max_{\{q_i\}, p} \sum_{i \in N_g} (p + \varphi_i) q_i - \sum_{i \in N_g} C_i(q_i)$$

subject to:

$$q_i \geq 0, \quad i \in N_g$$
$$q_i \leq \bar{q}_i, \quad i \in N_g$$

$$\sum_{i=1}^{N} q_i = \sum_{i=1}^{N} P_i^{-1}(p + \varphi_i)$$

$$r_i + q_i \geq 0, \quad i = 1 \ldots N$$

Because the firms observe the system-wide demand, their decision problems are not geographically separable and this model's solution is sensitive to whether or not a firm owns plants at different locations. Moreover, when the network constraints are not violated, the locational price premiums go to zero, and this model produces the same solution as a Cournot equilibrium calculated for a single node with the aggregated system demand. Unfortunately, this model has a shortcoming when applied to systems with multiple subnetworks. For instance, in the case of a two-node one-line network, reducing the line capacity to zero creates two local monopolies. However, this model will still yield a duopoly equilibrium with prices lower than the monopoly prices.

5.3.2   The Hybrid-Bertrand-Cournot Model

In the following, we introduce a model of Cournot competition among generators that is capable of separating the firms' decision making into strategic subnetworks. In doing so, we study the preceding two models, and observe that whether or not the firms' decision problems are separable depends on the slopes of the residual demand functions they face. In this new model, we assume that firms behave *a la* Cournot with respect to inter-subnetwork imports/exports, but *a la* Bertrand with respect to intra-subnetwork transmission costs. As a result, the firms' decisions in different subnetworks are essentially independent, but the generator ownership structures within individual subnetworks affect the solution.

5.3.2.1   The Firms' Problems   The first step to characterize the firms' problems is to quantify the aggregate demand function in each subnetwork. This is obtained by summing the inverse function of (5.2) for the node set $\tilde{N}_s$ of each subnetwork s (replacing p with p_s):

$$\sum_{i \in \tilde{N}_s} q_i + \sum_{i \in \tilde{N}_s} r_i = \sum_{i \in \tilde{N}_s} P_i^{-1}(p_s + \varphi_i), \quad s = 1 \ldots S$$

This equation allows the firms to compete for sales in each subnetwork. In mathematical terms, the decision problem for a firm g is

$$\max_{\{q_i\}_{i \in N_g}, \{p_s\}_{s=1}^S} \sum_{s=1}^{S} \sum_{i \in N_g \cap \tilde{N}_s} (p_s + \varphi_i) q_i - \sum_{i \in N_g} C_i(q_i)$$

subject to:

$$q_i \geq 0, \quad i \in N_g$$
$$q_i \leq \bar{q}_i, \quad i \in N_g$$
$$\sum_{i \in \tilde{N}_s} q_i + \sum_{i \in \tilde{N}_s} r_i = \sum_{i \in \tilde{N}_s} P_i^{-1}(p_s + \varphi_i), \quad s = 1 \ldots S$$
$$r_i + q_i \geq 0, \quad i = 1 \ldots N$$

Because this problem is parameterized by the total import/export $\sum_{i \in \tilde{N}_s} r_i$ in each subnetwork s and the locational price premiums $\{\varphi_i\}_{i=1}^{N}$, it can only be decomposed according to the structure of the subnetworks. If we let ρ_i^-, ρ_i^+, β_{gs}, and ξ_i be the Lagrange multipliers corresponding to the constraints, the KKT conditions for this problem are

- with respect to q_i:

$$p_s + \varphi_i - \beta_{gs} - C_i'(q_i) + \rho_i^- - \rho_i^+ + \eta_i = 0, \quad i \in \tilde{N}_s \cap N_g, s = 1 \ldots S$$

- with respect to p_s:

$$\beta_{gs} \sum_{i \in \tilde{N}_s} \frac{dP_i^{-1}(p_s + \varphi_i)}{dp_s} + \sum_{i \in \tilde{N}_s \cap N_g} q_i = 0, \quad s = 1 \ldots S$$

- with respect to β_{gs}

$$\sum_{i \in \tilde{N}_s} q_i + \sum_{i \in \tilde{N}_s} r_i = \sum_{i \in \tilde{N}_s} P_i^{-1}(p_s + \varphi_i), \quad s = 1 \ldots S$$

- with respect to ρ_i^-

$$0 \le \rho_i^- \perp q_i \ge 0, \quad i \in N_g$$

- with respect to ρ_i^+

$$0 \le \rho_i^+ \perp \bar{q}_i - q_i \ge 0, \quad i \in N_g$$

- with respect to ξ_i

$$0 \le \xi_i \perp r_i - q_i \ge 0, \quad i \in N_g$$

5.3.2.2 The Market Equilibrium Conditions

The market equilibrium conditions of the model are obtained by combining the KKT conditions for the ISO's and the firms' programs. In general, these conditions form a mixed nonlinear complementarity problem. When the nodal demand functions are linear and the cost functions are convex quadratic, that is,

$$P_i(q) = a_i - b_i q, \quad i = 1 \dots N$$

$$C_i(q) = c_i q + \frac{1}{2} d_i q^2, \quad i = 1 \dots N$$

the market equilibrium conditions become the following mixed linear complementarity problem (mixed LCP, see [5]):

$$p_s + \varphi_i - \beta_{gs} - c_i - d_i q_i + \rho_i^- - \rho_i^+ + \eta_i = 0, \quad i \in \tilde{N}_s \cap N_g, s = 1 \dots S, \quad g = 1 \dots G$$

$$-\beta_{gs} \sum_{i \in \tilde{N}_s} \frac{1}{b_i} + \sum_{i \in \tilde{N}_s \cap N_g} q_i = 0, \quad s = 1 \dots S, \quad g = 1 \dots G$$

$$\sum_{i \in \tilde{N}_s} q_i + \sum_{i \in \tilde{N}_s} r_i = \sum_{i \in \tilde{N}_s} \frac{a_i - p_s - \varphi_i}{b_i}, \quad s = 1 \dots S$$

$$0 \le \rho_i^- \perp q_i \ge 0, \quad i \in N$$

$$0 \le \rho_i^+ \perp \bar{q}_i - q_i \ge 0, \quad i \in N$$

$$0 \le \xi_i \perp r_i + q_i \ge 0, \quad i = 1 \dots N$$

$$P_i(r_i + q_i) - p + \sum_{l=1}^{L} (\lambda_l^- - \lambda_l^+) D_{li} + \eta_i = 0, \quad i = 1 \dots N$$

$$\sum_{i=1}^{N} r_i = 0$$

$$0 \le \lambda_l^- \perp \bar{k}_l + \sum_{i=1}^{N} D_{li} r_i \ge 0, \quad l = 1 \dots L$$

$$0 \le \lambda_l^+ \perp \bar{k}_l - \sum_{i=1}^{N} D_{li} r_i \ge 0, \quad l = 1 \dots L$$

$$\eta_i \perp r_i + q_i \ge 0, \quad i = 1 \dots N$$

This problem is not a square LCP because different Lagrange multipliers (η_i and ξ_i) are assigned to the common constraints (5.1) shared by the ISO's and the firms' problems. However, from an economic point of view, it is reasonable to assume that in equilibrium these common constraints should have the same shadow values for each entity, that is,

$$\eta_i = \xi_i, \quad i = 1 \ldots N$$

The practical importance of this, arguably strong assumption, is negligible, since the shadow prices are positive only if nodal prices are above the choke price on the demand curves, and the load is zero. This is a very unlikely occurrence in practice. Making this assumption is mathematically convenient, turning the market equilibrium conditions into the following square mixed LCP problem:

$$p_s + \varphi_i - \beta_{gs} - c_i - d_i q_i + \rho_i^- - \rho_i^+ + \eta_i = 0, \quad i \in \tilde{N}_s \cap N_g, \quad s = 1 \ldots S, \quad g = 1 \ldots G \tag{5.4}$$

$$-\beta_{gs} \sum_{i \in \tilde{N}_s} \frac{1}{b_i} + \sum_{i \in \tilde{N}_s \cap N_g} q_i = 0, \quad s = 1 \ldots S, \quad g = 1 \ldots G \tag{5.5}$$

$$\sum_{i \in \tilde{N}_s} q_i + \sum_{i \in \tilde{N}_s} r_i = \sum_{i \in \tilde{N}_s} \frac{a_i - p_s - \varphi_i}{b_i}, \quad s = 1 \ldots S \tag{5.6}$$

$$0 \le \rho_i^- \perp q_i \ge 0, \quad i = 1 \ldots N \tag{5.7}$$

$$0 \le \rho_i^+ \perp \bar{q}_i - q_i \ge 0, \quad i = 1 \ldots N \tag{5.8}$$

$$a_i - b_i(q_i + r_i) - p - \varphi_i = 0, \quad i = 1 \ldots N \tag{5.9}$$

$$\sum_{i=1}^{N} r_i = 0 \tag{5.10}$$

$$0 \le \lambda_l^- \perp \bar{k}_l + \sum_{i=1}^{N} D_{li} r_i \ge 0, \quad l = 1 \ldots L \tag{5.11}$$

$$0 \le \lambda_l^+ \perp \bar{k}_l - \sum_{i=1}^{N} D_{li} r_i \ge 0, \quad l = 1 \ldots L \tag{5.12}$$

$$0 \le \eta_i \perp r_i + q_i \ge 0, \quad i = 1 \ldots N \tag{5.13}$$

$$\varphi_i = -\sum_{l=1}^{L} (\lambda_l^- - \lambda_l^+) D_{li} - \eta_i = 0, \quad i = 1 \ldots N \tag{5.14}$$

In Subsection 5.3.1.1, we introduced the reference prices $\{p_s\}_{s=1}^{S}$ of the subnetworks to construct the aggregated subnetwork demand functions in the firms' problems. Because the firms have the full ability to influence these reference prices, our model will be incorrect if these prices aren't equal at the equilibrium.

Proposition 2: In the market equilibrium, all reference prices are equal, that is

$$p_s = p, \quad s = 1 \ldots S$$

Proof: Condition (6) implies

$$p_s = \frac{\displaystyle\sum_{i \in \tilde{N}_s} \frac{a_i - \varphi_i}{b_i} - \sum_{i \in \tilde{N}_s} q_i - \sum_{i \in \tilde{N}_s} r_i}{\displaystyle\sum_{i \in \tilde{N}_s} \frac{1}{b_i}}, \quad s = 1 \ldots S$$

Solving for $\{r_i\}_{i=1}^{N}$ from (9) and substituting the values into the above expression gives

$$p_s = \frac{\displaystyle\sum_{i\in\bar{N}_s}\frac{a_i-\varphi_i}{b_i} - \sum_{i\in\bar{N}_s}q_i - \sum_{i\in\bar{N}_s}\left(\frac{a_i-p-\varphi_i}{b_i}-q_i\right)}{\displaystyle\sum_{i\in\bar{N}_s}\frac{1}{b_i}} = p, \quad s=1\ldots S$$

5.3.2.3 *Computational Properties*   The preceding market equilibrium conditions represent a quasi-variational inequality problem due to the common constraints in the firms' and the ISO's programs. Next, we study its solution existence and the solution approach.

Lemma 1: Conditions (5.4)–(5.14) can be represented as a linear complementarity problem with a bisymmetric positive semi-definite matrix.

Proof: We group the parameters and variables as follows:

- $B \in R^{N\times N}$: A diagonal matrix where the (i, i) th is b_i,
- $D \in R^{L\times N}$: The PTDF matrix where the (l, i) th element is D_{li},
- $a = [a_i \ i = 1...N]$,
- $\bar{q} = [\bar{q}_i \ i = 1...N]$,
- $c = [c_i \ i = 1...N]$,
- $\bar{k} = [\bar{k}_l \ l = 1...L]$,
- $r = [r_i \ i = 1...N]$,
- $q = [q_i \ i = 1...N]$,
- $\rho_- = [\rho_i^- \quad i = 1...N]$,
- $\rho_+ = [\rho_i^+ \quad i = 1...N]$,
- $\lambda_- = [\lambda_l^- \quad l = 1...L]$,
- $\lambda_+ = [\lambda_l^+ \quad l = 1...L]$,
- $\eta = [\eta_i \ i = 1..N]$.

Further, let $e \in R^N$ be a vector of all 1's, and $H \in R^{N\times N}$ and $Q \in R^{N\times N}$ be two matrices such that

$$h_{ij} = \begin{cases} \dfrac{1}{\displaystyle\sum_{i=1}^{N}\frac{1}{b_i}} + \dfrac{1}{\displaystyle\sum_{i\in\bar{N}_S}\frac{1}{b_i}} + d_i & \text{if } i = j \\[2em] \dfrac{1}{\displaystyle\sum_{i=1}^{N}\frac{1}{b_i}} + \dfrac{1}{\displaystyle\sum_{i\in\bar{N}_S}\frac{1}{b_i}} & \begin{array}{l}\text{if } i \neq j, \text{ and nodes } i \text{ and } j \text{ belong to the same subnetwork}\\ \text{and the same firm}\end{array} \\[2em] \dfrac{1}{\displaystyle\sum_{i=1}^{N}\frac{1}{b_i}} & \text{Otherwise} \end{cases}$$

and

$$Q = B^{-1} - \frac{B^{-1}ee^T B^{-1}}{e^T B^{-1}e}.$$

Eliminating variables with free signs, we represent (5.4)–(5.14) as an LCP problem:

$$w = t + My \tag{5.15a}$$

$$0 \leq w \perp y \geq 0 \tag{5.15b}$$

where w and y are variable vectors, t and M are constants, such that

$$w = \begin{bmatrix} \rho_- \\ \bar{q}-q \\ \bar{k}+Dr \\ \bar{k}-Dr \\ r+q \end{bmatrix} \quad y = \begin{bmatrix} q \\ \rho_+ \\ \lambda_- \\ \lambda_+ \\ \eta \end{bmatrix} \quad t = \begin{bmatrix} c - \dfrac{ee^T B^{-1}}{e^T B^{-1}e}a \\ \bar{q} \\ \bar{k}+DQa \\ \bar{k}-DQa \\ Qa \end{bmatrix}$$

and

$$M = \begin{bmatrix} H & I & BQD^T & -BQD^T & BQ-I \\ -I & 0 & 0 & 0 & 0 \\ -DQB & 0 & DQD^T & -DQD^T & DQ \\ DQB & 0 & -DQD^T & DQD^T & -DQ \\ QB+I & 0 & QD^T & -QD^T & Q \end{bmatrix}$$

Note H and Q are both symmetric positive semi-definite and

$$\frac{M+M^T}{2} = \begin{bmatrix} H & 0 & 0 & 0 & 0 \\ 0 & 0 & 0 & 0 & 0 \\ 0 & 0 & 0 & 0 & 0 \\ 0 & 0 & 0 & 0 & 0 \\ 0 & 0 & 0 & 0 & 0 \end{bmatrix} + \begin{bmatrix} 0 \\ 0 \\ D \\ -D \\ I \end{bmatrix} Q \begin{bmatrix} 0 \\ 0 \\ D \\ -D \\ I \end{bmatrix}^T$$

we conclude that M is bisymmetric positive semi-definite.

Lemma 2: there exists at least one market equilibrium.

Proof: We observe that

$$w = \begin{bmatrix} c \\ \bar{q} \\ \bar{k} \\ \bar{k} \\ Qa \end{bmatrix} \quad \text{and} \quad y = \begin{bmatrix} 0 \\ 0 \\ 0 \\ 0 \\ a \end{bmatrix}$$

satisfy the linear conditions in (5.15). By Theorem 3.1.2 in [5], we conclude that conditions (5.4)–(5.14) have solutions.

Theorem 1: Assuming nondegeneracy, a solution to (5.4)–(5.14) is guaranteed by Lemke's algorithm [13].

Proof: This follows from Theorem 4.4.1 in [5] and lemmas 1 and 2.

It is worth noting that the models introduced in Section 5.3.1 are two special cases of (5.4)–(5.14) with N and 1 subnetworks, respectively. Their equilibrium conditions can also be presented as (5.15) where H is given by

$$
h_{ij} = \begin{cases} \dfrac{1}{\sum\limits_{i=1}^{N} \dfrac{1}{b_i}} + b_i + d_i & \text{if } i = j \\[2em] \dfrac{1}{\sum\limits_{i=1}^{N} \dfrac{1}{b_i}} & \text{Otherwise} \end{cases}
$$

for the extreme case where all N nodes are strategically decoupled and represent N subnetwork, and by

$$
h_{ij} = \begin{cases} \dfrac{2}{\sum\limits_{i=1}^{N} \dfrac{1}{b_i}} + d_i & \text{if } i = j \\[2em] \dfrac{2}{\sum\limits_{i=1}^{N} \dfrac{1}{b_i}} & \text{if } i \neq j, \text{ and nodes } i \text{ and } j \text{ belong to the same firm} \\[2em] \dfrac{1}{\sum\limits_{i=1}^{N} \dfrac{1}{b_i}} & \text{Otherwise} \end{cases}
$$

for the case where all generators are strategically coupled in a single subnetwork.

5.4 NUMERICAL EXAMPLE FOR THE SUBNETWORKS MODEL

We use the network in Figure 5.1 to illustrate the application and economics insights of the hybrid Bertrand-Cournot model. In this example, all eight lines are identical in terms of their electrical characteristics except that the two interfaces, lines 2–4 and 3–5, that have very low thermal limits of 2 MW. As a result, this network is separated into two strategic subnetworks where nodes 1 through 3 form one subnetwork and nodes 4 through 6 form the other. In addition, this system has six generators, each at one node (see Table 5.1). We assume four different hypothetical generator ownership structures (See Table 5.2) by assigning the generators to 2, 3,

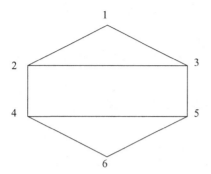

Figure 5.1   A six-bus network

TABLE 5.1   Generator information

Node	Capacity (MW)	Marginal cost ($/MWh)
1	120	15
2	80	20
3	25	30
4	80	20
5	25	30
6	120	15

TABLE 5.2   Generator ownership structures

Node	2 Firms	3 Firms	4 Firms	6 Firms
1	Firm #1	Firm #1	Firm #1	Firm #1
2	Firm #1	Firm #3	Firm #3	Firm #2
3	Firm #1	Firm #1	Firm #1	Firm #3
4	Firm #2	Firm #3	Firm #4	Firm #4
5	Firm #2	Firm #2	Firm #2	Firm #5
6	Firm #2	Firm #2	Firm #2	Firm #6

4, and 6 firms, respectively; such structures enable us to observe the sensitivity of the market equilibrium to market concentration.

We consider affine demand functions (see Table 5.3) which, together with the generator characteristics and the resource ownership patterns, lead to symmetric subnetworks. Therefore, as would be expected, the resulting nodal prices are uniform across all the nodes, and the flows on the interfaces are zero (see Table 5.4). In addition, our model predicts prices that are lower than the prices from the pure Cournot model (that treats each node as a subnetwork, so that generators believe that they face only the local price elasticity) and greater than those from the pure Bertrand model (that treats the entire network as a single subnetwork, resulting in generators believing that they can compete in all markets). The corresponding output quantities for each generator are summarized in Table 5.5. It should also be pointed

TABLE 5.3   Affine demand functions

Node	Price intercept ($/MWh)	Slope
1	100	1.0
2	100	0.8
3	100	1.2
4	100	0.8
5	100	1.2
6	100	1.0

TABLE 5.4   Nodal prices under symmetric subnetworks

Model	Node	2 Firms	3 Firms	4 Firms	6 Firms
Hybrid	1–6	60.00	45.00	45.00	42.30
Pure Cournot	1–6	60.63	60.63	60.63	60.63
Pure Bertrand	1–6	46.67	40.54	35.14	32.86

TABLE 5.5   Generator outputs under symmetric subnetworks

	Node	2 Firms	3 Firms	4 Firms	6 Firms
Hybrid Cournot-Bertrand	1	120.00	92.50	92.50	84.17
	6	120.00	92.50	92.50	84.17
	2	3.33	77.08	77.08	68.75
	4	3.33	77.08	77.08	68.75
	3	0	0	0	25.00
	5	0	0	0	25.00
Pure Cournot	1	45.63	45.63	45.63	45.63
	6	45.63	45.63	45.63	45.63
	2	50.78	50.78	50.78	50.78
	4	50.78	50.78	50.78	50.78
	3	25.00	25.00	25.00	25.00
	5	25.00	25.00	25.00	25.00
Pure Bertrand	1	120.00	120.00	120.00	110.12
	6	120.00	120.00	120.00	110.12
	2	44.44	63.33	80.00	79.29
	4	44.44	63.33	80.00	79.29
	3	0	0	0	17.62
	5	0	0	0	17.62

out that the hybrid Bertrand-Cournot model produces the same market outcomes for the three- and four-firm structures; this is because, in these structures, both subnetworks consist of duopoly firms.

Next, we create an asymmetric structure of the subnetworks by exchanging the cost functions at nodes 2 and 3. Such asymmetry is more likely to result in flow congestion on the interfaces and uneven nodal prices. Indeed, this is found true for most test scenarios except the duopoly structure (see Table 5.6). Again, the hybrid Bertrand-Cournot model leads to prices between those from the pure Cournot and Bertrand models. Similar to the symmetric case, our model produces identical market equilibria for the three- and four-firm structures.

5.5 BERTRAND MODEL WITH COMMON KNOWLEDGE CONSTRAINTS

In the pure Bertrand model described in Section 5.3.1.2, it was assumed that firms optimizing their profit are not aware of any transmission constraints and account for congestion only through the nodal price markups set by the ISO to which they respond as price takers. In this section we will allow firms to directly account to congestion on lines that are systematically capacitated and hence are designated as common knowledge constraints, while they still act at price takers to the portion of the nodal price markups that reflect congestion on lines that have not been so designated.

5.5.1 The Firm's Problems

When some line sets, $L_1 \subseteq L$ and $L_2 \subseteq L$, are constantly congested in the negative and positive directions, respectively, the ISO's KKT conditions take the form:

$$P_i(r_i + q_i) - p + \sum_{l \in L_1} D_{li}\lambda_l^- - \sum_{l \in L_2} D_{li}\lambda_l^+ + \sum_{l \in L\setminus L_1} (\lambda_l^- - \lambda_l^+)D_{li} = 0, \quad i \in N \quad (5.16)$$

$$\sum_{i \in N} r_i = 0 \quad (5.17)$$

$$\overline{k}_l + \sum_{i \in N} D_{li}r_i = 0, \quad \lambda_l^+ = 0, \quad l \in L_1 \quad (5.18)$$

$$\overline{k}_l - \sum_{i \in N} D_{li}r_i = 0, \quad \lambda_l^- = 0, \quad l \in L_2 \quad (5.19)$$

$$0 \le \lambda_l^- \perp \overline{k}_l + \sum_{i \in N} D_{li}r_i \ge 0, \quad l \in L\setminus L_1 \quad (5.20)$$

$$0 \le \lambda_l^+ \perp \overline{k}_l - \sum_{i \in N} D_{li}r_i \ge 0, \quad l \in L\setminus L_2 \quad (5.21)$$

Let

$$\tilde{\varphi}_i = - \sum_{l \in L\setminus L_1 \setminus L_2} (\lambda_l^- - \lambda_l^+) D_{li},$$

then condition (5.16) can be rewritten as

$$P_i(r_i + q_i) + p - \sum_{l \in L_1} D_{li}\lambda_l^- + \sum_{l \in L_2} D_{li}\lambda_l^+ + \tilde{\varphi}_i = 0, \quad i \in N \quad (5.22)$$

TABLE 5.6 Equilibria of asymmetric subnetworks

Model	Node	Nodal Prices				Generator outputs			
		2 Firms	3 Firms	4 Firms	6 Firms	2 Firms	3 Firms	4 Firms	6 Firms
Hybrid Cournot-Bertrand	1	60.00	47.88	47.88	45.18	120.00	101.39	101.39	93.05
	2	60.00	49.07	49.07	46.37	0	58.81	58.81	50.48
	3	60.00	46.69	46.69	43.99	3.33	0	0	25.00
	4	60.00	44.26	44.26	41.56	3.33	74.80	74.80	66.46
	5	60.00	46.64	46.64	43.94	0	0	0	25.00
	6	60.00	45.45	45.45	42.75	120.00	93.89	93.89	85.56
Pure Cournot	1	62.73	62.73	62.73	62.73	47.73	47.73	47.73	47.73
	2	63.11	63.11	63.11	63.11	41.39	41.39	41.39	41.39
	3	62.36	62.36	62.36	62.36	25.00	25.00	25.00	25.00
	4	60.48	60.48	60.48	60.48	50.61	50.61	50.61	50.61
	5	61.23	61.23	61.23	61.23	25.00	25.00	25.00	25.00
	6	60.86	60.86	60.86	60.86	45.86	45.86	45.86	45.86
Pure Bertrand	1	49.78	44.02	39.71	36.32	120.00	120.00	120.00	120.00
	2	54.78	47.50	40.60	38.20	17.32	44.43	65.34	50.56
	3	44.78	40.55	38.83	34.44	15.46	6.72	0	25.00
	4	45.23	37.50	34.43	31.30	34.03	63.46	80.00	69.70
	5	55.23	44.44	36.19	35.06	1.52	0	0	25.00
	6	50.23	40.97	35.31	33.18	120.00	120.00	120.00	112.11

Thus, the nodal prices are composed of three parts: the price p at the reference bus, the locational price premium $-\sum_{l \in L_1} D_{li} \lambda_l^- + \sum_{l \in L_2} D_{li} \lambda_l^+$ due to the systematically congested constraints and the locational price markups $\tilde{\varphi}_i$ that account for all the other constraints and taken by firms as ISO set parameters when valuating their residual demand. Solving for r_i in (5.22) and substituting into (5.17), (5.18), and (5.19) yields:

$$\sum_{i \in N} \left(P_i^{-1} \left(p - \sum_{l \in L_1} D_{li} \lambda_l^- + \sum_{l \in L_2} D_{li} \lambda_l^+ + \tilde{\varphi}_i \right) - q_i \right) = 0 \tag{5.23}$$

$$\bar{k}_l + \sum_{i \in N} D_{li} \left(P_i^{-1} \left(p - \sum_{m \in L_1} D_{mi} \lambda_m^- + \sum_{m \in L_2} D_{mi} \lambda_m^+ + \tilde{\varphi}_i \right) - q_i \right) = 0, \quad l \in L_1 \tag{5.24}$$

$$\bar{k}_l - \sum_{i \in N} D_{li} \left(P_i^{-1} \left(p - \sum_{m \in L_1} D_{mi} \lambda_m^- + \sum_{m \in L_2} D_{mi} \lambda_m^+ + \tilde{\varphi}_i \right) - q_i \right) = 0, \quad l \in L_2 \tag{5.25}$$

When the systematically congested lines are common knowledge, firms will account for conditions (5.18) and (5.19) in their profit maximization and hence the residual demand against which they maximize their profits is implicitly specified by (5.23), (5.24), and (5.25). Thus, in determining their profit maximizing output levels the firms try to influence the price at the reference bus and the shadow prices on the common knowledge constraints in the set $L_1 \cup L_2$, while behaving as price takers with respect to the ISO nodal price markup components reflecting congestion on all other transmission lines. Mathematically, each firm g solves the following profit-maximization problem:

$$\max_{\{q_i\}_{i \in N_g}, p, \{\lambda_l^-\}_{l \in L_1}, \{\lambda_l^+\}_{l \in L_2}} \sum_{i \in N_g} \left(p - \sum_{l \in L_1} D_{li} \lambda_l^- + \sum_{l \in L_2} D_{li} \lambda_l^+ + \tilde{\varphi}_i \right) q_i - \sum_{i \in N_g} C_i(q_i)$$

$$0 \le q_i \le \bar{q}_i, \quad i \in N_g$$

$$\sum_{i \in N} \left(P_i^{-1} \left(p - \sum_{l \in L_1} D_{li} \lambda_l^- + \sum_{l \in L_2} D_{li} \lambda_l^+ + \tilde{\varphi}_i \right) - q_i \right) = 0$$

$$\bar{k}_l + \sum_{i \in N} D_{li} \left(P_i^{-1} \left(p - \sum_{m \in L_1} D_{mi} \lambda_m^- + \sum_{m \in L_2} D_{mi} \lambda_m^+ + \tilde{\varphi}_i \right) - q_i \right) = 0, \quad l \in L_1$$

$$\bar{k}_l - \sum_{i \in N} D_{li} \left(P_i^{-1} \left(p - \sum_{m \in L_1} D_{mi} \lambda_m^- + \sum_{m \in L_2} D_{mi} \lambda_m^+ + \tilde{\varphi}_i \right) - q_i \right) = 0, \quad l \in L_2$$

If we let ρ_i^-, ρ_i^+, α_g, β_{gl}^-, and β_{gl}^+ be the Lagrange multipliers corresponding to the constraints, the KKT conditions for this problem are

- with respect to q_i

$$p - \sum_{l \in L_1} D_{li} \lambda_l^- + \sum_{l \in L_2} D_{li} \lambda_l^+ + \tilde{\varphi}_i - \alpha_g + \sum_{l \in L_1} D_{li} \beta_{gl}^- - \sum_{l \in L_2} D_{li} \beta_{gl}^+ - C'(q_i) + \rho_i^- - \rho_i^+ = 0, i \in N_g$$

- with respect to p

$$\alpha_g \sum_{i \in N} \frac{dP_i^{-1}\left(p - \sum_{m \in L_1} D_{mi}\lambda_m^- + \sum_{m \in L_2} D_{mi}\lambda_m^+ + \tilde{\varphi}_i\right)}{dp}$$

$$+ \sum_{l \in L_1} \beta_{gl}^- \sum_{i \in N} D_{li} \frac{dP_i^{-1}\left(p - \sum_{m \in L_1} D_{mi}\lambda_m^- + \sum_{m \in L_2} D_{mi}\lambda_m^+ + \tilde{\varphi}_i\right)}{dp}$$

$$- \sum_{l \in L_1} \beta_{gl}^+ \sum_{i \in N} D_{li} \frac{dP_i^{-1}\left(p - \sum_{m \in L_1} D_{mi}\lambda_m^- + \sum_{m \in L_2} D_{mi}\lambda_m^+ + \tilde{\varphi}_i\right)}{dp} + \sum_{i \in N_g} q_i = 0$$

- with respect to λ_l^-

$$\alpha_g \sum_{i \in N} \frac{dP_i^{-1}\left(p - \sum_{m \in L_1} D_{mi}\lambda_m^- + \sum_{m \in L_2} D_{mi}\lambda_m^+ + \tilde{\varphi}_i\right)}{d\lambda_l^-}$$

$$+ \sum_{k \in L_1} \beta_{gk}^- \sum_{i \in N} D_{ki} \frac{dP_i^{-1}\left(p - \sum_{m \in L_1} D_{mi}\lambda_m^- + \sum_{m \in L_2} D_{mi}\lambda_m^+ + \tilde{\varphi}_i\right)}{d\lambda_l^-}$$

$$- \sum_{k \in L_1} \beta_{gk}^+ \sum_{i \in N} D_{ki} \frac{dP_i^{-1}\left(p - \sum_{m \in L_1} D_{mi}\lambda_m^- + \sum_{m \in L_2} D_{mi}\lambda_m^+ + \tilde{\varphi}_i\right)}{d\lambda_l^-} + \sum_{i \in N_g} D_{li} q_i = 0$$

- with respect to λ_l^+

$$\alpha_g \sum_{i \in N} \frac{dP_i^{-1}\left(p - \sum_{m \in L_1} D_{mi}\lambda_m^- + \sum_{m \in L_2} D_{mi}\lambda_m^+ + \tilde{\varphi}_i\right)}{d\lambda_l^+}$$

$$+ \sum_{k \in L_1} \beta_{gk}^- \sum_{i \in N} D_{ki} \frac{dP_i^{-1}\left(p - \sum_{m \in L_1} D_{mi}\lambda_m^- + \sum_{m \in L_2} D_{mi}\lambda_m^+ + \tilde{\varphi}_i\right)}{d\lambda_l^+}$$

$$- \sum_{k \in L_1} \beta_{gk}^+ \sum_{i \in N} D_{ki} \frac{dP_i^{-1}\left(p - \sum_{m \in L_1} D_{mi}\lambda_m^- + \sum_{m \in L_2} D_{mi}\lambda_m^+ + \tilde{\varphi}_i\right)}{d\lambda_l^+} + \sum_{i \in N_g} D_{li} q_i = 0$$

- with respect to ρ_i^-

$$0 \le \rho_i^- \perp q_i \ge 0, \quad i \in N_g$$

- with respect to ρ_i^+

$$0 \le \rho_i^+ \perp \bar{q}_i - q_i \ge 0, \quad i \in N_g$$

- with respect to α_g

$$\sum_{i \in N}\left(P_i^{-1}\left(p - \sum_{m \in L_1} D_{mi}\lambda_m^- + \sum_{m \in L_2} D_{mi}\lambda_m^+ + \tilde{\varphi}_i\right) - q_i\right) = 0$$

- with respect to β_{gl}^-

$$\bar{k}_l + \sum_{i\in N} D_{li}\left(P_i^{-1}\left(p - \sum_{m\in L_1} D_{mi}\lambda_m^- + \sum_{m\in L_2} D_{mi}\lambda_m^+ + \tilde{\varphi}_i \right) - q_i \right) = 0, \quad l \in L_1$$

- with respect to β_{gl}^+

$$\bar{k}_l - \sum_{i\in N} D_{li}\left(P_i^{-1}\left(p - \sum_{m\in L_1} D_{mi}\lambda_m^- + \sum_{m\in L_2} D_{mi}\lambda_m^+ + \tilde{\varphi}_i \right) - q_i \right) = 0, \quad l \in L_2$$

5.5.2 The Market Equilibrium Conditions

An aggregation of the KKT conditions for the firms' and the ISO's problem leads to the market equilibrium conditions as a nonlinear mixed complementarity problem. In the sequel, we assume linear demand and linear marginal cost functions, and present these market equilibrium conditions as a mixed linear complementarity problem (mixed LCP, see [5]):

$$\alpha_i - (r_i + q_i)b_i - p + \sum_{l\in L}(\lambda_{li}^- - \lambda_{li}^+)D_{li} = 0, \quad i \in N$$

$$\sum_{i\in N} r_i = 0$$

$$0 \le \lambda_i^- \perp \bar{k}_l + \sum_{i\in N} D_{li}r_i \ge 0, \quad l \notin L_1 \cup L_2$$

$$0 \le \lambda_i^+ \perp \bar{k}_l - \sum_{i\in N} D_{li}r_i \ge 0, \quad l \notin L_1 \cup L_2$$

$$\bar{k}_l + \sum_{i\in N} D_{li}r_i = 0, \quad \lambda_i^+ = 0, \quad l \in L_1$$

$$\bar{k}_l - \sum_{i\in N} D_{li}r_i = 0, \quad \lambda_i^- = 0, \quad l \in L_2$$

$$p - \sum_{l\in L_1} D_{li}\lambda_l^- + \sum_{l\in L_2} D_{li}\lambda_l^+ + \tilde{\varphi}_i - \alpha_g + \sum_{l\in L_1} D_{li}\beta_{gl}^- - \sum_{l\in L_2} D_{li}\beta_{gl}^+ - c_i - d_i q_i + \rho_i^- - \rho_i^+ = 0,$$
$$i \in N$$

$$-\alpha_g \sum_{i\in N}\frac{1}{b_i} - \sum_{l\in L_1}\beta_{gl}^-\sum_{i\in N} D_{li}\frac{1}{b_i} + \sum_{l\in L_2}\beta_{gl}^+\sum_{i\in N} D_{li}\frac{1}{b_i} + \sum_{i\in N_g} q_i = 0,$$

$$\alpha_g \sum_{i\in N}\frac{D_{li}}{b_i} + \sum_{m\in L_1}\beta_{gl}^-\sum_{i\in N}\frac{D_{mi}D_{li}}{b_i} - \sum_{m\in L_2}\beta_{gl}^+\sum_{i\in N}\frac{D_{mi}D_{li}}{b_i} + \sum_{i\in N_g} D_{li}q_i = 0, \quad l \in L_1$$

$$\alpha_g \sum_{i\in N}\frac{D_{li}}{b_i} - \sum_{m\in L_1}\beta_{gl}^-\sum_{i\in N}\frac{D_{mi}D_{li}}{b_i} + \sum_{m\in L_2}\beta_{gl}^+\sum_{i\in N}\frac{D_{mi}D_{li}}{b_i} + \sum_{i\in N_g} D_{li}q_i = 0, \quad l \in L_2$$

$$0 \le \rho_i^- \perp q_i \ge 0, \quad i \in N$$

$$0 \le \rho_i^+ \perp \bar{q}_i - q_i \ge 0, \quad i \in N$$

$$\tilde{\varphi}_i = -\sum_{l\in L\backslash L_1\backslash L_2}(\lambda_l^- - \lambda_l^+)D_{li}$$

For linear demand and linear marginal cost functions, the ISO's and the firm's problems are concave-maximization problems, and the KKT conditions are necessary and sufficient for characterizing the global optimum of each agent.

5.6   NUMERICAL EXAMPLE OF EQUILIBRIUM WITH COMMON KNOWLEDGE CONSTRAINTS

This example employs the same six-bus network structure as shown in Figure 5.1, again assuming that all eight lines are identical in terms of their electrical characteristics but line 3–5 has a very low thermal limit of 2 MW. The supply and demand data is changed in order to highlight the features of this model. The demand functions are linear, with parameters specified in Table 5.7. This system has four generators; the generators at nodes 1 and 2 have a constant marginal cost of 30 $/MWh and the generators at nodes 4 and 6 have a relatively lower constant marginal cost of 10 $/MWh (see Table 5.8). In addition, these four generators are divided into a duopoly structure with the units at nodes 1 and 4 owned by firm #1 and the other two by firm #2.

We compute the market equilibrium of the model with a common knowledge constraint (the congestion on line 3–5) and compare it to that from the pure Bertrand model ignoring this constraint. In this case, the capacities of the generators are not binding in the market equilibrium. We find that, with the common knowledge constraint, both firms increase the output of the generators at nodes 1 and 2 and reduce the production of the units at nodes 4 and 6 (see Table 5.9). As a result, the prices at nodes 1 through 3 are reduced, but the prices at the other three nodes are raised (see Table 5.10). In addition, Table 5.11 illustrates greater profit for both firms. This suggests that, when recognizing common knowledge constraints, firms behave less competitively, which can be explained by the fact that capacitated lines only shift local demand horizontally but do not increase the elasticity of the residual demand functions.

TABLE 5.7   Demand functions

Node	Demand function	
	Price intercept ($/MWh)	Slope
1	50	1.00
2	50	0.82
3	50	1.13
4	50	1.10
5	50	0.93
6	50	0.85

TABLE 5.8   Generator information

Node	Capacity (MW)	Owner	Marginal cost ($/MWh)
1	100	Firm #1	30
2	100	Firm #2	30
4	100	Firm #1	10
6	100	Firm #2	10

TABLE 5.9 Generator output (MW)

Node	Model	
	Bertrand	Common knowledge constraints
1	0	8.4278
2	0	9.4984
3	0	0
4	79.4605	70.9568
5	0	0
6	48.7073	38.5607

TABLE 5.10 Prices ($/MWh)

Node	Model	
	Bertrand	Common knowledge constraints
1	42.3174	37.9236
2	37.4088	34.7477
3	47.2261	41.0994
4	22.6828	25.2201
5	12.8655	18.8683
6	17.7741	22.0442

TABLE 5.11 Firm's profit ($/h)

	Model	
	Bertrand	Common knowledge constraints
Firm #1	1007.7816	1146.7481
Firm #2	378.6554	509.5283
Total	1386.4371	1656.2764

Next, we study how generators' capacities affect the firms' ability to restrict the output from the low-cost generators. In doing so, we assume that all capacities are reduced by 50 percent so that, in the Bertrand model, both low-cost generators produce at the full capacity. Generators' outputs and corresponding prices for this case are summarized in Table 5.12 and Table 5.13. Unlike in the previous case with nonbinding generation capacities, the equilibrium corresponding to the common knowledge constraint now shows that the units at nodes 1 and 6 reduce their output. Thus, binding generation capacities lead to different behavior of the firms. Indeed, Table 5.14 reports that firm #1's profit decreases.

Finally, we point out that, due to loop flows in electricity networks, recognizing one particular common knowledge constraint might suffice to relieve congestion on other lines. For example, if the capacity of line 1–2 is 5 MW, the pure Bertrand

TABLE 5.12 Generation with constrained generation capacities (MW)

Node	Model	
	Bertrand	Common knowledge constraints
1	13.5491	13.2613
2	0	9.0971
4	50.0000	50.0000
6	50.0000	47.0237

TABLE 5.13 Prices with constrained generation capacities ($/MWh)

Node	Model	
	Bertrand	Public knowledge constraints
1	40.1432	37.9236
2	36.9665	35.2421
3	43.3199	40.6050
4	27.4365	27.1978
5	21.0832	21.8349
6	24.2598	24.5163

TABLE 5.14 Firm's profit ($/h)

	Model	
	Bertrand	Public knowledge constraints
Firm #1	1009.2562	964.9672
Firm #2	712.99	730.2980
Total	1722.2462	1695.2653

model produces an equilibrium where both lines 1–2 and 3–5 are congested, whereas the equilibrium with the common knowledge constraint indicates that only line 3–5 is capacitated while the flow on line 1–2 is 4.6 MW.

5.7 CONCLUDING REMARKS

This chapter examines two approaches for dealing with a limitation of existing Nash equilibrium models of congestion-prone electricity systems. These models either ignore the effects of joint ownership of generators and the effect of competitive interaction on the elasticity of the residual demand functions faced by the generators, or overestimate the effect of competitive interaction even when transmission capacity is limited or exausted. To address these shortcomings we first develop a hybrid Bertrand-Cournot model of electricity markets with multiple subnetworks. In this model, firms behave *a la* Cournot with respected to the ISO's inter-subnetwork transmission quantities, but *a la* Bertrand with respect to the intra-subnetwork transmission prices. This gives the modeler more flexibility as to how transmission price conjectures are represented in the model compared to pure Cournot or Bertrand models. When affine demand functions and quadratic cost functions are assumed, the market equilibrium conditions of this model become a linear complementarity problem with a bisymmetric positive semi-definite matrix. Numerical examples demonstrate that this model can lead to more realistic market equilibria.

In cases where the network cannot be partitioned into subnetworks as assumed by the hybrid Bertrand-Cournot approach, we propose a Bertrand type model where certain systematically congested lines are treated as common knowledge constraints and taken into consideration by the competing firms in assessing their residual demand and optimizing their output levels.

An important limitation of these approaches is that the definition of subnetworks as well as the designation of common knowledge constraints are exogenous. It is possible, for instance, that a Cournot conjecture about flows into a subnetwork is appropriate at some times (when congestion is more likely), but the Bertrand conjecture is preferable at other, less congested periods. Likewise a transmission interface may be congested most of the time but cannot be assumed to be congested all the time. Further research is needed to determine whether it is possible to endogenously determine the appropriate conjecture (perhaps in some iterative fashion). Empirical research is also desirable to determine what conjectures are actually held by firms in real markets.

ACKNOWLEDGMENTS

This research is supported in part by the National Science Foundation Grants ECS-0224779 and ECS-0224817, by the Power System Engineering Research Center (PSERC), by the University of California Energy Institute (UCEI), and by the U.S. Department of Energy.

REFERENCES

1. Barquin J, Vazquez M. Cournot equilibrium in power networks. Working Paper. Instituto de Investigacion Tecnologica (IIT), Universidad Pontificia Comillas, Madrid, Spain, 2005.
2. Borenstein S, Bushnell J, Stoft S. The competitive effects of transmission capacity in a deregulated electricity industry. *RAND Journal of Economics* 2000;31(2):294–325.
3. Cardell J, Hitt C, Hogan WW. Market power and strategic interaction in electricity networks. *Resource and Energy Economics* 1997;19:109–137.
4. Chao H-P, Peck SC. A market mechanism for electric power transmission. *Journal of Regulatory Economics* 1996;10(1):25–60.

5. Cottle RW, Pang JS, Stone RE. *The Linear Complementarity Problem*. Boston, MA: Academic Press; 1992.
6. Daxhelet O, Smeers Y. Variational inequality models of restructured electric systems. In: Ferris MC, Mangasarian OL, Pang J-S, editors. *Applications and Algorithms of Complementarity*. Boston MA: Kluwer; 2001.
7. Harker PT. Generalized Nash games and quasivariational inequalities. *European Journal of Operations Research* 1991;54(1):81–94.
8. Hobbs BF. Linear complementarity models of Nash-Cournot competition in bilateral and POOLCO power markets. *IEEE Transactions on Power Systems* 2001;16(2):194–202.
9. Hobbs BF, Metzler CB, Pang JS. Strategic gaming analysis for electric power systems: an MPEC approach. *IEEE Transactions on Power Systems* 2000;15(2):638–645.
10. Hobbs BF, Pang J-S. Nash-Cournot equilbria in electric power markets with piecewise linear demand functions and joint constraints. *Operations Research* 2007;55(1):113–127.
11. Hobbs BF, Rijkers FAM, Wals AF. Modeling strategic generator behavior with conjectured transmission price responses in a mixed transmission pricing system I: Formulation, II: Application. *IEEE Transactions on Power Systems* 2004;19(2):707–717; 872–879.
12. Hogan WW. Contract networks for electric power transmission. *Journal of Regulatory Economics* 1992;4(3):211–242.
13. Lemke CE. Bimatrix equilibrium points and mathematical programming. *Management Science* 1965;11(7):681–689.
14. Luo ZQ, Pang JS, Ralph D. *Mathematical Programs with Equilibrium Constraints*. Cambridge: Cambridge University Press; 1996.
15. Metzler C, Hobbs BF, Pang JS. Nash-Cournot equilibria in power markets on a linearized dc network with arbitrage: formulations and properties. *Networks and Spatial Economics* 2003;3(2):123–150.
16. Neuhoff K, Barquin J, Boots MG, Ehrenmann A, Hobbs BF, Rijkers FAM, Vázquez M. Network-constrained models of liberalized electricity markets: the Devil is in the details. *Energy Economics* 2005;27(3):495–525.
17. Oren SS. Economic inefficiency of passive transmission rights in congested electricity systems with competitive generation. *The Energy Journal* 1997;18(1):63–83.
18. Smeers Y, Wei J-Y. Spatial oligopolistic electricity models with Cournot firms and opportunity cost transmission prices. Center for Operations Research and Econometrics, Universite Catholique de Louvain, Louvain-la-Neuve, Belgium, 1997.
19. Sioshansi F, Pfaffenberger W, editors. *Electricity Market Reform: An International Perspective*. Global Energy Policy and Economics Series, Oxford: Elsevier; 2006.
20. Stoft S. Financial transmission rights meet Cournot: how TCCs curb market power. *Energy Journal* 1999;20(1):1–23.
21. Ventosa M, Baíllo Á, Ramos A, Rivier M. Electricity market modeling trends. *Energy Policy* 2005;33(7):897–913.
22. Wei J-Y, Smeers Y. Spatial oligopolistic electricity models with Cournot firms and regulated transmission prices. *Operations Research* 1999;47(1):102–112.
23. Yao J, Oren S, Adler I. Cournot equilibria in two-settlement electricity markets with system contingencies. *International Journal of Critical Infrastructure* 2007;3(1/2):142–160.
24. Yao J, Oren S, Adler I. Two settlements electricity markets with price caps and Cournot generation firms. *European Journal of Operations Research* 2007;181(3):1279–1296.
25. Yao J, Adler I, Oren S. Modeling and computing two-settlement oligopolistic equilibrium in congested electricity networks. *Operations Research* 2008;56(1):34–47.

CHAPTER 6

ELECTRICITY MARKET EQUILIBRIUM WITH REACTIVE POWER CONTROL

Xiao-Ping Zhang

6.1 INTRODUCTION

Electricity market equilibrium analysis has been considered a very important and interesting subject for research for a liberalized electricity market. However, due to the complexity of the nature of electricity industry, there may a limited number of generating firms participating in the market. By holding a large share in the electricity market, these generating firms may be able to control the market by acting in an uncompetitive manner by exercising market power. Basically, market power is usually exercised by strategic generating firms, either by asking for a higher price than the competitive levels or by holding the generating capacity below the competitive demand levels. Market power abuse by large strategic firms may be harmful to competition in the electricity market. It is therefore very useful to develop necessary analytical algorithms and computational tools in order to identify the levels of market power exercised by strategic firms and assess the market behavior. The objective of this chapter is to discuss computational tools for the analysis of the Nash supply function equilibrium using full AC network representation and considering network controls and constraints.

This chapter is organized as follows: In Section 6.2, there is an introduction to the AC power flow model in the rectangular coordinates. In Section 6.3, perfect electricity market modeling and analysis using AC optimal power flow (OPF) in the rectangular coordinates is presented. In Section 6.4, the concepts and assumptions of electricity market equilibrium analysis are discussed. In Section 6.5, electricity market equilibrium modeling with AC network representation is presented in detail and the new nonlinear interior point method with complementarity constraints for the calculation of Nash supply function equilibrium is derived. The algorithm involves a single-level procedure by combining the ISO problem for maximizing the social welfare with the generating firm's problem for maximizing individual profit with the implementation of a special nonlinear complementarity function to

Restructured Electric Power Systems: Analysis of Electricity Markets with Equilibrium Models,
Edited by Xiao-Ping Zhang
Copyright © 2010 Institute of Electrical and Electronics Engineers

handle the complementarity constraints. Section 6.6 discuses some implementation issues for the algorithm. In Section 6.7, numerical examples are presented to investigate the effects of reactive power and voltage control, transformer tap ratio control, as well as reactive power capabilities of synchronous generators on the electricity market equilibrium. In connection with the above, transmission congestion is also examined in the framework of supply function–based electricity market equilibrium analysis. In Section 6.8, general conclusions are drawn. In Section 6.9, some necessary derivatives for the deriving the algorithm and developing production grade computer code are given.

6.2 AC POWER FLOW MODEL IN THE RECTANGULAR COORDINATES

Following upon the discussions in Chapter 1, nodal voltage equation with N buses can be given as follows:

$$
\begin{bmatrix}
Y_{11} & Y_{12} & \cdots & Y_{1N} \\
Y_{21} & Y_{22} & \cdots & Y_{2N} \\
\cdots & \cdots & \cdots & \cdots \\
Y_{N1} & Y_{N2} & \cdots & Y_{NN}
\end{bmatrix}
\begin{bmatrix}
V_1 \\
V_2 \\
\cdots \\
V_3
\end{bmatrix}
=
\begin{bmatrix}
I_1 \\
I_2 \\
\cdots \\
I_N
\end{bmatrix}
\tag{6.1}
$$

where I_i $(i = 1, 2, \ldots, N)$ is the current injection at bus i; V_i $(i = 1, 2, \ldots, N)$ is nodal voltage at bus i; Y_{ii} $(i = 1, 2, \ldots, N)$ is the diagonal element of the system admittance matrix; and Y_{ij} $(i = 1, 2, \ldots, N; j = 1, 2, \ldots, N; i \neq j)$. The bus voltages and bus current injections are all in phasors. Due to the fact that, in load flow analysis, usually active and reactive power injections rather than current injections are given, and (6.1) is now changed into the following power equation using the definition of complex power $P_i + jQ_i = V_i I_i^*$:

$$
P_i + jQ_i = V_i I_i^* = V_i \sum_{j=1}^{N} (Y_{ij} V_j)^*
\tag{6.2}
$$

where * represents the conjugate operation. N is the total number of buses. Different from that in Chapter 1 using the polar coordinates, now in the rectangular coordinates, assuming that $V_i = e_i + jf_i$ $(i = 1, 2, \ldots, N)$ and $Y_{ij} = G_{ij} + jB_{ij}$, the active and reactive power injection equation (6.2) may be represented separately as

$$
P_i = \sum_{j=1}^{N} [e_i(G_{ij}e_j - B_{ij}f_j) + f_i(G_{ij}f_j + B_{ij}e_j)] \quad (i = 1, 2, \ldots, N)
\tag{6.3}
$$

$$
Q_i = \sum_{j=1}^{N} [f_i(G_{ij}e_j - B_{ij}f_j) - e_i(G_{ij}f_j + B_{ij}e_j)] \quad (i = 1, 2, \ldots, N)
\tag{6.4}
$$

With active and reactive power injections known, the following power mismatch equations can be obtained:

$$
\Delta P_i = P_i^{Spec} - P_i = 0 \quad (i = 1, 2, \ldots, N)
\tag{6.5}
$$

$$
\Delta Q_i = Q_i^{Spec} - Q_i = 0 \quad (i = 1, 2, \ldots, N)
\tag{6.6}
$$

where P_i^{Spec} and Q_i^{Spec} are the specified active and reactive power injections, respectively. P_i^{Spec} and Q_i^{Spec} are given by

$$P_i^{Spec} = Pg_i - Pd_i \quad (i = 1, 2, \dots, N) \tag{6.7}$$

$$Q_i^{Spec} = Qg_i - Qd_i \quad (i = 1, 2, \dots, N) \tag{6.8}$$

where Pg_i and Qg_i are the active power and reactive power of the generating unit at bus i, respectively. Pd_i and Qd_i are the active power and reactive power of the load at bus i, respectively. The objective of a load flow solution is to find a solution to (6.5) and (6.6) iteratively using the Newton-Raphson method. The elements of the Newton equation can be given by

$$\frac{\partial \Delta P_i}{\partial e_j} = -\frac{\partial P_i}{\partial e_j} = \begin{cases} -(G_{ij}e_i + B_{ij}f_i) & (j \neq i) \\ \displaystyle\sum_{j=1}^{N}(G_{ij}e_j - B_{ij}f_j) - (G_{ii}e_i + B_{ii}f_i) & (j = i) \end{cases} \tag{6.9}$$

$$\frac{\partial \Delta P_i}{\partial f_j} = -\frac{\partial P_i}{\partial f_j} = \begin{cases} B_{ij}e_i - G_{ij}f_i & (j \neq i) \\ \displaystyle\sum_{j=1}^{N}(G_{ij}f_j + B_{ij}e_j) + B_{ii}e_i - G_{ii}f_i & (j = i) \end{cases} \tag{6.10}$$

$$\frac{\partial \Delta Q_i}{\partial e_j} = -\frac{\partial Q_i}{\partial e_j} = \begin{cases} B_{ij}e_i - G_{ij}f_i & (j \neq i) \\ \displaystyle\sum_{j=1}^{N}(G_{ij}f_j + B_{ij}e_j) + B_{ii}e_i - G_{ii}f_i & (j = i) \end{cases} \tag{6.11}$$

$$\frac{\partial \Delta Q_i}{\partial f_j} = -\frac{\partial Q_i}{\partial f_j} = \begin{cases} G_{ij}e_i + B_{ij}f_i & (j \neq i) \\ -\displaystyle\sum_{j=1}^{N}(G_{ij}e_j - B_{ij}f_j) + G_{ii}e_i + B_{ii}f_i & (j = i) \end{cases} \tag{6.12}$$

6.3 ELECTRICITY MARKET ANALYSIS USING AC OPTIMAL POWER FLOW IN THE RECTANGULAR COORDINATES

6.3.1 Modeling of Power System Components in Optimal Power Flow

6.3.1.1 Modeling of Transmission Line The modeling of transmission line constraint in the rectangular coordinates representation can be derived from the equivalent circuit as follows. In Figure 6.1, $y_{ij} = g_{ij} + jb_{ij}$ is the series branch admittance of the transmission line, and y_c is the shunt admittance of the transmission line. The sending- and receiving-end powers of the transmission line are given by

$$S_{ij} = P_{ij} + jQ_{ij} = V_i I_{ij}^* = (e_i + jf_i)I_{ij}^* \tag{6.13}$$

$$S_{ji} = P_{ji} + jQ_{ji} = V_i I_{ji}^* = (e_i + jf_i)I_{ji}^* \tag{6.14}$$

Substituting (1.3) into (6.13), (6.14), we have:

$$P_{ij} = e_i(g_{ii}e_i - b_{ii}f_i) + f_i(g_{ii}f_i + b_{ii})e_i + e_i(G_{ij}e_j - B_{ij}f_j) + f_i(G_{ij}f_j + B_{ij}e_j) \tag{6.15}$$

$$Q_{ij} = f_i(g_{ii}e_i - b_{ii}f_i) - e_i(g_{ii}f_i + b_{ii}e_i) + f_i(G_{ij}e_j - B_{ij}f_j) - e_i(G_{ij}f_j + B_{ij}e_j) \tag{6.16}$$

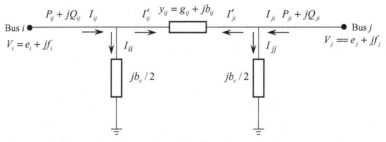

Figure 6.1   Equivalent π circuit model of transmission line

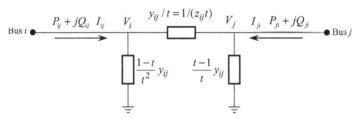

Figure 6.2   Equivalent π circuit model of transformer

$$P_{ji} = e_j(g_{jj}e_i - b_{jj}f_i) + f_j(g_{jj}f_i + b_{jj}e_j) + e_j(G_{ji}e_i - B_{ij}f_i) + f_j(G_{ji}f_i + B_{ji}e_i) \quad (6.17)$$

$$Q_{ji} = f_j(g_{jj}e_j - b_{jj}f_i) - e_j(g_{jj}f_j + b_{jj}e_j) + f_j(G_{ji}e_i - B_{ij}f_i) - e_j(G_{ji}f_i + B_{ji}e_i) \quad (6.18)$$

where $g_{ii} = g_{jj} = g_{ij}$, $b_{ii} = b_{jj} = b_c/2 + b_{ij}$, $G_{ij} = G_{ji} = -g_{ij}$, $B_{ij} = B_{ji} = -b_{ij}$.

The voltage equation is given by equation (1.3), which can be included in the system voltage equation according to Kirchoff's current law, while (6.15)–(6.21) can be included into bus power mismatch equations (6.5) and (6.6).

The transmission line capacity constraint is given by the following inequalities:

$$S_{ij}^2 = (P_{ij}^2 + Q_{ij}^2) \le (S_{ij}^{max})^2 \quad (6.19)$$

$$S_{ji}^2 = (P_{ji}^2 + Q_{ji}^2) \le (S_{ij}^{max})^2 \quad (6.20)$$

where S_{ij}^{max} is the rating of the transmission line, which reflects the thermal limit of the line.

6.3.1.2   Modeling of Transformer Control

In Figure 6.2, the equivalent of a transformer is given by where off nominal tap ratio is at the side of bus i. For the transformer equivalent circuit shown in Figure 6.2, the voltage equation is given by (1.4). However, for the sake of calculations, we can replace the tap ration t by a new parameter t_{ij} and t_{ij} is given by $t_{ij} = 1/t$. Then the voltage equation for the transformer shown in (1.4) can be re-written in terms of t_{ij} as follows:

The bus voltage equation of the transformer is given by

$$\begin{bmatrix} t_{ij}^2 y_{ij} & -t_{ij} y_{ij} \\ -t_{ij} y_{ij} & y_{ij} \end{bmatrix} \begin{bmatrix} V_i \\ V_j \end{bmatrix} = \begin{bmatrix} I_{ij} \\ I_{ji} \end{bmatrix} \tag{6.21}$$

The transformer current I_{ij} at bus i side in rectangular coordinates is given by:

$$I_{ij} = t_{ij}^2 (g_{ij} + jb_{ij})(e_i + jf_i) - t_{ij}(g_{ij} + jb_{ij})(e_j + jf_j) \tag{6.22}$$

According to the definition of the complex power at bus i side $S_{ij} = V_i I_i^*$, the active and reactive power flows at bus i side are given by:

$$P_{ij} = e_i(g_{ii}e_i - b_{ii}f_i) + f_i(g_{ii}f_i + b_{ii}e_i) + e_i(G_{ij}e_j - B_{ij}f_j) + f_i(G_{ij}f_j + B_{ij}e_j) \tag{6.23}$$

$$Q_{ij} = e_i(-g_{ii}f_i - b_{ii}e_i) + f_i(g_{ii}e_i - b_{ii}f_i) - e_i(G_{ij}f_j + B_{ij}e_j) + f_i(G_{ij}e_j - B_{ij}f_j) \tag{6.24}$$

where $g_{ii} = (t_{ij})^2 g_{ij}$, $b_{ii} = (t_{ij})^2 b_{ij}$, $G_{ij} = -g_{ij}t_{ij}$, $B_{ij} = -b_{ij}t_{ij}$

Similarly, the active and reactive power flows at bus j side are given by:

$$P_{ji} = e_j(g_{jj}e_j - b_{jj}f_j) + f_j(g_{jj}f_j + b_{jj}e_j) + e_j(G_{ij}e_i - B_{ij}f_i) + f_j(G_{ij}f_i + B_{ij}e_i) \tag{6.25}$$

$$Q_{ji} = e_j(-g_{jj}f_j - b_{jj}e_j) + f_j(g_{jj}e_j - b_{jj}f_j) - e_j(G_{ji}f_i + B_{ji}e_i) + f_j(G_{ji}e_i - B_{ji}f_i) \tag{6.26}$$

where $g_{jj} = g_{ij}$, $b_{jj} = b_{ij}$, $G_{ji} = -g_{ij}t_{ij}$, $B_{ji} = -b_{ij}t_{ij}$.

The voltage equation of the transformer shown in (6.21) can then be included in the system voltage equation according to Kirchoff's current law, while the power flow equations (6.23)–(6.26) then can be included into bus power mismatch equations (6.5) and (6.6).

The transformer capacity constraint is given by the following inequalities:

$$S_{ij}^2 = (P_{ij}^2 + Q_{ij}^2) \leq (S_{ij}^{\max})^2 \tag{6.27}$$

$$S_{ji}^2 = (P_{ji}^2 + Q_{ji}^2) \leq (S_{ij}^{\max})^2 \tag{6.28}$$

where $S_{ij}^{\max}$ is the rating of the transformer. In addition, for a tap-changing transformer, the tap ratio can be varied within its upper and lower limits:

$$t_{ij}^{\min} \leq t_{ij} \leq t_{ij}^{\max} \tag{6.29}$$

6.3.1.3 *Modeling of Generating Units* Basically, the reactive power and reactive power outputs can be varied within their operating limits:

$$Pg_i^{\min} \leq Pg_i \leq Pg_i^{\max} \tag{6.30}$$

$$Qg_i^{\min} \leq Qg_i \leq Qg_i^{\max} \tag{6.31}$$

where Pg_i and Qg_i are the active and reactive power generation outputs at bus i, respectively; $Pg_i^{\min}$ and $Pg_i^{\max}$ are the lower and upper active power generation limits, respectively; and $Qg_i^{\min}$ and $Qg_i^{\max}$ are the lower and upper reactive generation limits, respectively.

6.3.1.4 *Generator Reactive Power Capability* It should be pointed out that in (6.31), simple reactive generation limits are used to approximately describe the reactive power capability of a synchronous generator. For a more accurate modeling of reactive capability of a synchronous generator, voltage dependent reactive limits

should be used [1–5]. The capability diagram of a synchronous generator shows the allowed region of operation under steady state conditions.

The maximum and minimum limits for the reactive power generation with respective to the maximum stator current, $Ia_i^{\max}$, can be given as follows [2]:

$$Qg_i^{s,\min} = -\sqrt{V_i^2 \left(Ia_i^{\max}\right)^2 - Pg_i^2} \qquad (6.32)$$

$$Qg_i^{s,\max} = \sqrt{V_i^2 \left(Ia_i^{\max}\right)^2 - Pg_i^2} \qquad (6.33)$$

where the superscript s indicates the limits due to the maximum allowed current of the stator.

The maximum and minimum limits for the reactive power generation due to the maximum allowed rotor (which is corresponding to the maximum internal induced voltage $Eq_i^{\max}$) are given by [2]:

$$Qg_i^{r,\max} = -\frac{V_i^2}{Xd_i} + \sqrt{\left(\frac{V_i Eq_i^{\max}}{Xd_i}\right)^2 - \left(Pg_i\right)^2} \qquad (6.34)$$

where the superscript r indicates the limits due to the maximum allowed current of the rotor. Xd_i is the d axis synchronous reactance. It should be mentioned that (6.34) is only applicable to the rotor limiter with a cylindrical rotor while a more complicated expression for a salient pole generator is required.

The minimum reactive limit due to the minimum rotor current limit is given by

$$Qg_i^{r,\min} = -\frac{V_i^2}{Xd_i} + \frac{V_i Eq_i^{\min}}{Xd_i} \qquad (6.35)$$

where $Eq_i^{\min}$ is the minimum internal voltage of the synchronous generator corresponding to the minimum rotor current limit.

In addition, the under-excitation limiter limits the maximum load angle between the q axis and the voltage of the generator terminal.

$$Qg_i^{u,\min} = \frac{Pg_i}{\tan \delta_i^{\max}} - \frac{\left(V_i\right)^2}{Xq_i} \qquad (6.36)$$

where Xq_i is the d axis synchronous reactance.

Considering the above limits, the voltage and active dependent reactive power limits can be determined as follows:

$$Qg_i^{\min} = \max\left\{Qg_i^{s,\min}, Qg_i^{r,\max}, Qg_i^{u,\min}\right\} \qquad (6.37)$$

$$Qg_i^{\max} = \min\left\{Qg_i^{s,\max}, Qg_i^{r,\max}\right\} \qquad (6.38)$$

6.3.1.5 Modeling of Loads   Basically, the reactive power and reactive power demands may be varied within a certain range:

$$Pd_i^{\min} \leq Pd_i \leq Pd_i^{\max} \qquad (6.39)$$

$$Qd_i = Pd_i / Cd_i \qquad (6.40)$$

where Pd_i and Qd_i are the active and reactive power demands at bus i, respectively; $Pd_i^{\min}$ and $Pd_i^{\max}$ are the lower and upper active power demand limits, respectively;

Cd_i is usually a constant, which reflects a constant power factor of the load; and Qd_i is a dependent variable.

6.3.1.6 Bus Voltage Constraints Basically, each bus voltage should be operated with its limits:

$$\left(V_i^{\min}\right)^2 \le e_i^2 + f_i^2 \le \left(V_i^{\max}\right)^2 \tag{6.41}$$

where $V_i^{\min}$ and $V_i^{\max}$ are the lower and upper limits of the voltage at bus i, respectively.

6.3.2 Electricity Market Analysis

Basically, an optimal power flow (OPF) method is to determine a feasible steady state solution of an electrical power system involving physical, operating, and financial constraints while minimizing or maximizing a chosen objective function as discussed in Chapter 1. Here we start with OPF formulation in the rectangular coordinates briefly and build up foundations for the formulation and solution techniques for electricity market equilibrium problems. For instance, the optimization problem here is to maximise the social welfare for an electric power system and the objective is given as follows:

$$Maximize \quad f(x) = \left[\sum_i^{Nd}\left(\gamma_i Pd_i - \frac{1}{2}\delta_i Pd_i^2\right) - \sum_i^{Ng}\left(\alpha_{fi} Pg_{fi} + \frac{1}{2}\beta_{fi} Pg_{fi}^2\right)\right] \tag{6.42}$$

The objective function is equivalent to the following:

$$Minimize \quad f(x) = \left[\sum_i^{Ng}\left(\alpha_{fi} Pg_{fi} + \frac{1}{2}\beta_{fi} Pg_{fi}^2\right) - \sum_i^{Nd}\left(\gamma_i Pd_i - \frac{1}{2}\delta_i Pd_i^2\right)\right] \tag{6.43}$$

subject to the following constraints:

Nonlinear equality constraints

$$\begin{aligned} \Delta P_i(x) = Pg_i - Pd_i - P_i(t, e, f) = 0 \quad (i = 1, 2, \ldots, N) \\ \Delta Q_i(x) = Qg_i - Qd_i - Q_i(t, e, f) = 0 \quad (i = 1, 2, \ldots, N) \end{aligned} \tag{6.44}$$

$$\text{or} \tag{6.45}$$

$$\Delta Q_i(x) = Qg_i - Cd_i Pd_i - Q_i(t, e, f) = 0 \quad (i = 1, 2, \ldots, N)$$

Nonlinear inequality constraints

$$h_j^{\min} \le h_j(x) \le h_j^{\max} \quad (j = 1, 2, \ldots, M) \tag{6.46}$$

where $f(x)$ is the social welfare

$x = [Pg, Qg, t, e, f, Pd]^T$ the vector of variables

N the number of buses

M the number of double-sided nonlinear inequality constraints

Ng the number of generators

Nd	the number of load buses
Pg_{fi}	the vector of active power generation
Qg_{fi}	the vector of reactive power generation
f	generating firm $f = (1, 2, \ldots, N)$
Pd_i	the vector of active load power
t	the vector of transformer tap-ratios
e	the vector of real part of bus voltage
f	the vector of imaginary part of bus voltage
α_{fi}	linear production cost functions coefficient of generator
β_{fi}	quadratic production cost functions coefficient of generator
γ_i	linear load cost coefficient
δ_i	quadratic load cost coefficient
$\Delta P_i(x)$	active power mismatch equations (equalities)
$\Delta P_i(x)$	active power mismatch equations (equalities)
$h_j(x)$	functional inequality constraints including line flow and voltage magnitude constraints, simple inequality constraints of variables such as generator active power, generator reactive power, transformer tap ratio

In order to solve the nonlinear optimization problem as shown in (6.42)–(6.46), the nonlinear inequality constraints (6.46) can be transformed into equivalent equality constraints by introducing the nonnegative slack variables and including them into the barrier logarithmic functions [6] in the objective as discussed in Chapter 1.

$$Min\left\{ f(x) - \mu \sum_{j=1}^{M} \ln(sl_j) - \mu \sum_{j=1}^{M} \ln(su_j) \right\} \tag{6.47}$$

subject to the following constraints:

$$\Delta P_i(x) = 0 \tag{6.48}$$

$$\Delta Q_i(x) = 0 \tag{6.49}$$

$$h_j - sl_j - h_j^{\min} = 0 \tag{6.50}$$

$$h_j + su_j - h_j^{\max} = 0 \tag{6.51}$$

where $sl > 0$, $su > 0$, and $\mu > 0$; sl and su are the lower and upper nonnegative slack variables; and μ is the barrier parameter. The dual variables of (6.48), (6.49), (6.50), (6.51) are λp_i, λq_i, πl_j, and πu_j, respectively. The Lagrange function for the optimization problem (6.47)–(6.51) is given by:

$$L_\mu = f(x) - \mu \sum_{j=1}^{M} \ln(sl_j) - \mu \sum_{j=1}^{M} \ln(su_j) - \sum_{i=1}^{N} \lambda p_i \Delta P_i - \sum_{i=1}^{N} \lambda q_i \Delta Q_i$$

$$- \sum_{j=1}^{M} \pi l_j \left(h_j - sl_j - h_j^{\min} \right) - \sum_{j=1}^{M} \pi u_j \left(h_j + su_j - h_j^{\max} \right) \tag{6.52}$$

The first-order KKT conditions for (6.52) are given by:

$$\nabla_{x_\ell} L_\mu = \nabla_{x_\ell} f(x) - \sum_i \nabla_{x_\ell} \Delta P_i \lambda p_i - \sum_i \nabla_{x_\ell} \Delta Q_i \lambda q_i - \sum_j \nabla_{x_\ell} h_j \pi l_j - \sum_j \nabla_{x_\ell} h_j \pi u_j = 0$$

$$(6.53)$$

$$\nabla_{\lambda p_i} L_\mu = -\Delta P_i = 0 \tag{6.54}$$

$$\nabla_{\lambda q_i} L_\mu = -\Delta Q_i = 0 \tag{6.55}$$

$$\nabla_{\pi l_j} L_\mu = -\left(h_j - sl_j - h_j^{\min}\right) = 0 \tag{6.56}$$

$$\nabla_{\pi u_j} L_\mu = -\left(h_j + su_j - h_j^{\max}\right) = 0 \tag{6.57}$$

The Newton equation for the KKT conditions of (6.53)–(6.57) is given by:

$$
\begin{bmatrix}
Sl^{-1}\Pi l & 0 & I & 0 & 0 & 0 & 0 \\
0 & -Su^{-1}\Pi u & 0 & -I & 0 & 0 & 0 \\
0 & 0 & 0 & 0 & -\nabla h & 0 & 0 \\
I & -I & 0 & 0 & -\nabla h & 0 & 0 \\
0 & 0 & -\nabla h^T & -\nabla h^T & H & -J_p^T & -J_q^T \\
0 & 0 & 0 & 0 & -J_p & 0 & 0 \\
0 & 0 & 0 & 0 & -J_q & 0 & 0
\end{bmatrix}
\begin{bmatrix}
\Delta sl \\ \Delta su \\ \Delta \pi l \\ \Delta \pi u \\ \Delta x \\ \Delta \lambda p \\ \Delta \lambda q
\end{bmatrix}
=
\begin{bmatrix}
Sl^{-1}\nabla_{sl}L_\mu \\ Su^{-1}\nabla_{su}L_\mu \\ -\nabla_{\pi l}L_\mu \\ -\nabla_{\pi u}L_\mu \\ -\nabla_x L_\mu \\ -\nabla_{\lambda p_i}L_\mu \\ -\nabla_{\lambda q_i}L_\mu
\end{bmatrix}
$$

$$(6.58)$$

where

$$H = \nabla^2 f(x) - \sum \lambda p \nabla^2 \Delta P(x) - \sum \lambda q \nabla^2 \Delta Q(x) - \sum (\pi l + \pi u)\nabla^2 h(x),$$

$$Jp = \left[\frac{\partial \Delta P(x)}{\partial x}\right], \quad Jq = \left[\frac{\partial \Delta Q(x)}{\partial x}\right], \quad Sl = diag(sl_j),$$

$$Su = diag(su_j), \quad \Pi l = diag(\pi l_j), \Pi u = diag(\pi u_j).$$

The Hessian matrix for the Newton equation in (6.58) is given by:

$H =$

$Ht_{ij}t_{ij}$	0	0	0	0	0	0	$Ht_{ij}e_i$	$Ht_{ij}f_i$	$Ht_{ij}e_j$	$Ht_{ij}f_j$
	HPg_iPg_i	0	0	0	0	0	0	0	0	0
		HQg_iQg_i	0	0	0	0	0	0	0	0
			HPg_jPg_j	0	0	0	0	0	0	0
				HQg_jQg_j	0	0	0	0	0	0
					HPd_iPd_i	0	0	0	0	0
						HPd_jPd_j	0	0	0	0
							He_ie_i	He_if_i	He_ie_j	He_if_j
							Hf_if_i	Hf_ie_j	Hf_if_j	
								He_je_j	He_jf_j	
									Hf_jf_j	

$$(6.59)$$

The theoretic aspects of the interior points can be found in [7–13] while various applications of the interior point methods in solving power system problems can be found in [14–26]. Some of the elements of the Hessian matrix are derived in the Appendix in Section 6.9. In comparison to the OPF formulation in the polar coordinates, the OPF formulation in the rectangular coordinates can reduce the truncation error due to the elimination of the higher order terms in the Taylor series expansion in solving the KKT equations.

6.4 ELECTRICITY MARKET EQUILIBRIUM ANALYSIS

6.4.1 Nash Supply Function Equilibrium Model

In the past, many different electricity market equilibrium models have been proposed. Due to the unique nature and complexity of the electricity market, researchers have shown great interest in oligopolistic equilibrium analysis and especially the Nash supply function equilibrium (SFE) model. In the following, an interior point optimization algorithm based on complex AC network model for determining the SFE equilibrium of bid-based electricity markets is presented. Based on the full AC network formulation, the algorithm can provide sophisticated modeling of any network operating constraints and system controls. The main features of the algorithm proposed are:

- Full AC network model is used, which can take into consideration all the operating constraints such as the generation capacity limits, bus voltage limits, transmission line capacity constraints.
- System controls such as transformer tap-ratio control, reactive control, and even FACTS control can be fully considered.
- Active and reactive network losses can be fully considered.
- The effect of the reactive power on the market outcomes can be considered.

The SFE problem can be formulated as a bi-level optimization problem. At the lower level, an interior point OPF problem as discussion in Section 6.3, which is to maximize the social welfare. The KKT conditions of the problem at the lower level is reformulated as mixed non-linear complementarity constraints and included as equalities for the optimization problem at the higher level, the objective of which is to maximize the individual profit of each strategic generating firm participating in the electricity market. With a special nonlinear complementarity function, the nonlinear complementarity constraints are transformed into nonlinear algebraic expressions, and then the KKT conditions of the resulting combined problem can be derived. The final combined problem is then solved iteratively based on the solution techniques of the interior point optimization algorithm.

6.4.2 Assumptions for the Supply Function Equilibrium Electricity Market Analysis

In this chapter, the proposed analysis determines the Nash supply function equilibrium in bid-based pool generation electricity market with the full AC transmission network representation. We assume that there are N generating firms, where each one owns a number of generating units denoted by f_i. Each f_i unit is connected with bus i in the network. Each generating firm submits bids half-hourly to the ISO-governed pool. The bids are assumed to be linear SFE form and physically are given by linearly increasing functions related to the output power of the generating unit and the marginal generation cost. The inverse demand functions are also assumed to be linear form with negative gradient and consist of the load demand and the related prices. We suggest that the generation cost function is given by the following quadratic form:

$$C_{Pg_{fi}} = \alpha_{fi} Pg_{fi} + \frac{1}{2}\beta_{fi} Pg_{fi}^2 \qquad (6.60)$$

where

f generating firm, $f = (1, 2, \dots, N)$

i bus number $i = (1, 2, \dots, j)$

Pg_{fi} the vector of active power generation

α_{fi}, β_{fi} production cost function's coefficients of generator

The marginal cost $MC_{Pg_{fi}}$ for generating unit f_i at bus i is given by:

$$MC_{Pg_{fi}} = \alpha_{fi} + \beta_{fi} Pg_{fi} \qquad (6.61)$$

Note that, marginal cost of generation is defined as the cost for producing one more unit of power. The inverse load demand function is given by:

$$p_{d_i} = \gamma_i - \delta_i Pd_i \qquad (6.62)$$

where

p_{d_i} demand function

Pd_i load demand at bus i

γ_i, δ_i load cost coefficients

Note that, the inverse demand function describes the behavior of the consumers in terms of how much money are willing to pay in order to acquire an additional quantity of power.

The true marginal cost of generation is strategic information of each generating firm and is disclosed from public. Based on that assumption, the ISO does not know the true marginal cost of each unit, thus uses the bids for the calculation of the social welfare. The generating firms can exercise market power by increasing their bids and hence increasing their profits. In principle, in imperfect competition the optimal bid should be different than the marginal cost as shown in Figure 6.3. The linear supply function bid is given by:

$$p_{LSF_i} = a_{fi} + b_{fi} Pg_{fi} \qquad (6.63)$$

Once all the bids have been submitted to the pool, and the bidding process for the interval is closed, the ISO has to balance the market trying to maximise the social welfare. The ISO decides what amount of electrical power to buy from each generating unit and how much to supply to the consumers based on the SFE bid and the demand function so that the electricity market equilibrium is reached and the market clearing price is obtained. This is usually performed with the aid of an OPF algorithm. The equilibrium is reached when supply meets demand as shown in Figure 6.4.

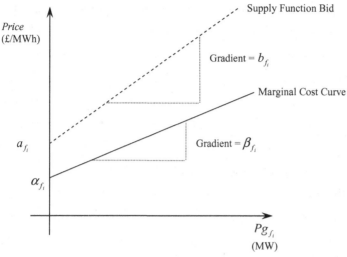

Figure 6.3   The relationship between the marginal cost and supply function bid

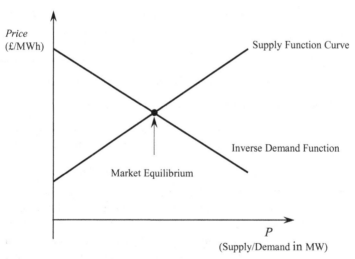

Figure 6.4   Supply function equilibrium

6.4.3   Parameterization Methods for Linear Supply Functions in Electricity Market Equilibrium Analysis

For the linear supply function shown in (6.63), there are four possible parameterization methods for the construction of the optimal linear supply function bids.

6.4.3.1   Intercept Parameterization   In this parameterization method, the strategic firms adjust the intercept a_{f_i} in the supply function in order to submit the profit-maximizing bids to the pool while keeping slope b_{f_i} in the linear supply function constant. Usually the bid slope b_{f_i} is fixed at the value of the marginal cost function slope. The intercept-parameterization has been studied in [27, 28, 30, 36, 38].

6.4.3.2 Slope Parameterization   The strategic firms adjust the slope b_{f_i} in the supply function while keeping the intercept a_{f_i} constant. Most models assume the bid intercept is equal to the true value of the marginal cost function intercept. The slope parameterization is naturally related to the original SFE model by Klemperer and Meyer [32]. A number of applications for this parameterization method can be found in [27, 28, 30, 33–37].

6.4.3.3 Slope-Intercept Parameterization   The slope-intercept parameter-ization method allows more degrees of freedom for the choice of the strategic supply function by arbitrarily parameterizing both slope b_{f_i} and intercept a_{f_i} independently.

The slope-intercept parameterization method has been demonstrated in [27–30] using small test system with the DC network model and in [31] using a small test system with the AC network model.

6.4.3.4 Linear Slope-Intercept Parameterization   In contrast to the slope-intercept parameterization method, in this parameterization method, the strategic firms adjust both the slope b_{f_i} and intercept a_{f_i} in the supply function while a_{f_i} and b_{f_i} have a fixed linear relationship. Such a parameterization method is shown in (6.64) by introducing the bidding parameter k_{f_i}. The marginal cost is multiplied by the factor k_{f_i} to form the linear supply function bid such as:

$$p_i = a_{f_i} + b_{f_i} Pg_{f_i} = k_{f_i} \cdot \frac{\partial C_{Pg_{f_i}}}{\partial Pg_{f_i}} = k_{f_i}(\alpha_{f_i} + \beta_{f_i} Pg_{f_i}) \tag{6.64}$$

where k_{f_i} is considered to be the strategic variable (bidding parameter) for generating firm f at bus i. The intercept and gradient are given by $a_{f_i} = k_{f_i}\alpha_{f_i}$ and $b_{f_i} = k_{f_i}\beta_{f_i}$, respectively.

The linear slope-intercept parameterization method has been used in market equilibrium analysis [28, 39–42]. In this chapter, the linear slope-intercept param-eterization as shown in (6.64) is used. It should be mentioned, as the modeling framework is general. The implementation other parameterization methods can be considered without difficulty.

6.5   COMPUTING THE ELECTRICITY MARKET EQUILIBRIUM WITH AC NETWORK MODEL

6.5.1   Objective Function for the Social Welfare for Imperfect Competition

Perfect competition cannot be applied in practice, since the electricity market is considered an oligopoly governed by a limited number of strategic firms that each try to maximize, their individual profits by exercising market power. In imperfect competition the supply bid is not equal to the generation marginal cost. In this situation, the objective function for the maximization of the social welfare, shown in (6.42), should be modified to reflect the strategic variable of the generating firm. Consequently, the social welfare is replaced by the quasi-social welfare as follows:

$$Max \text{ Social Welfare} = \left(\gamma_i Pd_i - \frac{1}{2}\delta_i Pd_i^2 \right) - k_{f_i} C_{Pg_{f_i}} \tag{6.65}$$

where k_{f_i} is the bidding parameter; Pd_i is the load demand at bus i; $C_{Pg_{f_i}}$ is the generation cost function; and γ_i, δ_i are the load cost coefficients.

6.5.2 Objective Function for the Maximization of Profit of the Generating Firm

Consider a generating firm f_i that owns a generator at bus i. The primary task of the firm is to try to maximize its profits from the sale of the electricity produced. The primary objective of the firm can be modeled as the difference between the income resulting from the sale of electricity and the generating cost for producing the amount of electric power, which is given by:

$$Max \quad \text{Profit} = \left[\lambda p_i Pg_{f_i} - C_{Pg_{f_i}} \right] \tag{6.66}$$

where Pg_{f_i} is the vector of active power generation; λp_i is the nodal price (£/kWh); and $C_{Pg_{f_i}}$ is the generation cost function.

6.5.3 Formulation of Market Equilibrium Model

6.5.3.1 ISO's Optimization Problem The formulation of the ISO optimization problem is based upon the interior point OPF for the maximization of the social welfare presented in Section 6.3.2. The major difference is the introduction of variable k_{f_i}, which represents the bidding strategy of firm f for its generator at bus i. The optimization problem of the ISO for maximizing the quasi social welfare is given by:

$$Max f_1(x) = \left[\sum_i^{NL} \left(\gamma_i * Pd_i - \frac{1}{2}\delta_i * Pd_i^2 \right) - \sum_i^{NG} k_{f_i} \left(\alpha_{f_i} * Pg_{f_i} + \frac{1}{2}\beta_{f_i} * Pg_{f_i}^2 \right) \right] \tag{6.67}$$

subject to the following constraints:

- Nonlinear equality constraints

$$\Delta P_i(x) = Pg_{f_i} - Pd_i - P_i(t, e, f) = 0 \tag{6.68}$$
$$\Delta Q_i(x) = Qg_{f_i} - Qd_i - Q_i(t, e, f) = 0 \tag{6.69}$$

- Nonlinear inequality constraints

$$h_j^{\min} \le h_j(x) \le h_j^{\max} \tag{6.70}$$

where $x = [Pg, Qg, t, e, f, Pd]^T$ is the vector of variables.

By applying Fiacco-McCormick barrier method, the ISO optimization problem can be transformed into the following equivalent optimization problem with equality constraints:

$$\text{Objective:}\quad Min\left\{-f_1(x)-\mu\sum_{j=1}^{M}\ln(sl_j)-\mu\sum_{j=1}^{M}\ln(su_j)\right\} \tag{6.71}$$

subject to the following constraints:

$$\Delta P_i = 0 \tag{6.72}$$
$$\Delta Q_i = 0 \tag{6.73}$$
$$h_j - sl_j - h_j^{\min} = 0 \tag{6.74}$$
$$h_j + su_j - h_j^{\max} = 0 \tag{6.75}$$

where $sl > 0$ and $su > 0$

Applying Lagrange's method for optimization with equality constraints, the following Lagrange function is obtained:

$$L_{1\mu} = -f_1(x)-\mu\sum_{j=1}^{M}\ln(sl_j)-\mu\sum_{j=1}^{M}\ln(su_j)-\sum_{i=1}^{N}\lambda p_i\Delta P_i - \sum_{i=1}^{N}\lambda q_i\Delta Q_i$$
$$-\sum_{j=1}^{M}\pi l_j\left(h_j-sl_j-h_j^{\min}\right)-\sum_{j=1}^{M}\pi u_j\left(h_j+su_j-h_j^{\max}\right) \tag{6.76}$$

Based on the Lagrange's function as shown in (6.76), the first-order KKT conditions are derived as follows:

$$\nabla_x L_{1\mu} = -\nabla_x f_1(x)-\sum_i\nabla_x\Delta P_i\lambda p_i - \sum_i\nabla_x\Delta Q_i\lambda q_i - \nabla_x h_j\pi l_j - \nabla_x h_j\pi u_j = 0 \tag{6.77}$$
$$\nabla_{\lambda p}L_{1\mu} = -\Delta P_i = 0 \tag{6.78}$$
$$\nabla_{\lambda q}L_{1\mu} = -\Delta Q_i = 0 \tag{6.79}$$
$$\nabla_{\pi l}L_{1\mu} = -\left(h_j - sl_j - h_j^{\min}\right) = 0 \tag{6.80}$$
$$\nabla_{\pi u}L_{1\mu} = -\left(h_j + su_j - h_j^{\max}\right) = 0 \tag{6.81}$$
$$\nabla_{sl}L_{1\mu} = \mu - sl_j\pi l_j = 0 \tag{6.82}$$
$$\nabla_{su}L_{1\mu} = \mu + su_j\pi u_j = 0 \tag{6.83}$$

For the sake of the following derivations, define $x_1 = [Pg, Qg, t, e, f]^T$ and $x_2 = [x_1, Pd]^T$. Based on this definition, (6.77) can be split into two expressions for x_1, Pd, respectively:

$$\nabla_{x_1}L_{1\mu} = -\nabla_{x_1}f_1(x_1)$$
$$-\nabla_{x_1}\left(\sum_i\Delta P_i\lambda p_i\right)-\nabla_{x_1}\left(\sum_i\Delta Q_i\lambda q_i\right)-\nabla_{x_1}\left(\sum_j h_j\pi l_j\right)-\nabla_{x_1}\left(\sum_j h_j\pi u_j\right)=0$$
$$\tag{6.84}$$

$$\nabla_{Pd_i}L_{1\mu} = -\nabla_{Pd_i}f_1(x_2)-\sum_i\nabla_{Pd_i}\Delta P_i\lambda p_i - \sum_i\nabla_{Pd_i}\Delta Q_i\lambda q_i = 0 \tag{6.85}$$

Noting the fact that $\dfrac{\partial f_1(x)}{\partial Pd_i}=\gamma_i - \delta_i Pd_i$, $\dfrac{\partial\Delta P_i}{\partial Pd_i}=-1$, and $\dfrac{\partial\Delta Q_i}{\partial Pd_i}=-Cd_i$, (6.85) can be simplified as follows:

$$\nabla_{Pd_i} L_{1\mu} = \lambda p_i + \lambda q_i Cd_i + \delta_i Pd_i - \gamma_i = 0 \tag{6.86}$$

where $Pd_i \geq 0$ and $\lambda p_i + \lambda q_i Cd_i + \delta_i Pd_i - \gamma_i \geq 0$.

6.5.3.2 Nonlinear Complementarity Constraints   The NCP function shown in (3.28), called the Fischer-Burmeister function, is nondifferentiable at the origin, and its Hessian is unbounded at the origin. It is a complementarity function for nonlinear complementarity problem (NCP), which is called a C-function or NCP function, that is,

$$\Psi(a,b) = 0 \Leftrightarrow a \geq 0, b \geq 0, ab = 0 \tag{6.87}$$

With the above relationship, the (6.86) can be transformed into complementarity constraint by setting $a = Pd_i$ and $b = \lambda p_i + \lambda q_i Cd_i + \delta_i Pd_i - \gamma_i$

$$Pd_i \cdot (\lambda p_i + \lambda q_i Cd_i + \delta_i Pd_i - \gamma_i) = 0, Pd_i \geq 0, (\lambda p_i + \lambda q_i Cd_i + \delta_i Pd_i - \gamma_i) \geq 0 \tag{6.88}$$

The constraint in (6.88) indicates that, as the algorithm will approach the optimal solution, the value of a and b should be either equal or greater than zero and their product should approach zero. The NCP function shown in (3.28) is given by:

$$\Psi(a,b) = a + b - \sqrt{a^2 + b^2} \tag{6.89}$$

With the property shown in (6.87) and the NCP function shown in (6.89), the first-order KKT condition equation for Pd_i, as shown in (6.86), can be transformed into the following NCP complementarity function while the nonnegative constraints $Pd_i \geq 0$ and $\lambda p_i + \lambda q_i Cd_i + \delta_i Pd_i - \gamma_i \geq 0$ are not required as they are implicitly contained in the NCP complementarity function:

$$\Psi(a,b) = Pd_i + (\lambda p_i + \lambda q_i Cd_i + \delta_i Pd_i - \gamma_i)$$
$$- \sqrt{Pd_i^2 + (\lambda p_i + \lambda q_i Cd_i + \delta_i Pd_i - \gamma_i)^2} = 0 \tag{6.90}$$

6.5.4 Formulation of the Optimization Market Equilibrium Problem as EPEC

The primary objective of each generating firm is to maximize its individual profits and the optimization problem of the generating firm is given by:

$$\text{Max } f_2(x_2) = \left[\sum_i^{Ng} \left(\lambda p_i * Pg_{f_i} - \alpha_{f_i} * Pg_{f_i} - \frac{1}{2} \beta_{f_i} * Pg_{f_i}^2 \right) \right] \tag{6.91}$$

subject to the following constraints:

$$k_{f_i \min} \leq k_{f_i} \leq k_{f_i \max} \tag{6.92}$$

where $k_{f_i \min}$, $k_{f_i \max}$ are the lower and upper limits of the bidding parameter k_{f_i}.

Note that the real generation Pg_{f_i} and the nodal price λp_i present in the optimization problem of the generating firm are produced by the nonlinear program given in (6.67)–(6.70). Since this is the case, these variables should satisfy the first-order KKT conditions of the ISO problem given in (6.67)–(6.70). Also, these variables can be expressed as implicit functions of all the generating firms' bidding strategies $\{k_f = k_{f_1}, k_{f_2}, \ldots, k_{f_N}\}$. Based on this assumption, the optimization problem of the generating firm can be transformed into the following optimization problem, which maximizes the profits of the generating firm and in addition it optimizes the social welfare of the consumers for the calculation of the SFE equilibrium. This is achieved with the addition of the first-order KKT condition equations of ISO optimization problem, (6.77)–(6.83), as equality constraints to the generating firm's optimization problem such that:

$$\text{Max } f_2(x_2) = \left[\sum_i^{Ng} \left(\lambda p_i * Pg_{f_i} - \alpha_{f_i} * Pg_{f_i} - \frac{1}{2}\beta_{f_i} * Pg_{f_i}^2 \right) \right] \qquad (6.93)$$

subject to the following constraints:

$$k_{f_i \min} \le k_{f_i} \le k_{f_i \max} \qquad (6.94)$$

$$\nabla_{x_{1k}} L_{1\mu} = -\nabla_{x_{1k}} f_1(x) - \sum_i \nabla_{x_{1k}} \Delta P_i \lambda p_i - \sum_i \nabla_{x_{1k}} \Delta Q_i \lambda q_i - \nabla_{x_{1k}} h_j \pi l_j - \nabla_{x_{1k}} h_j \pi u_j = 0 \qquad (6.95)$$

$$\nabla_{\lambda p} L_{1\mu} = -\Delta P_i = 0 \qquad (6.96)$$

$$\nabla_{\lambda q} L_{1\mu} = -\Delta Q_i = 0 \qquad (6.97)$$

$$\nabla_{\pi l} L_{1\mu} = -\left(h_j - sl_j - h_j^{\min} \right) = 0 \qquad (6.98)$$

$$\nabla_{\pi u} L_{1\mu} = -\left(h_j + su_j - h_j^{\max} \right) = 0 \qquad (6.99)$$

$$\nabla_{sl} L_{1\mu} = \mu - sl_j \pi l_j = 0 \qquad (6.100)$$

$$\nabla_{su} L_{1\mu} = \mu + su_j \pi u_j = 0 \qquad (6.101)$$

$$\nabla_{Pd_i} L_{1\mu} = Pd_i \cdot (\lambda p_i + \lambda q_i Cd_i + \delta_i Pd_i - \gamma_i) = 0$$

$$\Updownarrow$$

$$\Psi(a, b) = Pd_i + (\lambda p_i + \lambda q_i Cd_i + \delta_i Pd_i - \gamma_i)$$

$$- \sqrt{Pd_i^2 + (\lambda p_i + \lambda q_i Cd_i + \delta_i Pd_i - \gamma_i)^2} = 0 \qquad (6.102)$$

The optimization problem (6.93)–(6.102) is the optimization market equilibrium problem formulated as EPEC, which was discussed in Chapter 3. Basically, an EPEC is a mathematical program to find an equilibrium point that simultaneously solves a set of MPECs where each MPEC is parameterized by decision variables of other MPECs. The EPEC problems often arise from noncooperative games, for instance, multi-leader-follower games [46], where each leader is solving a Stackelberg game formulated as an MPEC [44]. In the past, a number of EPEC models have been proposed to investigate the strategic behavior of generating firms in electricity markets [5, 43, 45–51] while the necessary optimality conditions of EPECs via multiobjective optimization have been studied in [52].

6.5.5 Lagrange Function for the EPEC Optimization Problem

Applying the Fiacco-McCormick barrier method, the optimization market equilibrium problem shown in (6.93)–(6.102), which was formulated as EPEC, can be transformed into the optimization problem with equality constraints as follows:

$$Min\left\{-f_2(x_2)-\mu\sum_{j=1}^{Ng}\ln\left(sl_{k_{fi}}\right)-\mu\sum_{j=1}^{Ng}\ln\left(su_{k_{fi}}\right)\right\} \tag{6.103}$$

$$h_{k_{fi}}-sl_{k_{fi}}-h_{k_{fi}}^{\min}=0 \tag{6.104}$$

$$h_{k_{fi}}+su_{k_{fi}}-h_{k_{fi}}^{\max}=0 \tag{6.105}$$

KKT conditions for the ISO optimization problem:
Equalities given by (6.95)–(6.102)
$$\tag{6.106}$$

The Lagrange function of the transformed optimization problem of (6.103)–(6.106) is given by:

$$
\begin{aligned}
L_{2\mu} =&-f_2(x_2)-\mu\ln\left[sl_{k_{fi}}\right]-\mu\ln\left[su_{k_{fi}}\right]\\
&-\pi l_{k_{fi}}\left[h_{k_{fi}}-sl_{k_{fi}}-h_{k_{fi}}^{\min}\right]-\pi u_{k_f}\left[h_{k_{fi}}+su_{k_{fi}}-h_{k_{fi}}^{\max}\right]\\
&+\sum_k\omega_{x_{1k}}\left[-\nabla_{x_{1k}}f_1(x_1)-\sum_i\nabla_{x_{1k}}\Delta P_i\lambda p_i-\sum_i\nabla_{x_{1k}}\Delta Q_i\lambda q_i-\sum_j\nabla_{x_{1k}}h_j\pi l_j\right.\\
&\left.-\sum_j\nabla_{x_{1k}}h_j\pi u_j\right]+\sum_i\omega_{\lambda p_i}[-\Delta P_i]+\sum_i\omega_{\lambda q_i}[-\Delta Q_i]\\
&+\sum_j\omega_{\pi l_j}[-(h_j-sl_j-h_j^{\min})]+\sum_j\omega_{\pi u_j}[-(h_j+su_j-h_j^{\max})]+\sum_j\omega_{sl_j}[\mu-sl_j\pi l_j\\
&+\sum_j\omega_{su_j}[\mu+su_j\pi u_j]-\sum_i\omega_{\phi_i}[\Psi(P_{d_i},\lambda p_i+\lambda q_iCd_i+\delta_iPd_i-\gamma_i)]
\end{aligned}
$$

$$\tag{6.107}$$

where $\omega_{x_{1k}}$, $\omega_{\lambda p_i}$, $\omega_{\lambda q_i}$, $\omega_{\pi l_j}$, $\omega_{\pi u_j}$, ω_{sl_j}, ω_{su_j}, and ω_ϕ are dual variables that correspond to the KKT conditions equations of the ISO problem (6.95)–(6.102), respectively, while in particular ω_ϕ is the dual variable of the complementarity condition (6.102). Denote $x_1 = [Pg, Qg, t, e, f]^T$, $x_2 = [x_1, Pd]^T$, $k = 1, 2, \ldots, 2Ng + 2N + Nt$ and $m = 1, 2, \ldots, 2Ng + 2N + Nl + Nt$ where Ng is sum of generators, N is the total number of system buses, Nl is the number of loads, and Nt is the number of tap-changing transformers.

Based on the Lagrange function given by (6.107), the first order KKT conditions for the EPEC problem of (6.103)–(6.106) that calculates the SFE equilibrium point are given by:

$$
\begin{aligned}
\nabla_{x_{2m}}L_{2\mu}=&-\nabla_{x_{2m}}f_2(x_2)\\
&+\sum\omega_{x_{1k}}\left[-\nabla_{x_{2m}}\nabla_{x_{1k}}f_1(x_1)-\sum\nabla_{x_{2m}}(\nabla_{x_{1k}}\Delta P_i)\lambda p_i-\sum\nabla_{x_{2m}}(\nabla_{x_{1k}}\Delta Q_i)\lambda q_i\right]\\
&+\sum\omega_{x_{1k}}\left[-\sum\nabla_{x_{2m}}(\nabla_{x_{1k}}h_j\pi l_j)-\sum\nabla_{x_{2m}}\nabla_{x_{1k}}(h_j\pi u_j)\right]\\
&-\sum\omega_{\lambda p_i}\nabla_{x_{2m}}\Delta P_i-\sum\omega_{\lambda q_i}\nabla_{x_{2m}}\Delta Q_i-\sum\omega_{\pi l_j}\nabla_{x_{2m}}h_j-\sum\omega_{\pi u_j}\nabla_{x_{2m}}h_j\\
&-\sum\omega_\phi\nabla_{x_{2m}}\Psi(P_{d_i},\lambda p_i+\lambda q_iCd_i+\delta_iPd_i-\gamma_i)=0
\end{aligned}
$$

$$\tag{6.108}$$

$$\nabla_{\lambda p_i}L_{2\mu}=-\sum\nabla_{\lambda p_i}f_2(x_2)-\sum\omega_{x_{1k}}\nabla_{x_{1k}}\Delta P_i-\omega_\phi\nabla_{\lambda p_i}\Psi=0 \tag{6.109}$$

$$\nabla_{\lambda q_i}L_{2\mu}=-\sum\omega_{x_{1k}}\nabla_{x_{1k}}\Delta Q_i-\omega_\phi\nabla_{\lambda q_i}\Psi=0 \tag{6.110}$$

$$\nabla_{\lambda q_i}L_{2\mu}=-\sum\omega_{x_{1k}}\nabla_{x_{1k}}\Delta Q_i-\omega_\phi\nabla_{\lambda q_i}\Psi=0 \tag{6.111}$$

$$\nabla_{\pi l_j} L_{2\mu} = -\sum \omega_{x_{1k}} \nabla_{x_{1k}} h_j - \omega_{sl_j} sl_j = 0 \tag{6.112}$$

$$\nabla_{\pi u_j} L_{2\mu} = -\sum \omega_{x_{1k}} \nabla_{x_{1k}} h_j + \omega_{su_j} su_j = 0 \tag{6.113}$$

$$\nabla_{sl_j} L_{2\mu} = -\omega_{sl_j} \pi l_j + \omega_{\pi l_j} = 0 \tag{6.114}$$

$$\nabla_{su_j} L_{2\mu} = \omega_{su_j} \pi u_j - \omega_{\pi u_j} = 0 \tag{6.115}$$

$$\nabla_{\omega_{x_{1k}}} L_{2\mu} = \sum \left[-\nabla_{x_{1k}} f_1(x_1) - \sum \nabla_{x_{1k}} \Delta P_i \lambda p_i - \sum \nabla_{x_{1k}} \Delta Q_i \lambda q_i \right.$$
$$\left. - \sum \nabla_{x_{1k}} h_j \pi l_j - \sum \nabla_{x_{1k}} h_j \pi u_j \right] = 0 \tag{6.116}$$

$$\nabla_{\omega_{\lambda p_i}} L_{2\mu} = -\Delta P_i = 0 \tag{6.117}$$

$$\nabla_{\omega_{\lambda q_i}} L_{2\mu} = -\Delta Q_i = 0 \tag{6.118}$$

$$\nabla_{\omega_{\pi l_j}} L_{2\mu} = -\left[h_j - sl_j - h_j^{\min} \right] = 0 \tag{6.119}$$

$$\nabla_{\omega_{\pi u_j}} L_{2\mu} = -\left[h_j + su_j - h_j^{\max} \right] = 0 \tag{6.120}$$

$$\nabla_{\omega_{sl_j}} L_{2\mu} = \mu - sl_j \pi l_j = 0 \tag{6.121}$$

$$\nabla_{\omega_{su_j}} L_{2\mu} = \mu + su_j \pi u_j = 0 \tag{6.122}$$

$$\nabla_{\omega_{\phi_i}} L_{2\mu} = -\Psi(P_{d_i}, \lambda p_i + \lambda q_i C d_i + \delta_i P d_i - \gamma_i) = 0 \tag{6.123}$$

$$\nabla_{k_{fi}} L_{2\mu} = -\sum \nabla_{k_{fi}} \omega_{x_{1k}} \nabla_{x_{1k}} f_1(x) - \left(\pi l_{k_{fi}} + \pi u_{k_{fi}} \right) \nabla_{k_{fi}} h_{k_{fi}} = 0 \tag{6.124}$$

$$\nabla_{sl_{k_{fi}}} L_{2\mu} = \mu - sl_{k_{fi}} \pi l_{k_{fi}} = 0 \tag{6.125}$$

$$\nabla_{su_{k_{fi}}} L_{2\mu} = \mu + su_{k_{fi}} \pi u_{k_{fi}} = 0 \tag{6.126}$$

$$\nabla_{\pi l_{k_{fi}}} L_{2\mu} = -\left[h_{k_f} - sl_{k_{fi}} - h_{k_f \min} \right] = 0 \tag{6.127}$$

$$\nabla_{\pi u_{k_{fi}}} L_{2\mu} = -\left[h_{k_{fi}} + su_{k_{fi}} - h_{k_f \max} \right] = 0 \tag{6.128}$$

6.5.6 Newton Equation for the EPEC Problem

Applying the Taylor series expansion, the first-order KKT conditions equations of the EPEC problem shown in (6.108)–(6.128) can be linearized as follows:

$$-\nabla_{x_{2m}} L_{2\mu} = \left[-\sum \nabla_{x_{2m}} \left(\nabla_{x_{1k}} \Delta P_i \right) \lambda p_i - \sum \nabla_{x_{2m}} \left(\nabla_{x_{1k}} \Delta Q_i \right) \lambda q_i \right] \Delta \omega_{x_{1k}}$$
$$+ \left[-\sum \nabla_{x_{2m}} \left(\nabla_{x_{1k}} h_j \right) \pi l_j - \sum \nabla_{x_{2m}} \left(\nabla_{x_{1k}} h_j \right) \pi u_j \right] \Delta \omega_{x_{1k}}$$
$$- \sum \omega_{x_{1k}} \nabla_{x_{2m}}^2 \left(\nabla_{x_{1k}} h_j \right) \pi u_j \Delta x_{2m} - \sum \omega_{x_{1k}} \nabla_{x_{2m}} \left(\nabla_{x_{1k}} \Delta P_i \right) \Delta \lambda p_i$$
$$- \sum \omega_{x_{1k}} \nabla_{x_{2m}} \left(\nabla_{x_{1k}} \Delta Q_i \right) \Delta \lambda q_i - \sum \omega_{x_{1k}} \nabla_{x_{2m}} \left(\nabla_{x_{1k}} h_j \right) \Delta \pi l_j$$
$$- \sum \omega_{x_{1k}} \nabla_{x_{2m}} \left(\nabla_{x_{1k}} h_j \right) \Delta \pi u_j - \sum \nabla_{x_{2m}} \Delta P_i \Delta \omega_{\lambda p_i} - \sum \nabla_{x_{2m}} \Delta Q_i \Delta \omega_{\lambda q_i}$$
$$- \sum \nabla_{x_{2m}} h_j \Delta \omega_{\pi l_j} - \sum \nabla_{x_{2m}} h_j \Delta \omega_{\pi u_j} + \left[-\sum \omega_{\lambda p_i} \nabla_{x_{2m}}^2 \Delta P_i \right.$$
$$\left. - \sum \omega_{\lambda q_i} \nabla_{x_{2m}}^2 \Delta Q_i - \sum \omega_{\pi l_j} \nabla_{x_{2m}}^2 h_j - \sum \omega_{\pi u_j} \nabla_{x_{2m}}^2 h_j \right] \Delta x_{2m}$$
$$- \sum \nabla_{x_{2m}} \Psi \Delta \omega_\phi - \sum \omega_\phi \nabla_{x_{2m}}^2 \Psi \Delta x_{2m} - \sum \omega_\phi \nabla_{x_{2m}} \nabla_{\lambda p_i} \Psi \Delta \lambda p_i$$
$$- \sum \omega_\phi \nabla_{x_{2m}} \nabla_{\lambda q_i} \Psi \Delta \lambda q_i - \sum \nabla_{x_{2m}}^2 f_2(x_2) \Delta x_{2m} - \sum \nabla_{\lambda p_i} \nabla_{x_{2m}} f_2(x_2) \Delta \lambda p_i$$
$$- \sum \omega_{x_{1k}} \nabla_{x_{2m}}^2 \left(\nabla_{x_{1k}} \Delta P_i \right) \Delta \lambda p_i - \sum \omega_{x_{1k}} \nabla_{x_{2m}}^2 \left(\nabla_{x_{1k}} \Delta Q_i \right) \Delta \lambda q_i$$
$$- \sum \nabla_{x_{2m}} \nabla_{x_{1k}} f_1(x_1) \Delta \omega_{x_{1k}} - \sum \nabla_{k_{fi}} \nabla_{x_{2m}} \omega_{x_{1k}} \nabla_{x_{1k}} f_1(x_1) \Delta k_{fi} \tag{6.129}$$

$$-\nabla_{\lambda p_i} L_{2\mu} = -\sum_i \left(\nabla_{x_{1k}} \Delta P_i \right) \Delta \omega_{x_{1k}} - \sum_k \omega_{x_{1k}} \nabla_{x2} \left(\nabla_{x_{1k}} \Delta P_i \right) \Delta x_{2m}$$
$$- \sum_i \nabla_{\lambda p_i} \Psi \Delta \omega_\phi - \sum_i \omega_\phi \nabla^2_{\lambda p_i} \Psi \Delta p_i - \sum_i \omega_\phi \nabla_{\lambda p_i} \nabla_{\lambda q_i} \Psi \Delta q_i$$
$$- \sum_i \omega_\phi \nabla_{P d_i} \nabla_{\lambda p_i} \Psi \Delta P d_i \tag{6.130}$$

$$-\nabla_{\lambda q_i} L_{2\mu} = -\sum_i \left(\nabla_{x_{1k}} \Delta Q_i \right) \Delta \omega_{x_{1k}} - \sum_k \omega_{x_{1k}} \nabla_{x2} \left(\nabla_{x_{1k}} \Delta Q_i \right) \Delta x_{2m}$$
$$- \sum_i \nabla_{\lambda q_i} \Psi \Delta \omega_\phi - \sum_i \omega_\phi \nabla^2_{\lambda q_i} \Psi \Delta p_i - \sum_i \omega_\phi \nabla_{\lambda q_i} \nabla_{\lambda p_i} \Psi \Delta q_i$$
$$- \sum_i \omega_\phi \nabla_{P d_i} \nabla_{\lambda q_i} \Psi \Delta P d_i \tag{6.131}$$

$$-\nabla_{\pi l_j} L_{2\mu} = -\sum_j \nabla_{x_{1k}} h_j \Delta \omega_{x_{1k}} - \sum_k \omega_{x_{1k}} \nabla_{x2} (\nabla_{x1} h_j) \Delta x_{2m} - \omega_{sl_j} \Delta sl_j - sl_j \Delta \omega_{sl_j} \tag{6.132}$$

$$-\nabla_{\pi u_j} L_{2\mu} = -\sum_j \nabla_{x1} h_j \Delta \omega_{x1} - \sum_k \omega_{x_{1k}} \nabla_{x_{2m}} (\nabla_{x_{1k}} h_j) \Delta x_{2m} + \omega_{su_j} \Delta su_j + su_j \Delta \omega_{su_j} \tag{6.133}$$

$$-\nabla_{sl_j} L_{2\mu} = -\omega_{sl_j} \Delta \pi l_j - \pi l_j \Delta \omega_{sl_j} + \Delta \omega_{\pi l_j} \tag{6.134}$$

$$-\nabla_{su_j} L_{2\mu} = \omega_{su_j} \Delta \pi u_j + \pi u_j \Delta \omega_{su_j} - \Delta \omega_{\pi u_j} = 0 \tag{6.135}$$

$$-\nabla \omega_{x_{1k}} L_{2\mu} = \begin{bmatrix} -\sum_i \nabla_{x_{1k}} \left(\nabla_{x_{1k}} \Delta P_i \right) \lambda p_i - \sum_i \nabla_{x_{1k}} \left(\nabla_{x_{1k}} \Delta Q_i \right) \lambda q_i \\ -\sum_j \nabla_{x_{1k}} \left(\nabla_{x_{1k}} h_j \right) \pi l_j - \sum_j \nabla_{x_{1k}} \left(\nabla_{x_{1k}} h_j \right) \pi u_j \end{bmatrix} \Delta x_{1k}$$
$$- \sum_i \nabla_{x_{1k}} \Delta P_i \lambda p_i - \sum_i \nabla_{x_{1k}} \Delta Q_i \lambda q_i - \sum_i \nabla_{x_{1k}} h_j \Delta \pi l_j - \sum_i \nabla_{x_{1k}} h_j \Delta \pi u_j \tag{6.136}$$

$$-\nabla \omega_{\lambda p_i} L_{2\mu} = -\sum_k \left(\nabla_{x_{1k}} \Delta P_i \right) \Delta x_{1k} \tag{6.137}$$

$$-\nabla \omega_{\lambda q_i} L_{2\mu} = -\sum_k \left(\nabla_{x_{1k}} \Delta Q_i \right) \Delta x_{1k} \tag{6.138}$$

$$-\nabla \omega_{\pi l_j} L_{2\mu} = -\sum_k \nabla_{x_{1k}} h_j \Delta x_{1k} + \Delta sl_j \tag{6.139}$$

$$-\nabla \omega_{\pi u_j} L_{2\mu} = -\sum_k \nabla_{x_{1k}} h_j \Delta x_{1k} - \Delta su_j \tag{6.140}$$

$$-\nabla \omega_{sl_j} L_{2\mu} = -sl_j \Delta \pi l_j - \pi l_j \Delta sl_j \tag{6.141}$$

$$-\nabla \omega_{su_j} L_{2\mu} = su_j \Delta \pi u_j + \pi u_j \Delta su_j \tag{6.142}$$

$$-\nabla \omega_{\phi_i} L_{2\mu} = -\nabla_{P d_i} \Psi \Delta P d_i - \nabla_{\lambda p_i} \Psi \Delta \lambda_{p_i} - \nabla_{\lambda q_i} \Psi \Delta \lambda_{q_i} \tag{6.143}$$

$$-\nabla k_{f_i} L_{2\mu} = -\sum_k \nabla_{k_{f_i}} \omega_{x_{1k}} \nabla^2_{x_{1k}} f_1(x) \Delta x_{1k} - \sum_k \nabla_{k_{f_i}} \nabla_{x_{1k}} f_1(x) \Delta \omega_{x_{1k}} - \left(\Delta \pi l_{k_{f_i}} + \Delta \pi u_{k_{f_i}} \right) \tag{6.144}$$

$$-\nabla sl_{k_{f_i}} L_{2\mu} = -sl_{k_{f_i}} \Delta \pi l_{k_{f_i}} - \pi l_{k_{f_i}} \Delta sl_{k_{f_i}} \tag{6.145}$$

$$-\nabla su_{k_{f_i}} L_{2\mu} = su_{k_{f_i}} \Delta \pi u_{k_{f_i}} + \pi u_{k_{f_i}} \Delta su_{k_{f_i}} \tag{6.146}$$

$$-\nabla \pi l_{k_{fi}} L_{2\mu} = -\sum_m \left(\nabla_{x_{2m}} h_{k_f}\right)\Delta x_{2_m} + \Delta sl_{k_{fi}} \tag{6.147}$$

$$-\nabla \pi u_{k_{fi}} L_{2\mu} = -\sum_m \left(\nabla_{x_{2m}} h_{k_f}\right)\Delta x_{2_m} - \Delta su_{k_{fi}} \tag{6.148}$$

Combining the linearized first-order KKT conditions equations shown in (6.129)–(6.148), the Newton equation for the EPEC problem can be given as follows:

$$\begin{bmatrix} G_{11} & G_{12} & G_{13} \\ & G_{22} & G_{23} \\ & & G_{33} \end{bmatrix}\begin{bmatrix} y_1 \\ y_2 \\ y_2 \end{bmatrix} = -\begin{bmatrix} b_1 \\ b_2 \\ b_3 \end{bmatrix} \tag{6.149}$$

where:

$$G = \begin{bmatrix} G_{11} & G_{12} & G_{13} \\ & G_{22} & G_{23} \\ & & G_{33} \end{bmatrix} = \left[\nabla^2 L_{2\mu}\right] \tag{6.150}$$

$$y = \begin{bmatrix} y_1 \\ y_2 \\ y_2 \end{bmatrix} \tag{6.151}$$

$$b = \begin{bmatrix} b_1 \\ b_2 \\ b_3 \end{bmatrix} = \left[\nabla L_{2\mu}\right] \tag{6.152}$$

$$G_{11} = \begin{bmatrix} 0 & 0 & -\omega_{sl} & 0 & 0 & 0 & 0 \\ 0 & 0 & 0 & -\omega_{su} & 0 & 0 & 0 \\ & & 0 & 0 & -\omega_x \nabla_x^2 h & 0 & 0 \\ & & & 0 & -\omega_x \nabla_x^2 h & 0 & 0 \\ & & & & K & -Np^T & -Nq^T \\ & & & & & 0 & 0 \\ & & & & & & 0 \end{bmatrix} \tag{6.153}$$

$$G_{12} = \begin{bmatrix} -\pi l & 0 & I & 0 & 0 & 0 & 0 & 0 \\ 0 & \pi u & 0 & -I & 0 & 0 & 0 & 0 \\ -sl & 0 & 0 & 0 & -\nabla_x h & 0 & 0 & 0 \\ 0 & su & 0 & 0 & -\nabla_x h & 0 & 0 & 0 \\ \hline 0 & 0 & -\nabla_x h^T & -\nabla_x h^T & H & -Jp^T & -Jq^T & \nabla_x \Psi \\ 0 & 0 & 0 & 0 & -Jp & 0 & 0 & \nabla_{\lambda p}\Psi \\ 0 & 0 & 0 & 0 & -Jq & 0 & 0 & \nabla_{\lambda q}\Psi \end{bmatrix} \tag{6.154}$$

$$G_{13} = \begin{bmatrix} 0 \\ 0 \\ 0 \\ 0 \\ H_{xK} \\ 0 \\ 0 \end{bmatrix} \tag{6.155}$$

$$G_{22} = \begin{bmatrix} 0 & 0 & 0 & 0 & 0 & 0 & 0 & 0 \\ & 0 & 0 & 0 & 0 & 0 & 0 & 0 \\ & & 0 & 0 & 0 & 0 & 0 & 0 \\ & & & 0 & 0 & 0 & 0 & 0 \\ & & & & 0 & 0 & 0 & 0 \\ & & & & & 0 & 0 & 0 \\ & & & & & & 0 & 0 \\ & & & & & & & 0 \end{bmatrix} \tag{6.156}$$

$$G_{23} = \begin{bmatrix} 0 & 0 & 0 & 0 & 0 & 0 & 0 & 0 \end{bmatrix}^T \tag{6.157}$$

$$G_{33} = H_{K_f} \tag{6.158}$$

$$y_1 = \begin{bmatrix} \Delta sl & \Delta su & \Delta \pi l & \Delta \pi u & \Delta x & \Delta \lambda p & \Delta \lambda q \end{bmatrix}^T \tag{6.159}$$

$$y_2 = \begin{bmatrix} \Delta \omega_{sl} & \Delta \omega_{su} & \Delta \omega_{\pi l} & \Delta \omega_{\pi u} & \Delta \omega_x & \Delta \omega_{\lambda p} & \Delta \omega_{\lambda q} & \Delta \omega_\phi \end{bmatrix}^T \tag{6.160}$$

$$y_3 = K_f \tag{6.161}$$

$$b_1 = \begin{bmatrix} \nabla_{sl} L_{2\mu} & \nabla_{su} L_{2\mu} & \nabla_{\pi l} L_{2\mu} & \nabla_{\pi u} L_{2\mu} & \nabla_x L_{2\mu} & \nabla_{\lambda p} L_{2\mu} & \nabla_{\lambda q} L_{2\mu} \end{bmatrix}^T \tag{6.162}$$

$$b_2 = \begin{bmatrix} \nabla_{\omega_{sl}} L_{2\mu} & \nabla_{\omega_{su}} L_{2\mu} & \nabla_{\omega_{\pi l}} L_{2\mu} & \nabla_{\omega_{\pi u}} L_{2\mu} & \nabla_{\omega_x} L_{2\mu} & \nabla_{\omega_{\lambda p}} L_{2\mu} & \nabla_{\omega_{\lambda q}} L_{2\mu} & \nabla_{\omega_\phi} L_{2\mu} \end{bmatrix}^T \tag{6.163}$$

$$b_3 = K_f \tag{6.164}$$

where

$$K(x_2, \lambda, \pi l, \pi u) = -\nabla_{x_2}^2 f_2(x_2) - \sum \omega_\lambda \nabla_{x_2}^2 g(x) - \sum (\omega_{\pi l} + \omega_{\pi u}) \nabla_{x_2}^2 h(x)$$
$$- \sum \omega_{x_1} \nabla_{x_2}^2 (\nabla_{x_1} h_j)(\pi u_j + \pi l_j) - \sum \omega_{x_1} \nabla_{x_2}^2 (\nabla_{x_1} \Delta P) \lambda p$$
$$- \sum \omega_{x_1} \nabla_{x_2}^2 (\nabla_{x_1} \Delta Q) \lambda q - \omega_\phi \nabla_{x_2}^2 \Psi \tag{6.165}$$

$$H(x_1, \lambda, \pi l, \pi u) = -\nabla_{x_1}^2 f_1(x_1) - \sum \lambda \nabla_{x_1}^2 g - \sum (\pi l + \pi u) \nabla_{x_1}^2 h(x) \tag{6.166}$$

$$N_p(x) = \sum \omega_{x_{1k}} \nabla_{x_{2m}} (\nabla_{x_{1k}} \Delta P_i) + \sum \omega_\phi \nabla_{x_{2m}} \nabla_{\lambda p_i} \Psi \tag{6.167}$$

$$N_q(x) = \sum \omega_{x_{1k}} \nabla_{x_{2m}} (\nabla_{x_{1k}} \Delta Q_i) + \sum \omega_\phi \nabla_{x_{2m}} \nabla_{\lambda q_i} \Psi \tag{6.168}$$

$$g = \begin{bmatrix} \Delta P(x) \\ \Delta Q(x) \end{bmatrix} \tag{6.169}$$

$$J_p(x) = \left[\frac{\partial \Delta P(x)}{\partial x} \right] \tag{6.170}$$

$$J_q(x) = \left[\frac{\partial \Delta Q(x)}{\partial x} \right] \tag{6.171}$$

$$\lambda = \begin{bmatrix} \lambda p \\ \lambda q \end{bmatrix} \tag{6.172}$$

$$\omega_\lambda = \begin{bmatrix} \omega_{\lambda p} \\ \omega_{\lambda q} \end{bmatrix} \tag{6.173}$$

It should be pointed out that the system matrix of the Newton equation in (6.149) is symmetrical and it is a highly sparse matrix. The characteristics should be taken into consideration in solving the Newton equation. As elements $sl_{k_{fi}}$, $su_{k_{fi}}$, $\pi l_{k_{fi}}$, and $\pi u_{k_{fi}}$ are only coupled with k_{fi}, in the formulation of (6.149) they have been eliminated from the matrix using Gaussian elimination. The derivations of K, N_p, and N_q are shown in Section 6.9.

6.5.7 Modeling of Reactive Power and Voltage Control

In order to ensure efficient, secure, and reliable operation of electric power systems, the following operating objectives should be satisfied: (a) bus voltage magnitudes should be within acceptable limits; (b) system transient stability and voltage stability can usually be enhanced by proper voltage control and reactive power management; and (c) the reactive power flows should be minimized such that the active and reactive power losses can be reduced. In addition, the by-product of the minimized reactive power flows can actually reduce the voltage drop across transmission lines and transformers. In the vertically integrated electricity company including generation, transmission, and distribution, voltage control and reactive power support service is provided together with providing active power to customers. In electricity market environments, voltage control and reactive power support is considered as an ancillary service, which is unbundled from active power supply.

In electric power systems, there are different types of power system components such as synchronous generators, shunt reactors, shunt capacitors, series capacitors, synchronous condensers, FACTS devices, and tap-changing transformers, which can provide reactive power and voltage control. Except for tap-changing transformers, all other reactive control components can generate either inductive or capacitive reactive power. In particular, synchronous generators, synchronous condensers, FACTS devices such as SVC, STATCOM, SSSC, UPFC, and VSC HVDC can generate both inductive and capacitive reactive power and regulating speed of these devices is very fast. Hence, they can not only provide normal system reactive and voltage control but also provide dynamic reactive power and voltage control in terms of system disturbances. Synchronous generators, shunt reactors, shunt capacitors, and tap-changing transformers are very popular solutions to reactive power and voltage control. Synchronous generators are very important reactive sources, which can generate or absorb reactive power depending on excitation control. Equipped with modern excitation control systems, synchronous generators can provide both static and dynamic voltage control and reactive power support. Tap-changing transformers are used for voltage control in transmission and distribution systems. Transformers are usually used to control the voltages at buses to which they are connected. Transformers are not reactive power generation devices, however, tap-changing transformers can alternate reactive power distribution of the network by changing their tap ratios such that active and reactive losses of the network may be minimized and voltage profiles may be improved.

In this chapter, we mainly focus on reactive power and voltage control by generators and tap changing transformers since they are popular reactive power and voltage control components. Generator bus voltages can be controlled in two different methods, namely global method and local method as follows:

- *Global optimization method (GM):* In this method, generator bus voltages are optimized within ±5% from nominal value (1 p.u).

- *Local optimization method (LM):* In the second method, generator bus voltages are controlled to the fixed value, say the nominal voltage of 1.0 p.u.

6.6 IMPLEMENTATION ISSUES OF ELECTRICITY MARKET EQUILIBRIUM ANALYSIS WITH AC NETWORK MODEL

The resulting SFE problem given in equation (6.150) is then solved iteratively based on the solution techniques of the interior point OPF presented in Chapter 1.

6.6.1 Initialization of the Optimization Solution

For the optimization algorithm, the initialization of variables is very important as it helps the algorithm converge to the final solution quickly. It is assumed that the system is set for a flat start for voltage initial conditions of the system. In per unit systems, initial voltages are set to $e_i = 1.0$ and $f_i = 0.0$. The initial values for the remaining the variables are given as follows:

- Pg_{fi} and Qg_{fi} are set to the average between the maximum and minimum values.
- The transformer tap-ratio control t_{ij} is set to be equal to 1.0 p.u.
- The bidding parameter k_{fi} is set equal to 1.
- The slack variable sl_j is set to $h_j - h_{jmin}$ while the corresponding dual variable πl_j is set to μ/sl_j;
- The slack variable su_j is set to $h_{jmax} - h_j$ while the dual variable πu_j is set to $-\mu/su_j$.

- The dual variable λp_i is set to $k_{f_i} \times MC_{P_{g_i}}$ while λq_i is set to zero.
- All the auxiliary dual variables ω are set to zero.

6.6.2 Updating the Optimization Solution

By solving equation (6.150), Δsl, Δsu, $\Delta \pi l$, $\Delta \pi u$, Δx, $\Delta \lambda p$, $\Delta \lambda q$, $\Delta \omega_{sl}$, $\Delta \omega_{su}$, $\Delta \omega_{\pi l}$, $\Delta \omega_{\pi u}$, $\Delta \omega_x$, $\Delta \omega_{\lambda p}$, $\Delta \omega_{\lambda q}$, $\Delta \omega_\Phi$, and Δk_f, can be obtained. With these values being known, the primal variables Δsl, Δsu, Δx, and Δk_f can be updated using the following set of equations:

$$y_p^{(k+1)} = y_p^{(k)} + \sigma \alpha_p \Delta y_p^{(k+1)} \tag{6.174}$$

and the dual variables $\Delta \pi l$, $\Delta \pi u$, $\Delta \lambda p$, $\Delta \lambda q$, $\Delta \omega_{sl}$, $\Delta \omega_{su}$, $\Delta \omega_{\pi l}$, $\Delta \omega_{\pi u}$, $\Delta \omega_x$, $\Delta \omega_{\lambda p}$, $\Delta \omega_{\lambda q}$, and $\Delta \omega_\Phi$, can be updated using the following set of equations:

$$y_d^{(k+1)} = y_d^{(k)} + \sigma \alpha_d \Delta y_d^{(k+1)} \tag{6.175}$$

The step lengths α_p and α_d can be determined using the approach discussed in Chapter 1. The updating of barrier parameter μ and the calculations of the complementary gap $Cgap$ are also referred to in Chapter 1.

6.6.3 Solution Procedure

The solution procedure of the nonlinear interior point algorithm for determining the SFE electricity market equilibrium may be summarized as follows:

Step 0: Initialization of primal and dual variables sl, su, πl, πu, x, λp, λq, ω_{sl}, ω_{su}, $\omega_{\pi l}$, $\omega_{\pi u}$, ω_x, $\omega_{\lambda p}$, $\omega_{\lambda q}$, ω_Φ, and k_f

Step 1: Formulation of Newton equation (6.150)

Step 2: Solution of the Newton equation (6.150) using sparse matrix techniques

Step 3: Update variables using equations (6.174)–(6.175)

Step 4: Compute barrier parameter and complementary gap using equations (1.92) and (1.93), respectively

Step 5: Test for convergence:

$$\max(|\Delta P_i|, |\Delta Q_i|) \le \varepsilon_{PQ} \tag{6.176}$$

$$\mu \le \varepsilon_\mu \tag{6.177}$$

$$C_{gap} \le \varepsilon_{gap} \tag{6.178}$$

where ε_{PQ}, ε_μ, and ε_{gap} are convergence tolerances for bus power mismatches, barrier parameter, and complementarity gap, respectively.

If equations (6.176), (6.177), and (6.178) are satisfied, go to Step 6; otherwise go to Step 1 and continue the iterations.

Step 6: SFE electricity market equilibrium reached

Step 7: Output results

6.7   NUMERICAL EXAMPLES

6.7.1   Reactive Power and Voltage Control

6.7.1.1   Description of the Test Systems   Numerical results are presented in this section to show test results on a 3-bus system and the modified IEEE 14-bus system. The impact of the reactive power on the electricity market equilibrium is investigated using two different voltage regulation control modes at the generation buses. Generator bus voltages can be controlled in two different methods, the global method and the local method:

- *Global optimization method (GM):* In this method, generator bus voltages are optimized within ±5% from rated (1 p.u) value.

- *Local optimization method (LM):* In this method, generator bus voltages are controlled to the fixed value, say the nominal voltage of 1.0 p.u.

In addition, using the LM, further tests are to be presented with different settings of load power factor in between 0.7 and 1.0 to simulate different reactive power demand cases.

The 3-bus system consists of 3 buses where buses 1 and 2, buses 2 and 3, and buses 1 and 3 are connected to each other with a transmission line of equal impedance.

The IEEE 14-bus system consists of 14 buses, 20 transmission lines, and 5 generators. The generating units are grouped into three generating firms. Firm f_1 owns the units at buses 1 and 3, firm f_2 owns the units at buses 2 and 8, and firm f_3 owns the units at bus 6. The system data of the IEEE 14-bus system is referred to in [53].

6.7.1.2   Test Results of the 3-Bus System   For the 3-bus system, it is assumed that firm f_1 owns the generator at bus 1, and f_3 owns the generator at bus 3, while there are loads at buses 1 and 2, respectively. The generation marginal cost for the generator at bus 1 is $MC_{Pg_1} = 18 + 0.008Pg_1$ £/MWh and for the generator at bus 3 is $MC_{Pg_3} = 15 + 0.010Pg_3$ £/MWh. The load demand function at bus 1 is $D_{Pd_1} = 40 - 0.080Pd_1$ £/MWh and at bus 2 is equal to $D_{Pd_2} = 40 - 0.060Pd_2$ £/MWh.

Four case studies on the 3-bus test system are presented here. For these cases, the load power factor is set to 0.7 in order to simulate load conditions of high reactive power demand. The test results for nodal prices, social welfare, and the profits of the four cases are summarized in Table 6.1 and the results for the generating outputs, loads, and bidding parameters are given by Table 6.2.

Cases 1 and 2 are almost identical in terms of network conditions, where in Case 1, GM for voltage control at generator buses is utilized, while in Case 2, LM for voltage control is used. The results from Table 6.1 and Table 6.2 show that different voltage control methods, namely GM and LM, have a significant effect on the electricity market equilibrium point. The nodal prices in Case 2 are significantly higher than those in Case 1. Consequently, the increase in nodal prices resulted in a reduced (quasi) social welfare. It has also been noticed that the bidding strategies

TABLE 6.1   Nodal prices, firm's profits, and quasi social welfare for the 3-bus test system

Case no.	Transmission line capacity limit	Voltage control method	Nodal price λp_i(£/MWh)			Generating firm's profit (£/hour)		Social welfare (£/hour)
			λp_1	λp_2	λp_3	f_1	f_3	
1	None	GM	29.4	29.4	29.4	1117	2774	2085
2	None	LM	33.1	33.1	33.1	1211	1704	831.1
3	$S_{23max} = 0.3\,\text{p.u.}$	GM	33.1	37.3	28.9	1327	601.1	819.7
4	$S_{23max} = 0.3\,\text{p.u.}$	LM	31.5	34.3	28.8	1447	446.8	907.0

TABLE 6.2   System parameters for the 3-bus system test cases

Case no.	Pg_1 (p.u.)	Pg_3 (p.u.)	Qg_1 (p.u.)	Qg_3 (p.u.)	Pd_1 (p.u.)	Pd_2 (p.u.)	Qd_1 (p.u.)	Qd_2 (p.u.)	k_{f_1}	k_{f_3}
1	1.02	2.08	2.45	1.16	1.33	1.77	1.36	1.81	1.561	1.719
2	0.82	0.97	1.40	0.54	0.87	0.93	0.88	0.94	1.772	2.070
3	0.90	0.44	1.92	−0.49	0.86	0.48	0.88	0.49	1.767	1.871
4	1.11	0.33	1.28	0.20	1.06	0.37	1.08	0.38	1.668	1.879

(or parameters) k_{f_i} of the generating firms have also changed. In Case 2, the generation outputs have been reduced for both firms, while the increase in nodal prices helps firm f_1 to raise its profits. It appeared to be that the increase in nodal prices was not adequate to cover the reduction of firm f_3 production, and consequently the profits of firm f_3 have been significantly reduced and its generation output is less than half of that in Case 1. It should be mentioned here that for both Cases 1 and 2, there is no congestion in the network.

Cases 3 and 4 correspond to Cases 1 and 2, respectively, but there is congestion on transmission line 2–3 and the power flow of the transmission line is equal to its transmission line capacity limit, say, $S_{23max} = 0.3\,\text{p.u.}$ These cases have been performed to investigate the impact of voltage control at generation buses on the market equilibrium under stressed network conditions. The transmission constraint of line 2–3 is active in both cases. The impact of the transmission line constraint on the electricity market outcomes is as follows:

- Comparing Case 1 with Case 3, it has been found that the presence of the transmission line constraint has a significant effect on the production of firm f_3 and this has resulted in a large decrease in its profit.

- Comparing Cases 3 and 4, it can be seen that, due to the presence of the constraint, there is a significant variation between the nodal prices λp_i while the (quasi) social welfare shows a small increase. It has been noticed that the active generation outputs for firm f_1 increased but those of firm f_3 decreased.

TABLE 6.3 Electricity market equilibrium with different load power factor

Case no.	Power factor	Total Pg	Total Qg	Total Pd	Total Qd	k_{f_1}	k_{f_3}	Profit f_1	Profit f_3	Social welfare	λp_1	λp_2	λp_3
2	0.70	1.79	1.94	1.79	1.83	1.772	2.070	1211	1704	831.1	33.1	33.1	33.1
5	0.80	2.16	1.78	2.16	1.62	1.746	2.000	1211	2215	1018	32.6	32.6	32.6
6	0.90	3.17	1.84	3.17	1.52	1.541	1.711	1216	2663	2152	29.1	29.1	29.1
7	1.00	3.24	0.23	3.24	0.00	1.513	1.715	1430	2407	2208	28.9	28.9	28.9

TABLE 6.4 Test results for the IEEE 14-bus system

Case no.	Voltage control method	Firm's profit			Social welfare	Total Pg	Total Qg	Total Pd	Total Qd
		f_1	f_2	f_3					
8	GM	5434	1907	4112	6306	6.83	0.14	6.63	2.48
9	LM	5359	1895	4212	6782	6.30	−0.11	6.13	2.37

Cases 5, 6, and 7 were carried out to investigate the impact of load power factors on the electricity market equilibrium. Similar Case 2, these three cases also use the LM for voltage control at generator buses except that they have load power factors of 0.8, 0.9, and 1.0, respectively. The results are presented in Table 6.3.

Comparing the four cases as shown in Table 6.3, it can be seen that

- As the value of the power factor decreases, the nodal prices λp_i increase and the social welfare decreases. The individual profits of the two firms f_1 and f_3 have also been affected in proportion to the levels of production of each firm in each case.

- As the power factor decreases, the reactive generation levels increase, while the active power generation levels decrease. For the cases with lower power factor value, despite the increase in nodal prices λp_i, the profits of the firms are less than the cases with higher power factor values due to the decrease of active power generation levels. The value of the bidding parameter k_{f_i} is higher at a lower power factor.

6.7.1.3 The IEEE 14-Bus System
For the IEEE 14-bus system, it is assumed that firm f_1 owns the units at buses 1 and 3, firm f_2 owns the units at buses 2 and 8, and firm f_3 owns the units at bus 6. Two case studies are presented on this system. For Case 8, GM for voltage control at generator buses is applied and generator bus voltages are optimized within the 0.95–1.05 p.u. limits, while for Case 9, LM is used and generation voltage magnitudes are fixed at nominal voltage of 1 p.u. The test results for Cases 8 and 9 on the IEEE 14-bus system are presented in Table 6.4, where a load power factor of 0.93 was used for these two test cases. The results for

TABLE 6.5   Bidding parameters for Cases 8 and 9

Biding parameter Case no.	k_{f_1}	k_{f_2}	k_{f_3}	k_{f_6}	k_{f_8}
8	1.738	2.262	1.850	1.748	1.785
9	1.847	1.860	0.897	1.748	2.053

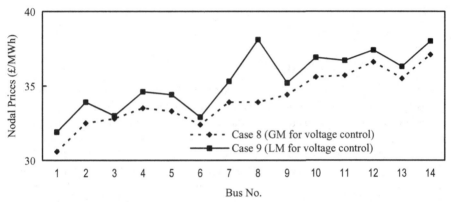

Figure 6.5   Nodal prices for the IEEE 14-bus system

the profits, the (quasi) social welfare, the bidding parameter k_{f_i} and the nodal prices λp_i, are shown in Table 6.4, Table 6.5, and Figure 6.5, respectively. Comparing the two cases, it can be seen from the results that the nodal prices λp_i have been affected and there is a variation of the distribution of profits among the three generating firms, and there is also an increase to the social welfare. The bidding strategies (parameters) k_{f_i} have been significantly changed; for instance, for the unit at bus 3 the value of k_{f_i} was reduced from 1.850 to 0.897. In connection with the changes of the bidding parameters, the total active power and reactive generation levels were decreased.

6.7.1.4   Discussions   From the test cases, it has been found that the incorporation of reactive power in the model provides the opportunity of investigating the bidding strategies of firms in excising market power while maximizing their benefit. The test results have demonstrated reactive power and voltage control has affected the supply and demand equilibrium point, consequently affecting the nodal prices, the profits, the social welfare, and the total outcome of the electricity market and hence affecting the electricity market efficiency. In addition to the reactive power and voltage control, the load power factor has also affected the nodal prices, the social welfare, and the profits of the firms.

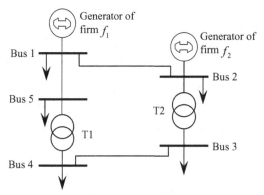

Figure 6.6 Five-bus test system

6.7.2 Transformer Control

6.7.2.1 Description of the Test Systems

In this section, test results on a 5-bus system and the IEEE 30-bus system are presented to investigate the impact of the transformer tap-ratio control on the market equilibrium. They demonstrate the performance and effectiveness of the proposed nonlinear interior point algorithm for computing the SFE of the electricity market.

As shown in Figure 6.6, the 5-bus system consists of five buses, three transmission lines, and two tap-changing transformers, two generators, and five loads. As shown in Figure 6.7, the IEEE 30-bus system consists of six generators, four tap-changing transformers, and 37 transmission lines. The IEEE 30-bus system data is referred to in [54].

The test cases in the following sections will show the impact of transmission line and generation capacity constraints, and transformer tap-ratio control on the electricity market equilibrium and market outcome.

6.7.2.2 Test Results on the 5-Bus System

For the 5-bus test system, it is assumed that generating firm f_1 owns the generating unit at bus 1, while firm f_2 owns the generating unit at bus 2. The generation marginal costs at bus 1 and bus 2 are equal to $MC_{Pg_1} = 11.0 + 0.009Pg_1$ £/MWh and $MC_{Pg_2} = 10.8 + 0.010Pg_2$ £/MWh, respectively. The inverse demand function for all buses is $D_{Pd_i} = 40 - 0.060Pd_i$ £/MWh.

In order to show the impact of the transformer tap-ratio control on the electricity market equilibrium, based on the 5-bus system, six cases are presented here. For Cases 10, 12, and 14, the tap-ratio control is assumed to be inactive and the two transformer tap ratios are set to the fixed value of 1.0 p.u. For Cases 11, 13, and 15, the tap-ratio control is considered to be active and the tap ratios can be optimized within ±10% of the nominal value of 1.0 p.u. The load power factor of the system was set to 0.9 for simulating normal operating conditions.

The description of the test cases is given by Table 6.6. The test results for Cases 10–17 on the 5-bus system are presented in Table 6.7, Table 6.8, and Table 6.9, which include nodal prices, bidding factors, firm profits, and social welfare.

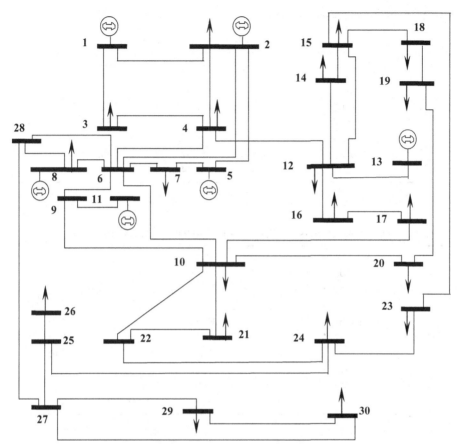

Figure 6.7   The IEEE 30-bus system

TABLE 6.6   Description of the tests cases

Case no.	Active con- straints (in p.u.)	With/without transformer control
10	None	Fixed
11		±10%
12	$Pg_2^{max} = 4.0$	Fixed
13		±10%
14	$S_{12}^{max} = S_{15}^{max} = 0.5$	Fixed
15		±10%

For both Cases 10 and 11, there is no active constraint in the system. These test cases have been examined in order to demonstrate the effect of transformer tap-ratio control on the electricity market equilibrium under normal operating conditions. In Case 10, there is no tap-ratio control, while in Case 11, the tap-ratios are

TABLE 6.7 Nodal prices for the 5-bus system

Case no.	Nodal prices λp_i (£/MWh)				
	λp_1	λp_2	λp_3	λp_4	λp_5
10	30.2	29.5	31.4	32.6	31.9
11	31.0	30.2	31.7	32.7	32.4
12	31.3	30.7	32.3	33.3	32.7
13	33.3	32.6	33.7	34.3	34.2
14	27.4	28.6	32.3	35.4	37.6
15	26.9	31.0	34.6	37.9	41.1

TABLE 6.8 Bidding parameters, profits, and social welfare for the 5-bus system

Case no.	Bidding strategies		Generating firm's profit (£/hour)		Social welfare (£/hour)
	kf_1	kf_2	f_1	f_2	
10	2.416	1.959	3085	7057	13640
11	2.481	1.923	3175	8357	14820
12	2.536	1.608	2948	7141	12940
13	2.751	1.788	2662	7924	12460
14	2.106	2.014	3430	5470	12650
15	1.944	2.449	4573	3580	11480

TABLE 6.9 System generation and transformer control parameters for the 5-bus system

Case no.	Pg_1 (p.u.)	Pg_2 (p.u.)	Qg_1 (p.u.)	Qg_2 (p.u.)	$\sum Pd_i$ (p.u.)	$\sum Qd_i$ (p.u.)	Tap-ratio (p.u.)	
							κ_1	κ_2
10	1.67	4.26	1.73	1.66	5.87	2.84	1.000	1.000
11	1.65	4.92	2.32	1.83	6.47	3.13	0.900	0.973
12	1.50	4.00	1.62	1.54	5.45	2.64	1.000	1.000
13	1.22	4.00	1.77	1.45	5.15	2.49	0.900	0.974
14	2.23	3.40	0.85	2.11	5.60	2.71	1.000	1.000
15	3.16	1.86	1.17	1.45	4.99	2.42	0.900	1.020

optimized. In comparison, in Case 11, there is a large increase in the total active power generation, and the total reactive power generation though the change in nodal prices between Cases 10 and 11 is small. Both firms also earn larger profits. It is interesting that, in Case 2, firm f_2 has decreased slightly its bidding parameter and increased its output power significantly by more than 20% of that of Case 10. It can also be seen that, for Case 11, with the tap-ratio control active, the social welfare is also increased.

Cases 12 and Case 13 are similar to Cases 10 and 11, respectively, except that there is the presence of the generation capacity constraint for the generator at bus 2 in these two cases. The purpose of these two cases is to investigate the interaction between the tap-ratio control and the generation capacity constraint and the impact on the electricity market equilibrium. Generation capacity limit applied here is $Pg_2^{max} = 4.0$ p.u. The generating bidding parameters of Case 13 with the tap-ratio control are higher than those of Case 12 without the tap-ratio control, which results in higher nodal prices. Although the capacity availability of firm f_2 is limited, the increase in nodal prices provides the opportunity to produce higher profits. On the other hand, the increase in nodal prices is not sufficient to increase the profits of firm f_1, and there is a lower profit for this firm. From this example, it can be seen that the transformer tap-ratio control gives incentives to the strategic firms to exercise market power in a different way than in the uncontrolled case, resulting in a different market outcome.

Cases 14 and 15, which correspond to Cases 10 and 11, respectively, are used to investigate the impact of transformer tap-ratio control on the electricity market with the presence of network congestion. It is assumed now that the transmission line capacity limits of lines 1–2 and 1–5 are set to $S_{12}^{max} = S_{15}^{max} = 0.5$ p.u. and hence there is congestion in these two transmission lines. In comparison with that in Case 14, with the tap-ratio control inactive, the bidding factors k_{f_i} of both firms in Case 15, with the tap-ratio control active and being optimized globally, have been altered significantly and a considerable variation in the nodal prices has been observed. The nodal prices at buses 2, 3, 4, and 5 of Case 15 are significantly increased, while the nodal price at bus 1 of Case 15 is slightly decreased. Due to the variation of nodal prices, there is a lower social welfare in Case 15 than in Case 14. In Case 15, the transformer tap-ratio control has provided incentives to the generating firms to change their bidding strategies. It has been found that in Case 15, despite the transmission congestion from both sides, firm f_1 submitted a lower bid, increased the active generation output, and subsequently gained higher profit, while firm f_2 has significantly increased its bidding parameter. In contrast, in Case 14, a large proportion of the demand at bus 1 is supplied by the unit of firm f_2 at bus 2. It should be mentioned that in Case 15 the power flows are from bus 1 to bus 2, while in Case 14 the power flows are reversed. The consequence is that firm f_2 has lower profits in Case 15 than in Case 14. By comparing Case 14 and Case 15, it can be concluded that the transformer tap-ratio control has a significant impact on the electricity market outcome in the presence of network congestion.

6.7.2.3 *Test Results on the IEEE 30-Bus System* For the IEEE 30-bus test system shown in Figure 6.7, the generating units are grouped into three generating firms. It is assumed that firm f_1 owns the units at buses 1, 2, and 13, firm f_2 owns the unit at bus 5, and firm f_3 owns the units at buses 8 and 11. For this system, Case 16 and Case 17 are presented here, where in Case 16 the tap-ratio control is inactive and the tap ratios are set to the fixed value of 1.0 p.u., while in Case 17 the tap-ratios

can be optimized globally within ±10% from the nominal value of 1.0 p.u. The load power factor at all load buses is set to 0.91.

The results for the bidding parameters k_{f_i} are presented in Table 6.10, while the results for the profits and social welfare and the system parameters are shown in Table 6.11. The nodal prices λp_i for both Cases 16 and 17 are shown in Figure 6.8.

TABLE 6.10   Bidding parameters of Case 16 and Case 17 for the IEEE 30-bus system

Bus no. of unit	Case 16 without tap-ratio control bidding parameter kf_i	Case 17 with tap-ratio control bidding parameter kf_i
1	2.805	2.618
2	3.424	1.274
5	2.009	2.011
8	1.922	1.881
11	1.678	1.667
13	1.690	1.622

TABLE 6.11   Profits and social welfare for Case 16 and Case 17 for the IEEE 30-bus system

Case no.	Firm's profit (£/MWh) f_1	f_2	f_3	Social welfare (£/MWh)	$\sum Pg_i$ (p.u.)	$\sum Qg_i$ (p.u.)	$\sum Pd_i$ (p.u.)	$\sum Qd_i$ (p.u.)
16	2079	4825	2680	11,990	5.62	2.53	5.55	2.50
17	4081	4278	2484	13,830	6.80	3.20	6.71	3.01

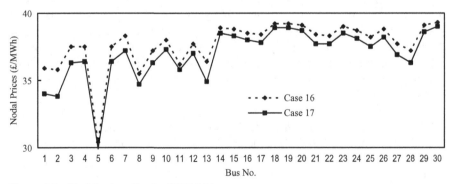

Figure 6.8   Nodal prices for the IEEE 30-bus system

Comparing the results of Cases 16 and 17, the following can be seen:

- The bidding strategies k_{f_i} have been significantly changed. For instance, k_{f_i} for the unit at bus 2 was reduced from 3.424 to 1.274. Total generation levels were also changed significantly. Consequently, the nodal prices λp_i were also changed, and a variation of the distribution of profits among the three generating firms was observed. In addition, the social welfare was also changed considerably.

- The transformer tap ratio control has a significant impact on the market outcome. It would therefore be useful to incorporate tap ratio control in electricity market equilibrium modeling and analysis.

6.7.3 Computational Performance

In order to demonstrate the computational performance of the algorithm proposed, the IEEE 118-bus system with 118 buses, 179 transmission lines, 54 generators, and 9 online tap-changing autotransformers was used. The convergence characteristics of the algorithm on the IEEE 118-bus system market behavior are shown in Figure 6.9. The specified tolerance for convergence corresponds to a value of ≤ 5 p.u. for both power mismatches. The algorithm converged within 40 iterations and the CPU time required was 6.5 seconds, which has the potential to be used in solving large-scale electricity market equilibrium calculations.

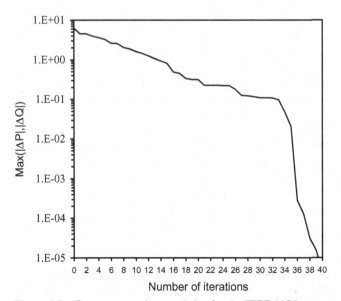

Figure 6.9   Convergence characteristics for the IEEE 118-bus test system

6.8 CONCLUSIONS

This chapter has presented a nonlinear interior point algorithm based on an AC network model for determining the Nash supply function equilibrium of bid-based electricity markets, which involves a single-level procedure by combining the ISO problem for maximizing the social welfare with the generating firm's problem for maximizing individual profit with the implementation of a special nonlinear complementarity function to handle the complementarity constraints.

Using the proposed nonlinear interior point algorithm, this chapter has investigated the effects of reactive power and voltage control, transformer tap ratio control, and reactive power capabilities of synchronous generators on the electricity market equilibrium. In connection with the above, transmission congestion has also been examined in the framework of supply function–based electricity market equilibrium analysis, which has been successfully implemented in the primal-dual nonlinear interior point algorithm with AC network modeling.

The numerical results have shown that reactive power and voltage control of generators in the network have a significant impact on the electricity market equilibrium, which affects nodal prices, social welfare, individual profits of the strategic generating firms, and power distribution in the network. In addition, variations of reactive power demand give strategic firms the opportunity of altering their strategic actions in order to exercise their market power, resulting in different market outcomes.

Transformer tap ratio control, a very popular voltage control method in the electricity network, has been examined in the framework of supply function–based electricity market equilibrium analysis. This algorithm has been used to show the significance of considering the transformer tap ratio control in electricity market equilibrium analysis. Preliminary study results have demonstrated that transformer tap ratio control gives incentives to strategic generating firms to exercise market power in a different way than that without transformer tap ratio control. Transformer control again has resulted in different market outcome and different power distribution in the network and should be considered from the electricity market equilibrium models in order to more accurately calculate the electricity market equilibrium. Tap ratio control can significantly affect the outcome of equilibrium analysis and have a direct impact on nodal prices, profits of the generating firms, and social welfare.

Numerical tests have also shown that the limitation of reactive power generation has an impact on overall market outcome. Consequently, generating firms choose different bidding strategies, resulting in withheld capacity, increased nodal prices, alterations in the profits, and reductions in the social welfare.

The incorporation of reactive power and voltage control and transformer tap ratio control as well as reactive power generation limitation in the electricity market analysis using the AC network model can provide a more accurate calculation of market equilibrium. In contrast to the DC network model that omits the presence of reactive power, the proposed AC power flow–based SFE market model can explicitly represent the reactive power flow–related physical characteristics and operating control. In principle, the full-featured AC network formulated can fully consider any network controls and constraints. This will give the opportunity to investigate the impact of different controls and constraints on the market equilibrium and hence also provide the possibility of the investigation of the interactions of different control actions and how the interactions would affect the market outcome. The proposed nonlinear interior point algorithm has shown superior computational performance and robustness.

6.9 APPENDIX

6.9.1 Second Derivatives for Power Mismatches in Rectangular Coordinates

First derivatives of power mismatches with respect to bus voltage have been given by (6.9)–(6.12). In the rectangular coordinates, the second derivatives of P_i and Q_i become constant. Some of the second derivatives are presented as follows:

$$\frac{\partial^2 \Delta P_i}{\partial e_i \partial e_j} = -G_{ij} \tag{6.179}$$

$$\frac{\partial^2 \Delta P_i}{\partial e_i \partial f_j} = B_{ij} \tag{6.180}$$

$$\frac{\partial^2 \Delta P_i}{\partial f_i \partial e_j} = -B_{ij} \tag{6.181}$$

$$\frac{\partial^2 \Delta P_i}{\partial f_i \partial f_j} = -G_{ij} \tag{6.182}$$

$$\frac{\partial^2 \Delta Q_i}{\partial e_i \partial e_j} = B_{ij} \tag{6.183}$$

$$\frac{\partial^2 \Delta Q_i}{\partial e_i \partial f_j} = G_{ij} \tag{6.184}$$

$$\frac{\partial^2 \Delta Q_i}{\partial f_i \partial e_j} = -G_{ij} \tag{6.185}$$

$$\frac{\partial^2 \Delta P_i}{\partial f_i \partial f_j} = B_{ij} \tag{6.186}$$

$$\frac{\partial^2 \Delta P_i}{\partial f_i \partial f_j} = -G_{ij} \tag{6.187}$$

$$\frac{\partial^2 \Delta P_i}{\partial e_i \partial f_j} = -B_{ij} \tag{6.188}$$

6.9.2 Second Derivatives for Transmission Line Constraints in Rectangular Coordinates

Some of the second derivatives of S_{ij}^2 are given as follows:

$$\frac{\partial^2 S_{ij}^2}{\partial^2 e_i} = 2(2g_{ii}P_{ij}) + 2(2g_{ii}e_i + G_{ij}e_j - B_{ij}f_j)^2 - 4Q_{ij}b_{ii} + 2(-2e_ib_{ii} - G_{ij}f_j - B_{ij}e_j)^2$$

(6.189)

$$\frac{\partial^2 S_{ij}^2}{\partial e_i \partial f_i} = 2(2g_{ii}e_i + G_{ij}e_j - B_{ij}f_j)(2g_{ii}f_i + G_{ij}f_j + B_{ij}e_j)$$
$$+ 2(-2b_{ii}e_i - G_{ij}f_j - B_{ij}e_j)(-2b_{ii}f_i + G_{ij}e_j - B_{ij}f_j)$$

(6.190)

$$\frac{\partial^2 S_{ij}^2}{\partial e_i \partial e_j} = 2G_{ij}P_{ij} + 2(2g_{ii}e_i + G_{ij}e_j - B_{ij}f_j)(G_{ij}e_i + B_{ij}f_i)$$
$$- 2B_{ij}Q_{ij} + 2(-2e_ib_{ii} - G_{ij}f_j - B_{ij}e_j)(G_{ij}f_i - B_{ij}e_i)$$

(6.191)

$$\frac{\partial^2 S_{ij}^2}{\partial e_i \partial f_j} = -2B_{ij}P_{ij} + 2(2g_{ii}e_i + G_{ij}e_j - B_{ij}f_j)(G_{ij}f_i - B_{ij}e_i)$$
$$- 2G_{ij}Q_{ij} - 2(-2e_ib_{ii} - G_{ij}f_j - B_{ij}e_j)(G_{ij}e_i + B_{ij}f_i)$$

(6.192)

$$\frac{\partial^2 S_{ij}^2}{\partial^2 f_i} = 4g_{ii}P_{ij} + 2(2g_{ii}f_i + G_{ij}f_j + B_{ij}e_j)^2 - 4Q_{ij}b_{ii} + 2(-2f_ib_{ii} + G_{ij}e_j - B_{ij}f_j)^2 \quad (6.193)$$

$$\frac{\partial^2 S_{ij}^2}{\partial f_i \partial e_j} = 2B_{ij}P_{ij} + 2(2g_{ii}f_i + G_{ij}f_j + B_{ij}e_j)(G_{ij}e_i + B_{ij}f_i)$$
$$+ 2G_{ij}Q_{ij} + 2(-2b_{ii}f_i + G_{ij}e_j - B_{ij}f_j)(G_{ij}f_i - B_{ij}e_i)$$

(6.194)

$$\frac{\partial^2 S_{ij}^2}{\partial f_i \partial f_j} = 2G_{ij}P_{ij} + 2(2g_{ii}f_i + G_{ij}f_j + B_{ij}e_j)(G_{ij}f_i - B_{ij}e_i)$$
$$- 2B_{ij}Q_{ij} - 2(-2b_{ii}f_i + G_{ij}e_j - B_{ij}f_j)(G_{ij}e_i + B_{ij}f_i)$$

(6.195)

$$\frac{\partial^2 S_{ij}^2}{\partial^2 e_j} = 2(G_{ij}e_i + B_{ij}f_i)^2 + 2(G_{ij}f_i - B_{ij}e_i)^2$$

(6.196)

$$\frac{\partial^2 S_{ij}^2}{\partial^2 f_j} = 2(G_{ij}f_i - B_{ij}e_i)^2 + 2(G_{ij}e_i + B_{ij}f_i)^2$$

(6.197)

6.9.3 Second Derivatives in Rectangular Coordinates

$$\frac{\partial^2 L_{2\mu}}{\partial^2 P_{gfi}} = -\nabla_{P_{gfi}}^2 f_2(x) = \beta_{fi}$$

(6.198)

$$\frac{\partial^2 L_{2\mu}}{\partial P_{gfi} \partial \lambda p_i} = -\nabla_{\lambda p_i} \nabla_{P_{gi}} f_2(x) = -1$$

(6.199)

$$\frac{\partial^2 L_{2\mu}}{\partial P_{gfi} \partial k_{fi}} = -\nabla_{k_{fi}} \nabla_{P_{gi}} \omega_{P_{gi}} \nabla_{P_{gi}} f_1(x) = \beta_{fi} \omega_{P_{gfi}}$$

(6.200)

$$\frac{\partial^2 L_{2\mu}}{\partial^2 P d_i} = -\nabla_{P d_i} \nabla_{P d_i} f_1(x) = d_i$$

(6.201)

$$\frac{\partial^2 L_{2\mu}}{\partial sl_{Pg} \partial \pi l_{Pg}} = -\omega_{slPg} \tag{6.202}$$

$$\frac{\partial^2 L_{2\mu}}{\partial sl_{Pg_{fi}} \partial \omega_{slPg}} = -\pi l_{Pg} \tag{6.203}$$

$$\frac{\partial^2 L_{2\mu}}{\partial sl_{Pg} \partial \omega_{\pi lPg}} = 1 \tag{6.204}$$

$$\frac{\partial^2 L_{2\mu}}{\partial sl_{Qg} \partial \pi l_{Qg}} = -\omega_{slQg} \tag{6.205}$$

$$\frac{\partial^2 L_{2\mu}}{\partial sl_{Qg_{fi}} \partial \omega_{slQg}} = -\pi l_{Qg} \tag{6.206}$$

$$\frac{\partial^2 L_{2\mu}}{\partial sl_{Qg} \partial \omega_{\pi lQg}} = 1 \tag{6.207}$$

$$\frac{\partial^2 L_{2\mu}}{\partial sl_V \partial \pi l_V} = -\omega_{slv} \tag{6.208}$$

$$\frac{\partial^2 L_{2\mu}}{\partial sl_V \partial \pi l_V} = -\omega_{slv} \tag{6.209}$$

$$\frac{\partial^2 L_{2\mu}}{\partial sl_V \partial \omega_{\pi lv}} = 1 \tag{6.210}$$

$$\frac{\partial^2 L_{2\mu}}{\partial e_i \partial \omega_{e_i}} = 2\lambda p_i G_{ii} - 2\lambda q_i B_{ii} - 2\pi l_{V_i} - 2\pi u_{V_i} \tag{6.211}$$

$$\frac{\partial^2 L_{2\mu}}{\partial e_i \partial \omega_{e_j}} = \lambda p_i G_{ij} - \lambda q_i B_{ij} + \lambda p_j G_{ji} - \lambda q_j B_{ji} \tag{6.212}$$

$$\frac{\partial^2 L_{2\mu}}{\partial e_i \partial \omega_{f_j}} = -\lambda p_i B_{ij} - \lambda q_i G_{ij} + \lambda p_j B_{ji} + \lambda q_j G_{ji} \tag{6.213}$$

$$\frac{\partial^2 L_{2\mu}}{\partial e_i \partial \lambda p_i} = 2\omega_{e_i} G_{ii} + \omega_{e_j} G_{ij} - \omega_{f_j} B_{ij} \tag{6.214}$$

$$\frac{\partial^2 L_{2\mu}}{\partial e_i \partial \lambda p_j} = \omega_{e_j} G_{ji} + \omega_{f_j} B_{ji} \tag{6.215}$$

$$\frac{\partial^2 L_{2\mu}}{\partial e_i \partial \lambda q_i} = -2\omega_{e_i} B_{ii} - \omega_{e_j} B_{ij} - \omega_{f_j} G_{ij} \tag{6.216}$$

$$\frac{\partial^2 L_{2\mu}}{\partial e_i \partial \lambda q_j} = -\omega_{e_j} B_{ji} + \omega_{f_j} G_{ji} \tag{6.217}$$

$$\frac{\partial^2 L_{2\mu}}{\partial e_i \partial \pi l_{V_i}} = -2\omega_{e_i} \tag{6.218}$$

$$\frac{\partial^2 L_{2\mu}}{\partial e_i \partial \pi u_{V_i}} = -2\omega_{e_i} \tag{6.219}$$

$$\frac{\partial^2 L_{2\mu}}{\partial e_i \partial \omega_{\lambda p_i}} = 2G_{ii}e_i + G_{ij}e_j - B_{ij}f_j \tag{6.220}$$

$$\frac{\partial^2 L_{2\mu}}{\partial e_i \partial \omega_{\lambda p_j}} = G_{ji}e_j + B_{ji}f_j \tag{6.221}$$

$$\frac{\partial^2 L_{2\mu}}{\partial e_i \partial \omega_{\lambda q_i}} = -2B_{ii}e_i - G_{ij}f_j - B_{ij}e_j \tag{6.222}$$

$$\frac{\partial^2 L_{2\mu}}{\partial e_i \partial \omega_{\lambda q_j}} = G_{ji}f_j - B_{ji}e_j \tag{6.223}$$

$$\frac{\partial^2 L_{2\mu}}{\partial e_i \partial \omega_{\pi l V_i}} = 2e_i \tag{6.224}$$

$$\frac{\partial^2 L_{2\mu}}{\partial e_i \partial \omega_{\pi u V_i}} = -2e_i \tag{6.225}$$

$$\frac{\partial^2 L_{2\mu}}{\partial^2 e_i} = 2\omega_{\lambda p_i}G_{ii} - 2\omega_{\lambda q_i}B_{ii} - 2\omega_{\pi l V_i} - 2\omega_{\pi u V_i} \tag{6.226}$$

$$\frac{\partial^2 L_{2\mu}}{\partial e_i \partial e_j} = \omega_{\lambda p_i}G_{ij} - \omega_{\lambda q_i}B_{ij} + \omega_{\lambda p_j}G_{ji} - \omega_{\lambda q_j}B_{ji} \tag{6.227}$$

$$\frac{\partial^2 L_{2\mu}}{\partial e_i \partial f_j} = -\omega_{\lambda p_i}B_{ij} - \omega_{\lambda q_i}G_{ij} + \omega_{\lambda p_j}B_{ji} + \omega_{\lambda q_j}G_{ji} \tag{6.228}$$

$$\frac{\partial^2 L_{2\mu}}{\partial Pd_i \partial \omega_{\lambda p_i}} = 1 \tag{6.229}$$

$$\frac{\partial^2 L_{2\mu}}{\partial Pd_i \partial \omega_{\lambda p_i}} = 1 \tag{6.230}$$

$$\frac{\partial^2 L_{2\mu}}{\partial Pd_i \partial \omega_{\lambda q_i}} = Cd_i \tag{6.231}$$

$$\frac{\partial^2 L_{2\mu}}{\partial Pd_i \partial \omega_{\phi_i}} = -\left[1 + d_i - \frac{Pd_i + d_i(\lambda p_i + \lambda q_i Cd_i + d_i Pd_i - c_i)}{\sqrt{Pd_i^2 + (\lambda p_i + \lambda q_i t Cd_i + d_i Pd_i - c_i)^2}}\right] \tag{6.232}$$

$$\frac{\partial^2 L_{2\mu}}{\partial^2 Pd_i} = \omega_\phi \frac{A}{Pd_i^2 + (\lambda p_i + \lambda q_i Cd_i + d_i Pd_i - c_i)^2} \tag{6.233}$$

where

$$A = (1 + d_i^2)\sqrt{Pd_i^2 + (\lambda p_i + \lambda q_i Cd_i + d_i Pd_i - c_i)^2}$$
$$- \frac{[Pd_i + d_i(\lambda p_i + \lambda q_i Cd_i + d_i Pd_i - c_i)]^2}{\sqrt{P_L^2 + (\lambda p_i + \lambda \lambda q_i Cd_i + d_i Pd_i - c_i)^2}} \tag{6.234}$$

$$\frac{\partial^2 L_{2\mu}}{\partial Pd_i \partial \lambda p_i} = \omega_\phi \frac{B}{Pd_i^2 + (\lambda p_i + \lambda q_i Cd_i + d_i Pd_i - c_i)^2} \tag{6.235}$$

where

$$B = d_i \sqrt{Pd_i^2 + (\lambda p_i + \lambda q_i Cd_i + d_i Pd_i - c_i)^2}$$
$$- \frac{(\lambda p_i + \lambda q_i Cd_i + d_i Pd_i - c_i)[Pd_i + d_i(\lambda p_i + \lambda q_i Cd_i + d_i Pd_i - c_i)]}{\sqrt{Pd_i^2 + (\lambda p_i + \lambda q_i Cd_i + d_i Pd_i - c_i)^2}} \tag{6.236}$$

$$\frac{\partial^2 L_{2\mu}}{\partial Pd_i \partial \lambda q_i} = \omega_\phi \frac{C}{Pd_i^2 + (\lambda p_i + \lambda q_i Cd_i + d_i Pd_i - c_i)^2} \tag{6.237}$$

where

$$C = d_i Cd_i \sqrt{Pd_i^2 + (\lambda p_i + \lambda q_i Cd_i + d_i Pd_i - c_i)^2}$$
$$- \frac{Cd_i(\lambda p_i + \lambda q_i Cd_i + d_i Pd_i - c_i)[Pd_i + d_i(\lambda p_i + \lambda q_i Cd_i + d_i Pd_i - c_i)]}{\sqrt{Pd_i^2 + (\lambda p_i + \lambda q_i Cd_i + d_i Pd_i - c_i)^2}} \tag{6.238}$$

$$\frac{\partial^2 L_{2\mu}}{\partial Pg_{f_i} \partial \omega_{\pi l Pg}} = -1 \tag{6.239}$$

$$\frac{\partial^2 L_{2\mu}}{\partial Pg_{f_i} \partial \omega_{\pi u Pg}} = -1 \tag{6.240}$$

$$\frac{\partial^2 L_{2\mu}}{\partial Pg_{f_i} \partial \omega_{\lambda p i}} = -1 \tag{6.241}$$

$$\frac{\partial^2 L_{2\mu}}{\partial \lambda p_i \partial \omega_{e_i}} = 2G_{ii}e_i + G_{ij}e_j - B_{ij}f_j \tag{6.242}$$

$$\frac{\partial^2 L_{2\mu}}{\partial \lambda p_i \partial \omega_{f_i}} = 2G_{ii}f_i + G_{ij}f_j + B_{ij}e_j \tag{6.243}$$

$$\frac{\partial^2 L_{2\mu}}{\partial \lambda p_i \partial \omega_{e_j}} = G_{ij}e_i + B_{ij}f_i \tag{6.244}$$

$$\frac{\partial^2 L_{2\mu}}{\partial \lambda p_i \partial \omega_{f_j}} = G_{ij}f_i - B_{ij}e_i \tag{6.245}$$

$$\frac{\partial^2 L_{2\mu}}{\partial \lambda q_i \partial \omega_{e_i}} = -2B_{ii}e_i - G_{ij}f_j - B_{ij}e_j \tag{6.246}$$

$$\frac{\partial^2 L_{2\mu}}{\partial \lambda q_i \partial \omega_{f_i}} = -2B_{ii}f_i + G_{ij}e_j - B_{ij}f_j \tag{6.247}$$

$$\frac{\partial^2 L_{2\mu}}{\partial \lambda q_i \partial \omega_{e_j}} = -B_{ij}e_i + G_{ij}f_i \tag{6.248}$$

$$\frac{\partial^2 L_{2\mu}}{\partial \lambda q_i \partial \omega_{f_j}} = -G_{ij}e_i - B_{ij}f_i \tag{6.249}$$

6.9.4 Second Derivatives of Transmission Line Constraints in Rectangular Coordinates

$$\frac{\partial^2 L_{2\mu}}{\partial su_\ell \partial \pi u_\ell} = \omega_{su_\ell} \tag{6.250}$$

$$\frac{\partial^2 L_{2\mu}}{\partial su_\ell \partial \omega_{\pi u_\ell}} = -1 \tag{6.251}$$

$$\frac{\partial^2 L_{2\mu}}{\partial \pi u_\ell \partial su_\ell} = \omega_{su_\ell} \tag{6.252}$$

where ℓ denotes transmission line constraints and $h_\ell = S_{ij}^2$.

$$\frac{\partial^2 L_{2\mu}}{\partial \pi u_\ell \partial e_i} = -\omega_{e_i} \nabla_{e_i} \nabla_{e_i} h_\ell - \omega_{f_i} \nabla_{e_i} \nabla_{f_i} h_\ell - \omega_{e_j} \nabla_{e_i} \nabla_{e_j} h_\ell - \omega_{f_j} \nabla_{e_i} \nabla_{f_j} h_\ell \tag{6.253}$$

$$\frac{\partial^2 L_{2\mu}}{\partial \pi u_\ell \partial f_i} = -\omega_{e_i} \nabla_{f_i} \nabla_{e_i} h_\ell - \omega_{f_i} \nabla_{f_i} \nabla_{f_i} h_\ell - \omega_{e_j} \nabla_{f_i} \nabla_{e_j} h_\ell - \omega_{f_j} \nabla_{f_i} \nabla_{f_j} h_\ell \tag{6.254}$$

$$\frac{\partial^2 L_{2\mu}}{\partial \pi u_\ell \partial e_j} = -\omega_{e_i} \nabla_{e_j} \nabla_{e_i} h_\ell - \omega_{f_i} \nabla_{e_j} \nabla_{f_i} h_\ell - \omega_{e_j} \nabla_{e_j} \nabla_{e_j} h_\ell - \omega_{f_j} \nabla_{e_j} \nabla_{f_j} h_\ell \tag{6.255}$$

$$\frac{\partial^2 L_{2\mu}}{\partial \pi u_\ell \partial f_i} = -\omega_{e_i} \nabla_{f_j} \nabla_{e_i} h_\ell - \omega_{f_i} \nabla_{f_j} \nabla_{f_i} h_\ell - \omega_{e_j} \nabla_{f_j} \nabla_{e_j} h_\ell - \omega_{f_j} \nabla_{f_j} \nabla_{f_j} h_\ell \tag{6.256}$$

$$\frac{\partial^2 L_{2\mu}}{\partial \pi u_\ell \partial f_i} = -\omega_{e_i} \nabla_{f_j} \nabla_{e_i} h_\ell - \omega_{f_i} \nabla_{f_j} \nabla_{f_i} h_\ell - \omega_{e_j} \nabla_{f_j} \nabla_{e_j} h_\ell - \omega_{f_j} \nabla_{f_j} \nabla_{f_j} h_\ell \tag{6.257}$$

$$\frac{\partial^2 L_{2\mu}}{\partial \pi u_\ell \partial \omega_{su_\ell}} = su_\ell \tag{6.258}$$

$$\frac{\partial^2 L_{2\mu}}{\partial \pi u_\ell \partial \omega_{e_i}} = -\nabla_{e_i} h_\ell \tag{6.259}$$

$$\frac{\partial^2 L_{2\mu}}{\partial \pi u_\ell \partial \omega_{f_i}} = -\nabla_{f_i} h_\ell \tag{6.260}$$

$$\frac{\partial^2 L_{2\mu}}{\partial \pi u_\ell \partial e_j} = -\nabla_{e_j} h_\ell \tag{6.261}$$

$$\frac{\partial^2 L_{2\mu}}{\partial \pi u_\ell \partial f_j} = -\nabla_{f_j} h_\ell \tag{6.262}$$

6.9.5 Third Derivatives of Power Mismatches with Transformer Control

$$\frac{\partial^3 \Delta P_i}{\partial z_1 \partial z_2 \partial z_3} = 0, \quad \frac{\partial^3 \Delta P_j}{\partial z_1 \partial z_2 \partial z_3} = 0 \tag{6.263}$$

$$\frac{\partial^3 \Delta Q_i}{\partial z_1 \partial z_2 \partial z_3} = 0, \quad \frac{\partial^3 \Delta Q_j}{\partial z_1 \partial z_2 \partial z_3} = 0 \tag{6.264}$$

where z_1, z_2, $z_3 \in [Pg, Qg, e, f]$ and $\notin t_{ij} \cdot t_{ij}$ is the tap-changing variable of the transformer connected between bus i and bus j and it is assumed that the tap ratio control is at the side of bus i.

For z_1, z_2, $z_3 \in [t_{ij}, e_i, f_i, e_j, f_j]$, the third derivatives

$$\frac{\partial^3 \Delta P_i}{\partial z_1 \partial z_2 \partial z_3}, \frac{\partial^3 \Delta P_j}{\partial z_1 \partial z_2 \partial z_3}, \frac{\partial^3 \Delta Q_i}{\partial z_1 \partial z_2 \partial z_3}, \frac{\partial^3 \Delta Q_j}{\partial z_1 \partial z_2 \partial z_3}$$

may not be equal to zero. Some of the third derivatives are presented as follows:

$$\frac{\partial^3 \Delta P_i}{\partial e_i \partial e_j \partial t_{ij}} = -G_{ij} / t_{ij} \tag{6.265}$$

$$\frac{\partial^3 \Delta P_i}{\partial e_i \partial f_j \partial t_{ij}} = B_{ij} / t_{ij} \tag{6.266}$$

$$\frac{\partial^3 \Delta P_i}{\partial f_i \partial e_j \partial t_{ij}} = -B_{ij} / t_{ij} \tag{6.267}$$

$$\frac{\partial^3 \Delta P_i}{\partial f_i \partial f_j \partial t_{ij}} = -G_{ij} / t_{ij} \tag{6.268}$$

$$\frac{\partial^3 \Delta Q_i}{\partial e_i \partial e_j \partial t_{ij}} = B_{ij} / t_{ij} \tag{6.269}$$

$$\frac{\partial^3 \Delta Q_i}{\partial e_i \partial f_j \partial t_{ij}} = G_{ij} / t_{ij} \tag{6.270}$$

$$\frac{\partial^3 \Delta Q_i}{\partial f_i \partial e_j \partial t_{ij}} = -G_{ij} / t_{ij} \tag{6.271}$$

$$\frac{\partial^3 \Delta P_i}{\partial f_i \partial f_j \partial t_{ij}} = B_{ij} / t_{ij} \tag{6.272}$$

$$\frac{\partial^3 \Delta P_i}{\partial f_i \partial f_j \partial t_{ij}} = -G_{ij} / t_{ij} \tag{6.273}$$

$$\frac{\partial^3 \Delta P_i}{\partial e_i \partial f_j \partial t_{ij}} = -B_{ij} / t_{ij} \tag{6.274}$$

The contribution of the power mismatches to submatrix K shown in (6.153) and (6.165) is given as follows:

$$-\sum \omega_{x_1} \nabla_{x_2}^2 (\nabla_{x_1} \Delta P) \lambda p - \sum \omega_{x_1} \nabla_{x_2}^2 (\nabla_{x_1} \Delta Q) \lambda q \tag{6.275}$$

6.9.6 Third Derivatives of Transmission Line Constraints

$$\frac{\partial^3 S_{ij}^2}{\partial^3 e_i} = 12 g_{ii} (2 g_{ii} e_i + G_{ij} e_j - B_{ij} f_j) - 12 b_{ii} (-2 b_{ii} e_i - G_{ij} f_j - B_{ij} e_j) \tag{6.276}$$

$$\frac{\partial^3 S_{ij}^2}{\partial e_i \partial e_i \partial f_i} = 4 g_{ii} (2 g_{ii} f_i + G_{ij} f_j + B_{ij} e_j) - 4 b_{ii} (-2 b_{ii} f_i + G_{ij} e_j - B_{ij} f_j) \tag{6.277}$$

$$\frac{\partial^3 S_{ij}^2}{\partial e_i \partial e_i \partial e_j} = 4g_{ii}(G_{ij}e_i + B_{ij}f_i) - 4b_{ii}(G_{ij}f_i - B_{ij}e_i)$$
$$+ 4G_{ij}(2g_{ii}e_i + G_{ij}e_j - B_{ij}f_j) - 4B_{ij}(-2b_{ii}e_i - G_{ij}f_j - B_{ij}e_j) \quad (6.278)$$

$$\frac{\partial^3 S_{ij}^2}{\partial e_i \partial e_i \partial f_j} = 4g_{ii}(G_{ij}f_i - B_{ij}e_i) - 4b_{ii}(-G_{ij}e_i - B_{ij}f_i)$$
$$- 4B_{ij}(2g_{ii}e_i + G_{ij}e_j - B_{ij}f_j) - 4G_{ij}(-2b_{ii}e_i - G_{ij}f_j - B_{ij}e_j) \quad (6.279)$$

$$\frac{\partial^3 S_{ij}^2}{\partial^3 f_i} = 12g_{ii}(2g_{ii}f_i + G_{ij}f_j + B_{ij}e_j) - 12b_{ii}(-2b_{ii}f_i + G_{ij}e_j - B_{ij}f_j) \quad (6.280)$$

$$\frac{\partial^3 S_{ij}^2}{\partial f_i \partial f_i \partial e_i} = 4g_{ii}(2g_{ii}e_i + G_{ij}e_j - B_{ij}f_j) - 4b_{ii}(-2b_{ii}e_i - G_{ij}f_j - B_{ij}e_j) \quad (6.281)$$

$$\frac{\partial^3 S_{ij}^2}{\partial f_i \partial f_i \partial e_j} = 4g_{ii}(G_{ij}e_i + B_{ij}f_i) - 4b_{ii}(G_{ij}f_i - B_{ij}e_i)$$
$$+ 4B_{ij}(2g_{ii}f_i + G_{ij}f_j + B_{ij}e_j) + 4G_{ij}(-2b_{ii}f_i + G_{ij}e_j - B_{ij}f_j) \quad (6.282)$$

$$\frac{\partial^3 S_{ij}^2}{\partial f_i \partial f_i \partial f_j} = 4g_{ii}(G_{ij}f_i - B_{ij}e_i) - 4b_{ii}(-G_{ij}e_i - B_{ij}f_i)$$
$$+ 4G_{ij}(2g_{ii}f_i + G_{ij}f_j + B_{ij}e_j) - 4B_{ij}(-2b_{ii}f_i + G_{ij}e_j - B_{ij}f_j) \quad (6.283)$$

$$\frac{\partial^3 S_{ij}^2}{\partial e_i \partial e_j \partial e_i} = 4G_{ij}(G_{ij}e_i + B_{ij}f_i) - 4B_{ij}(G_{ij}f_i - B_{ij}e_i) \quad (6.284)$$

$$\frac{\partial^3 S_{ij}^2}{\partial e_i \partial e_j \partial f_i} = 4B_{ij}(G_{ij}e_i + B_{ij}f_i) + 4G_{ij}(G_{ij}f_i - B_{ij}e_i) \quad (6.285)$$

$$\frac{\partial^3 S_{ij}^2}{\partial e_j \partial e_j \partial f_i} = -4B_{ij}(G_{ij}f_i - B_{ij}e_i) + 4G_{ij}(G_{ij}e_i + B_{ij}f_i) \quad (6.286)$$

$$\frac{\partial^3 S_{ij}^2}{\partial f_j \partial f_j \partial f_{ii}} = 4G_{ij}(G_{ij}f_i - B_{ij}e_i) + 4B_{ij}(G_{ij}e_i + B_{ij}f_i) \quad (6.287)$$

$$\frac{\partial^3 S_{ij}^2}{\partial e_i \partial f_i \partial e_j} = 2G_{ij}(2g_{ii}f_i + G_{ij}f_j + B_{ij}e_j) + 2B_{ij}(2g_{ii}e_i + G_{ij}e_j - B_{ij}f_j)$$
$$- 2B_{ij}(-2b_{ii}f_i + G_{ij}e_j - B_{ij}f_j) + 2G_{ij}(-2b_{ii}e_i - G_{ij}f_j - B_{ij}e_j) \quad (6.288)$$

$$\frac{\partial^3 S_{ij}^2}{\partial e_i \partial f_i \partial f_j} = -2B_{ij}(2g_{ii}f_i + G_{ij}f_j + B_{ij}e_j) + 2G_{ij}(2g_{ii}e_i + G_{ij}e_j - B_{ij}f_j)$$
$$- 2G_{ij}(-2b_{ii}f_i + G_{ij}e_j - B_{ij}f_j) - 2B_{ij}(-2b_{ii}e_i - G_{ij}f_j - B_{ij}e_j) \quad (6.289)$$

$$\frac{\partial^3 S_{ij}^2}{\partial e_i \partial e_j \partial f_j} = 2G_{ij}(G_{ij}f_i - B_{ij}e_i) - 2B_{ij}(G_{ij}e_i + B_{ij}f_i) \quad (6.290)$$

The contribution of the transmission line constraints to submatrix K shown in (6.153) and (6.165) is given as follows:

$$-\sum(\omega_{\pi l_\ell} + \omega_{\pi u_\ell})\nabla_{x_2}^2 h_\ell - \sum \omega_{x_1}\nabla_{x_2}^2(\nabla_{x_1}h_\ell)(\pi u_\ell + \pi l_\ell) \quad (6.291)$$

where $h_\ell = S_{ij}^2$. The second term in (6.291) involves the third derivatives of S_{ij}^2 with respect to voltage variables x_1 and x_2. $x_1, x_2 \in [e_i, f_i, e_j, f_j]$ and $\omega_{x_1} \in [\omega_{e_i}, \omega_{f_i}, \omega_{e_j}, \omega_{f_j}]$.

$$\frac{\partial^3 S_{ij}^2}{\partial e_i \partial e_j \partial f_j} = 2G_{ij}(G_{ij}f_i - B_{ij}e_i) - 2B_{ij}(G_{ij}e_i + B_{ij}f_i) \tag{6.292}$$

ACKNOWLEDGMENTS

The inputs from S.G. Petoussis and A.G. Petoussis are appreciated.

REFERENCES

1. Adibi MM, Milaniciz DP. Reactive capability limitation of synchronous machines. *IEEE Transactions on Power Systems* 1994;9(1):29–40.
2. Löf P-A, Andersson G, Hill DJ. Voltage dependent reactive power limits for voltage stability studies. *IEEE Transactions on Power Systems* 1995;10(1):221–228.
3. Adibi MM, Milaniciz DP, Volkmann TL. Optimizing generator reactive resources. *IEEE Transactions on Power Systems* 1995;10(1):221–228.
4. IEEE Task Force on Excitation Limiters. Under excitation limiter models for power system stability studies. *IEEE Transactions on Energy Conversion* 1995;10(3):524–531.
5. Gargiulo G, Mangoni V, Russo M. Capability charts for combined cycle power plants. *IEE Proceeding Generation, Transmission and Distribution* 2002;149(4):407–415.
6. Fiacco AV, McCormick GP. *Non-linear Programming: Sequential Unconstrained Minimisation Techniques.* New York: John Wiley & Sons; 1968.
7. Karmarkar N. A new polynomial time algorithm for linear programming. *Combinatorica* 1984;4:373–395.
8. Gill PE, Murray W. On projected Newton barrier methods for linear programming and an equivalence to Karmarkar's projective method. *Mathematical Programming* 1986;36:183–209.
9. Marsten R, Subramanian R, Saltzman M, Lustig I, Shanno D. Interior point methods for linear programming: just call Newton, Lagrange, and Fiacco and McCormick. *Interfaces* 1990;20(4): 105–116.
10. Lustig IJ, Marsten RE, Shanno DF. Interior point methods for linear programming: computational state of art. *ORSA Journal of Computing* 1994;6(1):1–14.
11. Mehrotra S. On the implementation of a primal-dual interior point method. *SIAM Journal of Optimization* 1992;2(4):575–601.
12. Astfalk G, Lustig I, Marsten RE, Shanno DF. The interior-point method for linear programming. *IEEE Software* 1992;4(4):61–68.
13. El-Bakry AS, Tapia RA, Tsuchiya T, Zhang Y. On the formulation and theory of the newton interior-point method for nonlinear programming. *Journal of Optimization Theory and Applications* 1996;89(3):507–541.
14. Vargas LS, Quintana VH, Vannelli A. A tutorial description of an interior point method and its applications to security-constrained economic dispatch. *IEEE Transactions on Power Systems* 1993;8(3):1315–1323.
15. Lu CN, Unum and MR. Network constrained security control using an interior point algorithm. *IEEE Transactions on Power Systems* 1993;8(3):1068–1076.
16. Yan X, Quintana V. An efficient predictor/corrector interior point method for security constrained economic dispatch. *IEEE Transactions on Power Systems* 1997;12(2):803–810.
17. Momoh JA, Guo SX, Ogbuobiri EC, Adapa R. The quadratic interior point method solving power system optimization problems. *IEEE Transactions on Power Systems* 1994;9(3):1327–1336.
18. Wei H, Sasaki H, Yokoyama R. An interior point quadratic programming algorithm to power system optimization problems. *IEEE Transactions on Power Systems* 1996;11(1):260–266.

19. Granville S. Optimal reactive power dispatch through interior point methods. *IEEE Transactions on Power Systems* 1994;9(1):136–146.

20. Wu YC, Debs AS, Marsten RE. A direct non-linear predictor-corrector primal-dual interior point algorithm for optimal power flows. *IEEE Transactions on Power Systems* 1994;9(2):876–883.

21. Irisarri GD, Wang X, Tong J, Mokhtari S. Maximum loadability of power systems using interior point nonlinear optimization method. *IEEE Transactions on Power Systems* 1997;12(1):167–172.

22. Wei H, Sasaki H, Yokoyama R. An interior point nonlinear programming for optimal power flow problems within a novel data structure. *IEEE Transactions on Power Systems* 1998;13(3):870–877.

23. Torres GL, Quintana VH. An interior point method for non-linear optimal power flow using voltage rectangular coordinates. *IEEE Transactions on Power Systems* 1998;13(4):1211–1218.

24. Zhang X-P, Hanschin EJ. Advanced implementation of UPFC in a non-linear interior point OPF. *IEE Proceeding Generation, Transmission & Distribution* 2001;148(5):489–496.

25. Zhang X-P, Hanschin EJ, Yao M. Modeling of the generalized unified power flow controller (GUPFC) in a nonlinear interior point OPF. *IEEE Transactions on Power Systems* 2001;16(3): 367–373.

26. Zhang X-P, Petoussis SG, Godfrey KR. Novel nonlinear interior point optimal power flow (OPF) method based on current mismatch formulation. *IEE Proceeding Generation, Transmission & Distribution* 2005;152(6):795–805.

27. Baldick R. Electricity market equilibrium models: the effect of parameterization. *IEEE Transactions on Power Systems* 2002;17(4):1170–1176.

28. Chen H, Wong KP, Chung CY, Nguyen DHM. A coevolutionary approach to analyzing supply function equilibrium model. *IEEE Transactions on Power Systems* 2006;21(3):1019–1028.

29. Haghighat H, Seifi H, Kian AR. Gaming analysis in joint energy and spinning reserve markets. *IEEE Transactions on Power Systems* 2007;22(4):2074–2085.

30. Hu X, Ralph D. Using EPECs to model bilevel games in restructured electricity markets with locational prices. *Operations Research* 2007;55(5):809–827.

31. Bautista G, Anjos MF, Vannelli A. Numerical study of affine supply function equilibrium in AC network-constrained markets. *IEEE Transactions on Power Systems* 2007;22(3):1174–1184.

32. Klemperer PD, Meyer MA. Supply function equilibria in oligopoly under uncertainty. *Econometrica* 1989;57(6):1243–1277.

33. Green R. Increasing competition in the British electricity spot market. *Journal of Industrial Economics* 1996;44(2):205–216.

34. Ferrero RW, Rivera JF, Shahidehpour SM. Application of games with incomplete information for pricing electricity in deregulated power pools. *IEEE Transactions on Power Systems* 1998; 13(1):184–189.

35. Wen F, David AK. Optimal bidding strategies and modeling of imperfect information among competitive generators. *IEEE Transactions on Power Systems* 2001;16(1):15–21.

36. Berry CA, Hobbs BF, Meroney WA, O'Neil RP, Stewart Jr WR. Understanding how market power can arise in network competition: a game theoretic approach. *Utilities Policy* 1999;8:139–158.

37. Correia PF, Overbye TJ, Hiskens IA. Searching for noncooperative equilibria in centralized electricity markets. *IEEE Transactions on Power Systems* 2003;18(4):1417–1424.

38. Hobbs BF, Metzler CB, Pang J-S. Strategic gaming analysis for electric power systems: an MPEC approach. *IEEE Transactions on Power Systems* 2000;15(2):638–645.

39. Younes Z, Ilic M. Generation strategies for gaming transmission constraints: will the deregulated electric power market be an oligopoly? *Decision Support Systems* 1999;24:207–222.

40. Weber JD, Overbye TJ. An individual welfare maximisation algorithm for electricity markets. *IEEE Transactions on Power Systems* 2002;17(3):590–596.

41. Petoussis SG, Zhang XP, Godfrey KR. Electricity market equilibrium analysis based on nonlinear interior point algorithm with complementarity constraints. *IET Generation, Transmission and Distribution* 2007;1(4):603–612.

42. Petoussis SG, Petoussis AG, Zhang XP, Godfrey KR. Impact of the transformer tap-ratio control on the electricity market equilibrium. *IEEE Transactions on Power Systems* 2008;23(1):65–75.

43. Hu X, Ralph D. Using EPECs to model bi-level games in restructured electricity markets with locational prices. *Operations Research* 2007;55(5):809–827.

44. Luo Z-Q, Pang J-S, Ralph D. *Mathematical Programs with Equilibrium Constraints*. Cambridge: Cambridge University Press; 1996.

45. Su C-L. A sequential NCP algorithm for solving equlibrium problems with equilibrium constraints. Technical report, Department of Management Science and Engineering, Stanford University, 2004.

46. Pang J-S, Fukushima M. Quasi-variational inequalities, generalized nash equilibrium, and multi-leader-follower games. *Computational Management Science* 2005;2(1):21–56.

47. Leyffer S, Munson TS. Solving multi-leader-follower games. Preprint ANL/MCS-P1243-0405, Argonne National Laboratory, Mathematics and Computer Science Division, April 2005, Revised March 2007.

48. Bautista G, Anjos MF, Vannelli A. Formulation of oligopolistic competition in AC power networks: an NLP approach. *IEEE Transactions on Power Systems* 2007;22(1):105–115.

49. Petoussis S, Zhang X-P, Godfrey K. Electricity market equilibrium analysis based on nonlinear interior point algorithm with complementary constraints. *IET Generation, Transmission and Distribution* 2007;1(4):603–612.

50. Yao J, Adler I, Oren SS. Modeling and computing two-settlement oligopolistic equilibrium in a congested electricity network. *Operations Research* 2008;56(1):34–47.

51. Zhang D, Xu H, Wu Y. A Stochastic EPEC model for electricity markets with two way contracts. www.optimization-online.org, February 2008.

52. Mordukhovich BS. Equilibrium problems with equilibrium constraints via multiobjective optimization. *Optimization Methods and Software* 2004;19(5):479–492.

53. IEEE 14-bus system. http://www.ee.washington.edu/research/pstca/pf14/pg_tca14bus.htm

54. Alsac O, Stott B. Optimal load flow with steady state security. *IEEE Transactions on Power Apparatus and Systems* 1974;93(3):745–751.

USING MARKET SIMULATIONS FOR ECONOMIC ASSESSMENT OF TRANSMISSION UPGRADES: APPLICATION OF THE CALIFORNIA ISO APPROACH

Mohamed Labib Awad, Keith E. Casey, Anna S. Geevarghese, Jeffrey C. Miller, A. Farrokh Rahimi, Anjali Y. Sheffrin, Mingxia Zhang, Eric Toolson, Glenn Drayton, Benjamin F. Hobbs, and Frank A. Wolak

7.1 INTRODUCTION

The need for new transmission planning processes that respond to the demands of a restructured power industry is widely acknowledged [1–10]:

> *"There is a need for complex models that will take into account bidding strategies, the expansion and location of new merchant power plants, volatility and uncertainty factors, and an accurate representation of the network system" [11].*

> *"ISOs are challenged when asked to develop a business case justifying a market economics project and lack the necessary market models to adequately forecast and 'prove' their need" [12].*

Unlike the previous vertically integrated regime in which a single regulated utility was responsible for serving its load, the restructured wholesale electric market is comprised of a variety of parties independently making decisions that affect the use of transmission. A new approach to evaluate the economic benefits of transmission expansion is therefore needed. Specifically, the approach must anticipate how a transmission expansion would affect (a) transmission users' access to customers and generation, (b) bidding and operating behavior of existing generation, and (c) incentives for new generation investment. The approach must also account for

Restructured Electric Power Systems: Analysis of Electricity Markets with Equilibrium Models,
Edited by Xiao-Ping Zhang
Copyright © 2010 Institute of Electrical and Electronics Engineers

uncertainty associated with key market factors such as hydro conditions, fuel prices, and demand growth. The California ISO's (CAISO's) response to this challenge has been to develop a planning approach called the Transmission Economic Assessment Methodology (TEAM) [13–15].

TEAM was developed because the CAISO is responsible for evaluating the need for transmission upgrades that California ratepayers may be asked to fund. These include construction of transmission projects needed either to promote economic efficiency or maintain reliability. The CAISO has clear standards for evaluating reliability-based projects. TEAM will help the CAISO fulfill its responsibility to identify economic projects that encourage efficient use of the grid.

The goal of TEAM is to streamline the evaluation process for economic projects, improve the accuracy of the evaluation, and add greater predictability to the evaluations of transmission need conducted by various agencies. In several previous cases, the CAISO has seen the same project receive multiple reviews of project need by various agencies, each carrying out its individual mandate. This has caused redundancies and inefficiencies [16, 17]. We believe that accepting the TEAM methodology as the standard for project evaluation will reduce redundant efforts and lead to faster and more widely supported decisions on transmission investment projects.

The TEAM methodology is based upon five principles for quantifying benefits. It represents the state-of-the-art in the area of transmission planning in terms of its simultaneous consideration of the network, market power, uncertainties, and multiple evaluation perspectives. This framework is a template defining the basic components that any transmission study in California should address, providing standards for the minimum functionality that modeling software should have. TEAM is intended to provide market participants, policy-makers, and permitting authorities with the information they need to make informed decisions.

This chapter summarizes the elements of the TEAM methodology for assessing the economic benefits of transmission expansions for wholesale market environments (Section 7.2). To illustrate its use, we summarize its application to a proposed transmission upgrade (Palo Verde-Devers 2, PVD2) (Section 7.3). We describe particular modeling procedures we used for the risk and market power analyses, which are new in transmission planning practice. We also summarize some issues that arise in applying TEAM to evaluating renewable-focused transmission (Section 7.4).

7.2 FIVE PRINCIPLES

The valuation methodology we propose here enhances traditional transmission evaluations in five ways, which we call "principles." With the exception of the market-based (market power) pricing principle, none of these individual principles is entirely novel, in that each has been considered previous transmission planning studies. However, no previous studies, to our knowledge, have considered all of the principles.

Although how the principles are applied will vary from study to study, the CAISO requires that the principles be considered in any economic evaluation of proposed upgrades presented to the CAISO for review. The TEAM report [13] suggests specific procedures that can be used to implement each principle. The study type and initial results will dictate the level of application. Our PVD2 study experience indicates that about 12 person-months of effort over three months is needed to fully apply TEAM, including analysis and public participation.

We note that the methodology was developed in collaboration with stakeholders in an open process. Further, its application to any particular project is subject to public review before submitting a project for approval to the CAISO Board of Governors. Finally, the TEAM results are reviewed in California Public Utilities Commission hearings. At any time during this process, stakeholders can propose alternatives for consideration by TEAM. This open process is intended to make the method's assumptions and procedures transparent to all interests involved.

7.2.1 First Principle: Benefit Framework

A benefit-cost analysis framework should enable users to clearly identify the beneficiaries and expected benefits of any kind of transmission project.

TEAM divides the total benefits due to a transmission expansion into three parts—changes in consumer surplus, producer surplus, and transmission owner (congestion revenue) benefits. For a vertically integrated utility, benefits arise from three sources—direct reductions in wholesale power costs, increases in net revenue for utility-owned generation, and increases in utility-derived congestion revenue.

The quantified benefits can be aggregated for individual subregions or groups of market participants (e.g., California ratepayers), as well as for the entire Western interconnection. A key policy question is which perspective should be used to evaluate projects. The answer depends on the viewpoint of the entity that the network is intended to benefit. If the network is operated to benefit ratepayers who have paid for the network, then the ratepayer perspective might be argued to be most appropriate. But in the long run, financially healthy utility generation and private supply may be needed to maximize ratepayer benefits. In this view, the network is operated to benefit all market participants and, thus, benefits to CAISO participants or the Western Electricity Coordinating Council (WECC) may be the relevant test. (The WECC includes 11 states, two Canadian provinces, and northwest Mexico.)

TEAM does not specify a single test as being the "right" test, nor any specific numerical threshold as being "do or die" for a project. Rather, each perspective provides important information to policy-makers [6]. If the benefit-cost ratio of an upgrade passes the CAISO participant test, but fails the WECC test of economic efficiency, then it may indicate that the expansion will mainly transfer benefits from one region to another. In contrast, if the project passes the societal test but fails the CAISO participant test, this implies that other project beneficiaries should help fund the project.

An additional consideration in weighing various perspectives is how to treat the loss of market power–derived rents by generation owners when the grid is expanded. Since market power reduces efficiency and harms consumers, it can be argued that it is reasonable to exclude the loss of those rents in benefit calculations. (These rents are distinguished from scarcity rents that arise in competitive markets.) This is the difference between the societal test and the modified societal test (based on societal benefits minus market power rents) used in the PVD2 study.

The basic calculations of cost-to-load and profits earned by market parties are given below; from these building blocks, the various benefit-cost metrics can be calculated. For simplicity, we here disregard the complications of long-run power purchase contracts, as well as ownership of and payments to transmission interfaces between different markets (e.g., California and Arizona). We also consider only the short-run, assuming capital stock is fixed; so we can ignore payments for fixed capital costs (e.g., customer wire charges paid to the transmission owner, or financing costs for generation), which are unaffected by operating decisions. Demand elasticity is zero (fixed load). Only one hour is considered. Of course, most or all these assumptions are relaxed in the actual calculations in any particular TEAM application.

Let L_i be the power consumed at bus i; g_{ui} the amount of utility-owned power produced at i; g_{mi} the amount of independent (merchant) power produced at i; and pi the price (LMP) at i. The function $C_{ui}(g_{ui})$ is the production cost associated with utility-owned generation, while $C_{mi}(g_{mi})$ is the production cost of merchant generation. The function $P_{imp}(imports_r, p)$ is the cost of imports to the transmission owner which, in general, depends on the level of imports to r as well as the vector of prices in all locations. For exporting regions, $imports_r$ will be negative and, generally, will be their "cost" (i.e., revenue will be earned). $I(r)$ is the set of buses in region r.

Net cost-to-load in region $r = CTL_r = \Sigma_{i \in I(r)} p_i L_i - \Pi_{ur} - MS_r$

= payments for power minus utility and transmission profits (7.1)

(since utility-owned generation and transmission are assumed to be regulated on a cost-of-service basis)

Operating profit (gross margin) to utility-owned generation in r

$= \Pi_{ur} = \Sigma_{i \in I(r)} [p_i g_{ui} - C_{ui}(g_{ui})]$

= revenue minus cost (7.2)

Operating profit (gross margin) to merchant-owned generation in r

$= \Pi_{mr}$

$= \Sigma_{i \in I(r)} [p_i g_{mi} - C_{mi}(g_{mi})]$ (7.3)

Transmission owner merchandising surplus in r

$= MS_r$

$= \Sigma_{i \in I(r)} p_i [L_i - g_{ui} - g_{mi}] - P_{imp}(\Sigma_{i \in I(r)} [L_i - g_{ui} - g_{mi}], p)$

= revenue from load minus payments to generation and for imports. (7.4)

Assuming that $\Sigma_r P_{imp}(\Sigma_{i \in I(r)} [L_i - g_{ui} - g_{mi}], p) = 0$ (i.e., payments by one region for imports equal receipts to all other regions for exports to that region), then the total benefit to all parties $= \Sigma_r [\Pi_{ur} + \Pi_{mr} + MS_r - CTL_r]$ simplifies to $-\Sigma_r \Sigma_{i \in I(r)} [C_{ui}(g_{ui}) + C_{mi}(g_{mi})]$, the sum of all production costs. This is because all the p_i terms cancel. (One party's expenditure is another's revenue.)

7.2.2   Second Principle: Full Network Representation

It is important to accurately model physical transmission flows to correctly forecast the impact of an upgrade. Models based on contract paths may suffice for some types of resource studies, but that approach is generally deficient when analyzing a network modification that impacts regional transmission flows and locational marginal prices (LMPs).

We have recently seen how critical an accurate network representation is to making correct decisions. A utility proposed a transmission addition and justified its economic viability using a contract-path model. However, the CAISO found the line to be uneconomic due to adverse physical impacts on other parts of the transmission system that the contract-path model disregarded. The CAISO's full network model showed far more flow into California from a particular direction because the proposed line reduced the impedance of the system in that direction. Thus, the CAISO experienced an actual reduction in transfer capability, and additional upgrades were needed to get the benefits projected by the utility [18].

It is possible that, with careful tuning, aggregate path-based models that disregard parallel flows can be adequate in many circumstances. Indeed, this was the most controversial issue in the California regulatory review of the TEAM methodology [18]. But obtaining such approximations is challenging and invites criticism in regulatory proceedings; using a full network model avoids criticisms about equivalences. A useful research direction would be a systematic comparison of the results of path-based and full network (DC and AC) models at various levels of aggregation to more fully understand when they differ, and the implications of such differences. This could lead to a fuller understanding of what simplifications can be safely made without distorting the results of economic studies.[1]

There are many different techniques for modeling physical transmission networks. More accurate techniques may also increase computational and data burdens. Recognizing these tradeoffs, the CAISO identified the need to model the correct network representation provided in WECC base cases. Any production cost program that utilizes this network model should include the ability to model the following:

- Either a DC or AC optimal power flow (OPF) that correctly represents thermal and other constraints upon physical power flows for high-voltage transmission facilities and interfaces resulting from specific hourly load and generation patterns. Use of a full AC load flow model to represent hourly conditions in a large market over a planning horizon is not presently possible. Several production costing models are available (e.g., GE-MAPS [19] and PLEXOS [20]) that include a linearized DC load flow.

- Individual facility thermal or surge impedance loading-based constraints, linear nomograms resulting from stability and other limits, and path limits.

- Flow limits that depend on variables such as area load, facility loading, or generator availability.

- Phase shifters, DC lines, and other controllable devices.

---

[1]For an example of a study comparing the accuracy of load-flow simulation methods, see [37].

- LMPs.
- Hourly flows on individual facilities, paths, or nomograms.

It is also desirable to model transmission losses.

While the TEAM approach recommends use of a network model, a simplified analysis (contract path or transportation models) can also be utilized if desired to screen a large number of cases for the purpose of identifying system conditions that may result in large benefits from a transmission expansion. Also, if the project proponent can convincingly demonstrate that a simpler model can estimate costs and market impacts as accurately as a full network model, it is permissible to use the simpler model; thus, TEAM is making a rebuttable presumption that a full network is necessary. Of course, in applying any transmission model, it is important to verify that results are not unduly affected by constraints that in the real world can be readily modified.

7.2.3 Third Principle: Market Prices

Historically, resource plans have relied on production cost simulations to quantify economic benefits of proposed upgrades. Such an approach made sense when utilities were vertically integrated and recovered costs through regulated rates. But naïvely assuming that profit-maximizing suppliers bid at marginal cost in a market environment may distort benefit estimates. Instead, suppliers are likely to optimize bidding strategies in response to system conditions or behavior of other market participants.

Modeling such bidding is important because transmission expansion can benefit consumers by improving market competitiveness. A project can enhance competitiveness of the wholesale market by increasing the number of independent generation owners that can supply energy at various locations. However, in theory, the presence of imperfect competition can either decrease or increase the benefits of transmission upgrades, depending on the situation [21].

Thus, strategic bidding can impact societal benefits of an upgrade, as well as transfers of benefits among participants. Because of this, forecasting market prices is critical.

There are two approaches to modeling strategic bidding in transmission valuation studies. The first involves use of game-theoretic models to simulate strategic bidding [e.g., 22]. Such a model typically represents several strategic suppliers, each seeking to maximize its profits by altering its bids or production in response to the strategies of other players. The second approach involves the use of estimated historical relationships between market structure and measures of market power such as bid-cost mark-ups or the difference between market prices and hypothetical competitive prices [23].[2]

---

[2]Several empirical studies have gauged the extent of unilateral market power exercised in a wholesale electricity market by computing the mark-up of the actual price over a counterfactual competitive benchmark price [24–26]. However, none of these studies have estimated predictive statistical models relating hour-by-hour mark-ups to shifting market conditions. The strength of the approach we use in the PVD2 case study is that it relies on California's experience with markets over the past seven years to estimate a stable predictive relationship between the mark-up of the actual market price over a counterfactual competitive price and key variables that measure system supply/demand conditions that influence mark-ups.

Each approach has advantages [13]. In our experience in California and else-where, we have found that game-theoretic models can be extraordinarily useful for providing general insights on how proposals for changes in market designs or industry structure might affect the ability to exercise market power. However, they have been less useful for predicting specific prices under particular supply and demand circumstances. In assessing these alternative approaches, we believe that an empirical approach to modeling strategic bidding is preferable to a game theoretic approach if relevant data are available and can be adapted to a detailed transmission network representation. On the other hand, game theoretic methods are advantageous in unprecedented situations or where data is lacking.

To the best of our knowledge, no one has successfully developed and implemented a market simulation model based on strategic supply bids that dynamically respond to supply conditions while incorporating a detailed physical transmission modeling capability. However, we acknowledge that much research and development remains to be done in this area, and that approaches other than the empirical bid mark-up method we use below may be more useful in other circumstances. TEAM does not specify the process to be used for forecasting market power. Rather, at this point, the CAISO asks only that a credible and comprehensive approach for forecasting market prices be utilized in the evaluation. We consider the empirical bidding model we use in the PVD2 analysis below to be one of several useful methods for deriving market prices.

7.2.4 Fourth Principle: Explicit Uncertainty Analysis

Decisions on whether to build new transmission are complicated by uncertainty. Future load growth, fuel costs, additions and retirements of generation capacity, exercise of market power, and availability of hydropower are among the many uncertainties that impact decision making. Some of these risks and uncertainties are readily quantified, but others are not.

There are two reasons why we must consider uncertainty. First, changes in system conditions can significantly affect transmission benefits and the relationship between benefits and underlying system conditions is nonlinear. (This is true in the case study; see Table 7.1 below.) Thus, evaluating an upgrade based just on average future system conditions might greatly under- or overestimate the expected project benefits and lead to a suboptimal decision. To capture all project impacts, we must examine a wide range of possible system conditions.

Second, historical evidence suggests that transmission upgrades have been particularly valuable during extreme conditions. A hypothetical interconnection between WECC and the eastern US that would have been able to convey many gigawatts of power during the 2000–2001 period would have been worth tens of billions of dollars, based on differences between the regions' prices. Had such a significant inter-connection been in place, western prices would not have risen to levels that they did during that period. (Such an interconnection could be analyzed by the TEAM approach, but has not since it would not be under CAISO jurisdiction.)

There are several approaches for assessing the impact of uncertainty on transmission expansion [e.g., 3, 4]. A complete evaluation process should incorporate

TABLE 7.1 Seventeen market cases considered in 2008 expected benefits analysis (all benefits are in millions of $2008, and are the difference between "with PVD2" and "without PVD2" simulations)

Case i	LD	GP	HY	MU	Pr_i	Societal	Modified societal	CAISO ratepayer (LMP only)	CAISO R.P. (LMP + contract path)
1	B	B	B	B	0.11	45.3	58.9	37.9	98.7
2	B	B	B	H	0.05	47	71.1	54.8	124.5
3	B	B	D	B	0.099	50.5	66.6	34.5	115.7
4	B	B	W	B	0.131	24.3	26.2	29.1	72.8
5	B	H	B	B	0.023	90	113.1	76.7	185.9
6	B	H	B	H	0.018	92.5	133.9	104.8	229.1
7	H	B	B	H	0.033	45.3	120.8	70.9	199.8
8	H	H	D	B	0.018	119.9	237	85.2	317.5
9	B	H	D	H	0.018	106	151.6	80.7	257.3
10	B	B	B	L	0.15	42.5	41.5	17	68.5
11	L	B	B	B	0.127	29.9	31.6	35.6	83.3
12	B	L	B	B	0.101	8.8	18.5	8	36.6
13	H	H	B	H	0.015	93.8	235.2	143.2	371.1
14	H	L	B	B	0.049	4.4	23.7	2.2	41
15	L	H	B	B	0.023	56.9	59.5	74.1	155.4
16	H	H	D	H	0.015	135.8	387.7	234.9	568.5
17	H	H	W	B	0.019	19.1	21.5	5.6	119.7
	Expected Value					41	61	39	110

Key: LD = load level; GP = gas price level; HY = hydro level; MU = mark-up

probabilistic analysis or scenario analysis. The probabilistic approach models uncertainties associated with parameters that affect project benefits, and assigns probabilities to, for example, scenarios of future loads, gas prices, and generating unit availabilities.

Unless the proposed project economics are overwhelmingly favorable when using "expected" input assumptions, we need to perform sensitivity studies using a range of input assumptions. We do this to compute the following risk measures:

• Expected value

• Range

• Values under specified rare but potentially important contingencies, such as loss of a major transmission link

Much of the economic value of an upgrade is realized when unusual or unexpected situations occur. Such situations may include high load growth, high gas prices, or extreme hydrological years. The "expected value" of a transmission upgrade should be based on both the usual or expected conditions as well as on the unusual, but plausible, situations. These are not combined mechanistically into a single index of project desirability or risk. Rather, the various measures provide a fuller picture of the advantages and disadvantages of a proposal.

A transmission upgrade can also be viewed as a type of insurance against extreme events. Providing the additional capacity incurs a capital and operating cost, but the benefit is that the impact of extreme events is reduced. The events considered could include physical contingencies such as extended transmission outages, as in the PVD2 analysis below. They could also include drastic changes in regulation (e.g., CO_2 caps).

An extension of risk analysis would assess the value of waiting for more information before committing to construction [27]. This so-called "option value" could be quantified by constructing decision trees representing the defer option and later construction possibilities, along with changes in scenario probabilities that could result from the information ("posterior probabilities").

7.2.5   Fifth Principle: Interactions with Other Resources

The economic value of a proposed upgrade directly depends on the cost of resources that could be added or implemented in lieu of the upgrade. We consider the following resource options singly and in combination:

- Central station, renewable, and distributed generation
- Demand-side management
- Modified operating procedures
- Additional remedial action schemes
- Alternative transmission upgrades

Examining such alternatives must recognize that an alternative can either complement the upgrade or substitute for it.

In addition to considering resource alternatives, another important issue to consider is the decision where to site new resources. One perspective is that the transmission should be sited after the siting of new generation. Another point of view is that the transmission should be planned anticipating how generation investment would react. (Sauma and Oren [21] carefully analyze these different perspectives.)

We believe the latter perspective will yield the greatest long-run societal benefits. Transmission additions have planning horizons that require decisions a decade in advance of the line being placed in service. A new combined cycle natural gas-fired generation unit can easily be built in half this time. Consequently, we believe it is best to plan the grid anticipating the entry decisions of new generation as a result of the upgrade [21]. In this way, the transmission planner influences generation decision making, rather than accounting for it after the fact.

The ideal means to account for private investment decisions is to model the profitability of generation investment [21]. We suggest a "what if" framework. As an example, if a new line was to be built, what would be the most likely resulting outcomes in the profitability of private generation decisions? Profitability should consider energy and ancillary service revenues, as well as markets for capacity or long-term energy contracts created in response to resource adequacy requirements. Comparing this to a case where we did not build the line, how much would the profitability of generation investments differ? The methodology can then optimize generation additions for both the with- and without upgrade cases, adding generation when its revenues can cover its fixed and variable costs. (As a less preferable alternative, fixed entry scenarios could be considered.) The difference in costs between the scenarios, including both the fixed and variable costs of the new resources, will be the value of the upgrade.

7.3 PALO VERDE-DEVERS NO. 2 STUDY

No other ISO, to our knowledge, has included all five of the above principles in their planning studies [28]. PJM, for example, includes multiple scenarios in their regional transmission expansion process [29], but not market-based pricing. The Italian ISO proposes a market simulation method based on the statistical methods we used [30], but does not consider the interaction of transmission and generation investment.

The purpose of our case study is to illustrate the application of the above principles to a market-driven upgrade. Below we summarize the project, assumptions of the analysis, results for each category of benefits, and resource alternatives to the project. We focus on identifying quantifiable economic benefits that can be attributed to PVD2. These include:

- Energy cost savings
- Operational benefits
- Capacity benefits
- System-loss reduction
- Emission reductions

Energy cost savings are estimated using the market simulation model PLEXOS [20], an optimal power flow model based on a linearized DC-load flow [31].[3] In theory, such a market simulator could also calculate the other categories of benefits, but as explained below, either data or software limitations preclude such calculations at this time; we recommend that such capabilities be developed for future analyses.

7.3.1 Market Model: PLEXOS

PLEXOS simulates hour-by-hour bid-based dispatch by minimizing as-bid costs, and yields dispatch quantities, flows, costs, and LMPs. The general formulation can be summarized as follows:

---

[3]PLEXOS simulates hour-by-hour bid-based dispatch by minimizing as-bid costs, and yields dispatch quantities, flows, costs, and LMPs. The model has the capability to include individual facility limits, path limits, and linearized nomograms capturing stability constraints on operations. Although PLEXOS has the capability of optimally shaping non-pumped hydropower output over time, we took hydro schedules over the day and year as varying over time but not changeable, reflecting historical operating patterns. The amount and timing of pumped storage is optimized by simulating 24 hours simultaneously.

MIN   Sum of hourly generation and ancillary services costs (as bid) over 24 hours

subject to:

- Generation limits, including multi-fuel constraints, ramp rate limits, and random plant outages (using multiple Monte Carlo runs)
- Pump storage constraints, including environmental restrictions
- Spin and non-spin ancillary services
- Transmission limits, including thermal and SIL limits, interface (multiline) limits, phase shifters, and linearized nomograms capturing stability constraints on operations
- PDTF representations of line flows, based on line reactances[4]

Demand is assumed to be perfectly inelastic (fixed).

Thus, PLEXOS simulates a market in which ancillary services and energy are in equilibrium (or, equivalently, are co-optimized by an ISO). Oligopolistic behavior is simulated using exogenous bid adders, as described below, that are calculated as a function of system supply-demand conditions. Theoretically, an alternative is to calculate market power endogenously using PLEXOS' Cournot modeling capabilities [20], but that is not practically possible for a system with tens of thousands of buses, as in the western US. If bid adders are zero (cost-based bidding), then PLEXOS is equivalent to a perfectly competitive market equilibrium model in which generators are price takers. This is, in essence, an implementation of the famous Samuelson principle [39]: a perfectly competitive market (with no market failures) can be simulated by maximizing the sum of consumer and producer surpluses or equivalently, in the case of zero price elasticity of demand, minimizing the sum of production costs.

The WECC implementation of PLEXOS included:

- Calculation of flows on 17,450 lines
- Constraints upon flows on 3 DC lines, 284 high voltage (500 kV) AC lines, and 129 interfaces
- Calculation of prices at 13,383 buses
- Representation of 57 phase shifters (7 optimized, 50 fixed)
- Hourly dispatch of 760 generators over a 24-hour day
- Bids of California plants based on empirical *RSI*-based mark-ups, with other plants bid competitively
- Optimal operation of 8 pumped storage plants, and predetermined output schedules from 117 hydro plants

With so many power plants and line flows to simulate over 24 hours, PLEXOS constitutes a very large linear program which, however, can be solved using standard linear programming solvers.

---

[4]PLEXOS has the capability of simulating quadratic resistance losses [38], but this capability was not used in this analysis.

7.3.2 Project Description

The PVD2 project is a proposed 500 kV line that would provide additional interconnection between southern California and Arizona. If approved, the project could come online by 2009, increasing California's import capability from the southwest by at least 1200 MW. This is important because California depends on imports for more than 20% of its power needs. The CAISO recently used the TEAM methodology to identify and quantify the economic benefits of this line [32].

The idea for the PVD2 project originated in a regional planning process called the Southwest Transmission Expansion Plan (STEP) [33]. PVD2 is the third of fourth major project recommended by that process. In parallel with the STEP process, the Southern California Edison Company (SCE) determined that PVD2 was cost effective and filed a report requesting that the CAISO approve the project addition. The CAISO then undertook an independent economic study of PVD2 applying TEAM.

The location of the PVD2 project is shown in Figure 7.1. It includes the following facilities:

- A new 230 mile 500 kV overhead line between Harquahala Generating Company's Harquahala Switchyard (near Palo Verde) and SCE's Devers 500 kV Substation

- Rebuilding and reconductoring of four 230 kV lines west of the Devers substation

- Voltage support facilities in southern California

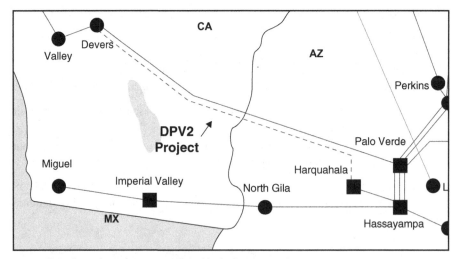

Figure 7.1   Location of proposed Palo Verde-Devers project

7.3.3 Input Assumptions

We conducted the energy benefits analysis for two future years, 2008 and 2013, using PLEXOS. Each hour of the year is simulated in PLEXOS, although a smaller number of runs could have been made while still spanning the range of possible system conditions. (The number of runs required to obtain an accurate estimation of production costs is an empirical question, and should be determined by experimentation for each particular system as a part of the study design.) We chose the years 2008 and 2013 because those were the only years for which vetted network and resource data were available from the Seams Steering Group (SSG) for WECC. It is common in long-range planning and market modeling studies to estimate benefits in five-year time steps because the planning cycle is often on the order of a half decade, and using a multiyear time step makes computation times more reasonable. Additional years could have been simulated if a simpler, unvetted network had been used; we decided that such an analysis would have raised other issues without significantly improving our understanding of the time distribution of benefits. All benefits are expressed in year 2008 dollars.

7.3.3.1 Transmission Consistent with the second TEAM principle ("full network modeling"), we studied the impact of the proposed PVD2 upgrade using a detailed transmission network model of the WECC (Figure 7.2). The model computed physical transmission flows, associated transmission charges, and nodal prices for each hour of 2008 and 2013 for the high-voltage WECC network. Constraints on flows were imposed for 284 500kV lines, two DC lines, and 124 interfaces (involving 468 lines), while flows were calculated for lower voltage lines. Flows were calculated for 17,500 lines of different voltage levels, allowing LMPs to be calculated for 13,400 buses in the system.

Consistent with the market design implemented in California in 2008, prices to California loads are based on zonal averages, while prices received by generators are bus-based.

7.3.3.2 Loads For loads outside California, WECC forecasts were used, and were disaggregated into hourly chronological load shapes for 21 regions and about 5700 locations (nodes). For California loads, we used the California Energy Commission (CEC) March 2003 forecast. From 2008 to 2013, overall energy growth in WECC is predicted to be about 1.7%/yr for the base case, and 1.4%/yr for the CAISO area. In 2013, the CAISO peak is 33% of the WECC peak.

7.3.3.3 Generation We obtained most of the system resource data from the SSG database. Their WECC database has about 800 thermal, hydro, pumped storage, and renewable generators with a total capacity of about 196,000MW in 2008 and 213,000MW in 2013. We added resources to the SSG database to reflect renewable portfolio standards in each of the states. Renewable resource additions included wind, solar, biomass, geothermal, and digester gas. We also added new gas-fired generation, primarily combined cycle plants, in each WECC area to attain a 15% planning reserve margin. The California gas-fired resources that we added on top of

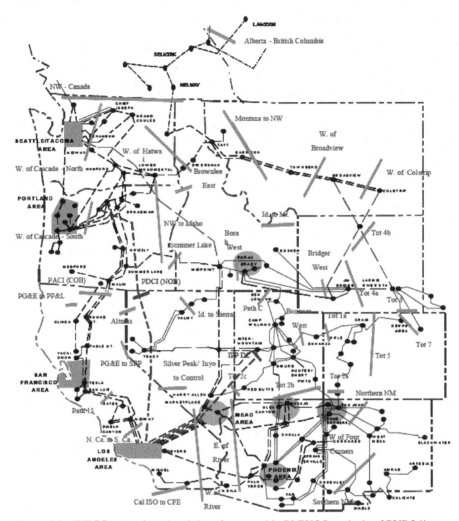

Figure 7.2   WECC network and path interfaces used in PLEXOS analysis of PVD2 line

the SSG additions were those that appeared to have a high likelihood of completion based on information compiled by the CEC.

The total CAISO resource capacity is 59,204 MW in 2008, and 64,447 MW in 2013. The WECC area planning reserve margin is 18% in both 2008 and 2013, although some regions are more resource rich than others; California in particular is projected to have lower reserves than most other regions.

The base gas price case was based on CEC forecasts [17], revised to reflect the gas price differential that existed at various city gates and gas pricing hubs as of August 2004.

7.3.3.4 Uncertainty Cases Consistent with fourth TEAM principle ("explicit uncertainty analysis"), the benefits of the line must be considered in the context of uncertainties that will unfold over the life of the project. We quantified the impact of this uncertainty by developing cases with different levels of input assumptions for load, gas prices, hydro conditions, and the exercise of market power. We believe that these cases cover a reasonable range of possibilities. We then calculated expected benefits across these cases taking into account their probabilities. In addition, we consider the line's "insurance benefit" by calculating benefits under various possible contingencies. Sixteen combinations of transmission and/or generation outages were considered as contingencies.

In the expected benefit calculation, we focused on the four key variables just mentioned, defining 17 combinations for each year. For the cases where we varied load, gas price, and market power, we examined three levels: very high (H), base (B), or very low (L). For the hydro cases, we also examined three levels: wet (W), base (B), or dry (D) year.

We determined the values of the demand and gas price cases by analyzing the historical accuracy of predictions of those variables, comparing CEC forecasts of loads and prices over the past 20 years [17] to their actually realized levels. Load distributions are characterized using normal distributions fitted to the historical forecast errors, while gas prices follow a log-normal distribution. The L and H levels used in the load and gas sensitivity cases are based on 90% confidence intervals from their distributions. For loads, those levels vary only slightly from the base case, while for gas and mark-up, the differences are large (Figure 7.3).

We took hydro ranges from 80 years of historical hydro production records. Derivation of the bid mark-up uncertainty cases is discussed in the subsection on market pricing, below.

The $3 \times 3 \times 3 \times 3 = 81$ possible combinations of values for the four uncertain variables are too many to simulate. Therefore, we considered a small but representative subset of the cases in the expected benefits calculations:

1. Base values for all four variables (one case).

2. Base values for three of the four variables, and the low value for the fourth variable (four cases).

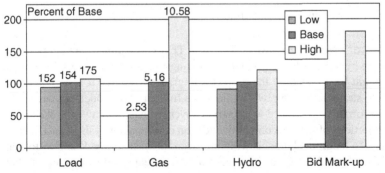

Figure 7.3 Comparison of very low, base, and very high assumptions, 2013 (WECC Peak Load in GW; Gas price, WECC Annual Average in $/MMBTU)

3. Base values for three of the four variables, and the high value for the fourth variable (three cases; the high load case with base values for other variables is not considered).

4. Additional cases representing plausible combinations of extreme scenarios such as a high stress condition (high load, high gas price, dry hydro, high market mark-up), economic boom (high load and gas prices), or recession induced by high fuel prices (low load, high gas price). Another consideration in selecting these cases was to make it possible for probabilities to be chosen so that the means and standard deviations of each of the individual variables matched the assumptions, and for correlations to be reasonable (for instance, we expect a positive correlation between dry conditions and high demand due to warm temperatures) (nine cases). Latin-hypercube sampling, which has been used in another TEAM application [13], could also have been used in the PVD2 study to select additional scenarios, but the 17 selected scenarios were considered to be sufficient.

Table 7.1 shows the selected 17 cases for 2008.

After choosing the cases, it is necessary to determine the probability that each will occur in the future. Each case is a realization of the various dimensions of uncertainty in future system conditions. However, the input data described above only provides an estimate of the marginal distribution of each of these dimensions. For example, we have information on the marginal distributions of future hydro conditions and gas prices, but not their joint density. Consequently, we must pick values for the joint probability of each set of future system conditions. We choose these probabilities using a nonlinear program that maximizes the logarithm of likelihood (the sum of the logarithm of the joint probabilities) of observing the 17 scenarios subject to the constraint that the joint probabilities replicate the first two moments of the marginal distribution of each variable. Mathematically, we choose the Pr_i for cases $i = 1, 2, \ldots, 17$ to maximize $\Sigma_i \ln(Pr_i)$ subject to the constraints:

- $\Sigma_i Pr_i = 1$, and
- the mean and standard deviation for each variable implied by these joint probabilities match the assumed values for the marginal distribution of each variable.

Table 7.1 shows the resulting probabilities.

7.3.3.5 Market Price Derivation

The third TEAM principle ("market prices") requires that energy prices be projected considering the potential for market power and how it might be affected by the proposed upgrade. Although it is a great challenge to model strategic bidding by suppliers in a full network model, we were able to rely on California's experience with markets over the past seven years. We chose to ground projections of market competitiveness on empirical analysis of past behavior, as opposed to theoretical models with unproven forecast ability. Using historical data, we were able to demonstrate a stable predictive relationship between market price-competitive price mark-ups and key variables that measure system supply/demand conditions. This regression approach may lack the rigorous foundation in economic theory that characterize other studies [24–26], but its simplicity and robustness together with its ability to capture the impact of system conditions and competitor behavior on prices on an hourly basis make it a useful tool here.

We estimated this mark-up relationship from observed data during two critical periods: from 1999 to 2000 when suppliers had few long-term commitments to supply energy to load, and the year 2003 when some suppliers had large long-term contractual commitments.[5] We estimated regressions predicting how hourly prices are marked up over the variable cost of the highest variable cost unit operating during that hour for every hour in each of three California regions (south, central, north), based on the amount of supply relative to demand, accounting for potential import quantities into that zone. These estimated relationships allow us to build a dynamic bid mark-up mechanism into PLEXOS in which suppliers' price bids are determined by their variable costs and the mark-up over these costs implied by the relationship relevant for this generation unit. More importantly, because this mechanism varies the bid mark-up with hourly system conditions, we can capture the impact of major transmission upgrades, such as PVD2, on import capability into the CAISO control area, thus reducing the ability of suppliers in the CAISO control area to bid above their variable cost. After incrementing bids by the mark-ups implied by these estimated relationships, we then ran PLEXOS to obtain market prices, which were then used in our assessment of energy benefits.

In the mark-up and system and market conditions relationship, mark-ups were expressed as a Lerner index $(P_a - P_c)/P_a$, where P_a represents the actual observed price and P_c is the price that would result from price-taking behavior by suppliers. The *RSI* is the variable in this relationship that can change as a result of a transmission upgrade. The *RSI* is defined as the ratio of total market supply minus the supply from the largest firm, divided by the load. Only flexible supplies were included, netting out obligations to one's own load and contractual obligations. Likewise, the denominator excluded such obligations from the load.[6] *RSI* < 1 indicates that the largest supplier is pivotal because system demand cannot be met without this supplier producing some energy regardless of the amount of energy produced by its competitors. When these circumstances occur, the pivotal supplier can name the price at which it would like to supply this electricity and be assured that it will receive this price. CAISO experience indicates that values of *RSI* less than 1.2 are associated with significant mark-ups [23].

7.3.3.5.1 *An Example*   An example of a regression relationship used is:

$$(P_a - P_c)/P_a = 0.14 - 0.53\text{RSI} + 0.65\text{LUH} + 0.086\text{D}_{\text{peak}} + 0.15\text{D}_{\text{sum}}$$
$$(0.013)(0.0073)\ (0.0092)\quad (0.0036)\quad (0.0031) \tag{7.5}$$

where LUH is the fraction of the load that is unhedged, D_{peak} is a binary variable indicating whether the hour occurs during the peak period (1 = yes, 0 = no), and

---

[5]Even though these were very divergent periods, the relationships were stable over time, giving us confidence in their usefulness.

[6]There are must-run and must-take generators in California that are required to run regardless of market prices because of local reliability constraints or contractual obligations that predate the start of the California market.

D_{sum} is a binary variable indicating whether it is summer. All of the parameters estimates are very large relative to their standard errors, shown in parentheses under the coefficients. The data used to estimate the regressions consisted of 31,333 hourly observations from November 1999 to October 2000, and from January to December 2003. The fit ($R^2 = 0.46$) is close to that of models of the Italian market (e.g., $R^2 = 0.61$ for ENEL's mark-ups in Sicily) [30].

Because our regression specification is used to derive future market prices for all the various scenarios considered, it is important to test the model's validity. For this purpose, we estimated several different specifications (linear, nonlinear, and with different sets of variables) and compared their predictive ability using an out-of-sample test. First, we divided the entire sample into two parts: an in-sample data set and an out-of-sample data set. The out-of-sample set consists of hourly data for a total of 60 days in 2003 (5 days for each month in 2003). The in-sample set consists of the remaining 2003 data along with the 1999–2000 data. Using the in-sample data set, we generated regression estimates for each regression specification. The specifications differed in terms of which variables were considered and the inclusion of nonlinear terms for RSI. Then, for each specification, we computed the projected Lerner Index for the out-of-sample data. Finally, we compared the projection results from each specification with the actual Lerner Index, and chose the one that generated the best out-of-sample fit. The linear specification (1) performed best. Thus, on the basis of both predictive power and simplicity, the simple model is preferred here; however, in other circumstances, more complex specifications may perform better.

The estimated relationship (7.5) was used obtain bid mark-ups for use in PLEXOS by inserting the appropriate values for the independent variables for each hour and each zone into the equations, rescaling them so that larger suppliers had higher mark-ups.[7] The PVD2 addition of 1200 MW in each direction increased estimated total market supply in Southern California, yielding a higher RSI for that region and, as a result, lower values of $(P_a - P_c)/P_a$ because of the negative coefficient for the RSI variable in (7.5).

To account for uncertainly in mark-ups implicit in our regression, we used ranges of mark-ups derived from the distribution of the error term in (7.5). In particular, we calculated the mark-ups used in particular scenarios as follows:

$$(P_a - P_c)/P_a = \text{MAX}[0, f(RSI, LUH, D_{peak}, D_{sum}) + t_{value}S] \qquad (7.6)$$

where $f()$ is the function in (7.5); S is the standard deviation of the error term in (7.5); and t_{value} is chosen to represent a particular mark-up scenario. For the L mark-up scenario, a t_{value} corresponding to the lower 90% confidence interval (-1.645) was

---

[7] Instead of applying the same bid-cost mark-ups to all strategic suppliers in the same region, we used a "proportional mark-up" approach, assuming that the largest supplier had the highest bid-cost mark-up in the region. According to the supply function equilibrium model [34], the price mark-up of a supplier is proportional to the quantity it supplies and inversely proportional to the sum of residual supply elasticity and absolute value of demand elasticity. This indicates that the largest supplier has more incentive than other suppliers to mark-up its bid. The same implication can be also drawn from Cournot-type models [22]. Thus, we scaled the result of (2) by the ratio of each supplier's uncontracted capacity to the uncontracted capacity of the largest supplier.

used, while for the H mark-up case, the upper 90% limit (+1.645) was applied. For the B mark-ups, $t_{value} = 0$.

7.3.3.5.2 Project Costs SCE estimated the capital cost of the PVD2 upgrade to be $680 million, including allowance for funds used during construction, assuming an in-service date of early 2009. In 2008 dollars, this was $667 M, based upon a 2% inflation rate. This is about $2.5 M/mile. These capital costs were then converted to an equivalent stream of annual revenue requirements. We estimate that the levelized revenue requirement for the PVD2 project will be $71 million per year for 50 years, assuming a real carrying charge of 10.43%/yr, accounting for taxes and administrative costs and adding fixed operating costs. This is the value that we compare the benefits to in order to determine the economic viability of the project.

7.3.4 Results

As noted at the start of this section, we made estimates of five benefit components: (1) energy savings; (2) operational benefits; (3) capacity savings; (4) system loss reductions; and (5) emission reductions. We derived the energy savings using the PLEXOS market simulation model. We estimated operational benefits, capacity savings, system losses, and emission benefits separately, outside of the market modeling process. Detailed results are available in [32].

7.3.4.1 Benefit Category 1: Energy Savings Energy savings are based on differences between generation costs and prices calculated with and without the proposed PVD2 upgrade. For market-based pricing scenarios, PLEXOS was solved by inserting bid functions for California independent power producers (constructed using the supplier's variable cost and the bid mark-ups implied by (1), (2)) and production (variable O&M) costs for everyone else into the objective function. However, costs for the purposes of the societal benefits calculations are based on assumed fuel costs, not as-bid costs.

To perform the expected benefits calculation, we evaluated the benefits for 17 different cases for each of the years 2008 and 2013 (Table 7.1). Each case is composed of two simulations, "without" and "with" the proposed PVD2 upgrade. As mentioned, we also considered a set of 16 contingency cases, representing extreme events for which it is difficult to assign a probability.

Consistent with the first TEAM principle ("benefit framework"), we quantified the benefits from four perspectives:

- *Societal.* Represents the WECC production cost savings resulting from adding the transmission upgrade. The total WECC benefit is also equal to the sum of the consumer, producer, and transmission owner benefits.
- *Modified Societal.* Represents the enhancement to overall market competitiveness in the WECC resulting from the upgrade. This is the same as societal benefits, except that producer benefit includes the net generator revenue from competitive prices only, and excludes generator net revenue from uncompetitive market conditions (i.e., bid mark-ups).

- *CAISO Ratepayer (LMP Only)*. Demonstrates whether benefits outweigh costs for CAISO ratepayers. This perspective is used to decide whether ISO ratepayers should fund the transmission expansion. This calculation is based on locational marginal pricing, and the congestion revenues that such pricing would imply throughout the WECC.

- *CAISO Ratepayer (LMP + Contract Path)*. Same perspective as above but the flow-based or LMP market is modified to reflect actual transmission pricing rules for selected contractual paths between CAISO and the Southwest region, rather than congestion pricing.

PLEXOS' geographic detail makes finer breakdowns possible, for example, by individual generating company or state. The focus here, however, is on the breakdown between California and the rest of the western US.[8]

The CAISO Ratepayer (LMP Only) analysis is performed assuming congestion revenue is based on WECC physical flows. An important assumption is that locational marginal pricing will be uniformly implemented by all WECC entities. However, this pricing mechanism may not be implemented in the immediate future. At present, most of the WECC instead operates based on contract path scheduling.

The distinction between LMP and contract path-based pricing is important. The CAISO Ratepayer (LMP Only) computes transmission congestion revenue for each line in the WECC. In some cases, this congestion revenue can be very high; the PLEXOS simulations show that the upgrade would lower those revenues. However, today some congestion is actually managed in real-time, resulting in uplift charges to load rather than congestion revenue to transmission owners. The net result is that the LMP method as applied to the CAISO Ratepayer perspective exaggerates the amount of congestion revenue that California transmission owners would receive, which turn inflates the loss of congestion revenue in today's environment due to the upgrade. This means that the LMP Only approach understates the net benefits to California consumers, since lower congestion revenue means that transmission owners must recover more of their fixed costs from load.

The CAISO Ratepayer (LMP + contract path) perspective corrects this problem by adjusting transmission congestion revenue both before and after the upgrade. The net impact of the adjustment was usually an increase in transmission upgrade benefits for the CAISO ratepayers, more closely reflecting the upgrade benefits that ratepayers would receive under present WECC scheduling rules.

Table 7.2 summarizes the energy benefits for 2008 and 2013 from these four perspectives. (Table 7.1, above, presented values for individual uncertainty cases for

---

[8] It is crucial in any benefit-cost analysis to avoid double-counting of benefits. For instance, consumer expenditures on energy need to be adjusted downwards for any increases in congestion revenues as a result of the transmission change because such charges are refunded as decreases in transmission portions of consumer bills. Such adjustments are also made for changes in the transmission loss surplus (which is also returned to consumers) and for changes in profits earned by regulated utility-owned generation (which, under average-cost regulation, are, in effect, returned to consumers). If demand is perfectly inelastic, then the decrease in WECC production costs should equal the sum of producer and consumer benefits, properly accounting for these refunds; this check was made to ensure that double-counting did not occur.

TABLE 7.2   **Estimated energy benefits (millions Per Year, 2008 dollars)**

Perspective	Expected value, 2008	Range across cases, 2008	Expected value, 2013	Range across cases, 2013
Societal	$41	$4–$200	$54	$20–$200
Modified Societal	$61	$6–$400	$81	$20–$600
CAISO Ratepayer (LMP)	$39	–$3–$300	$56	–$3–$400
CAISO R.P. (LMP + contract path)	$110	$10–$600	+$200	$50–$1,000

2008.) For perspective, the values shown in these tables can be compared to total power costs for the CAISO system. For 2013, we estimate the total wholesale energy costs to be about $12 B, about two orders of magnitude larger than these benefit estimates.

The table shows several interesting results. Consider for instance the 2008 benefits. Societal benefits (cost savings throughout the west), by coincidence, almost precisely equal CAISO ratepayer benefits (LMP). Societal benefits are $20 M higher if decreases in "market power"-based profits are disregarded. However, considering how transmission of imports to California is actually priced, CAISO ratepayer benefits ($110 M) are almost three times the societal benefit of $41 M. This means that independent generators in California along with ratepayers and generators in other states appear to suffer a decrease of $69 M in their benefits.

The benefits in Table 7.2 cannot be directly compared to the annual costs since they have not been levelized over the 50-year project life. Nor do they include the other benefits described later in this paper. To obtain a levelized annual benefit, we need to assume a discount rate and to extrapolate benefits beyond 2013 through the remainder of the 50-year project life. A real discount rate of 7.16% was used based on SCE's weighted cost of capital. A 1%/yr real escalation rate for benefits was selected for the period after 2013. The main reason is that most of the commodity costs that are a factor in setting market-clearing prices are likely to escalate in real terms in the long run (natural gas, labor, steel, concrete, land, emission offsets, etc.). The resulting levelized energy benefits are shown in Table 7.3. Assuming zero rather 1% escalation decreases both Societal and CAISO Ratepayer Benefits (LMP only) by about 5 $M/yr.

7.3.4.2 Uncertainty in Energy Benefit Estimates
The ranges of benefits shown in Table 7.2 provide perspective on how uncertain the benefits are for the four perspectives, but they provide no information regarding the relative likelihood of different levels of benefits. Since we assigned probabilities to many of the cases (e.g., Table 7.1), we can use that information to characterize the distribution of benefits. In Figure 7.4, we illustrate the relative probabilities of various benefit ranges for the CAISO Ratepayer (LMP Only) perspective in 2013. The highest benefits resulted from those cases where several adverse events occur simultaneously, such as high load, gas price, and market power together with dry hydro (Table

TABLE 7.3   Derivation of PVD2 benefit-cost ratios (expected levelized value, millions per year, 2008 dollars)

Component of B-C ratio	Societal	Modified societal	CAISO ratepayer (LMP only)	CAISO ratepayer (LMP + contract path)
Levelized benefits				
1. Energy	$56	$84	$57	$198
2. Operational	$20	$20	$20	$20
3. Capacity	$12	$12	$6	$6
4. System Loss	$2	$2	$1	$1
5. Emissions	$1	$1	$1	$1
Total	$91	$119	$84	$225
Levelized costs	$71	$71	$71	$71
Benefit-cost ratio	1.3	1.7	1.2	3.2

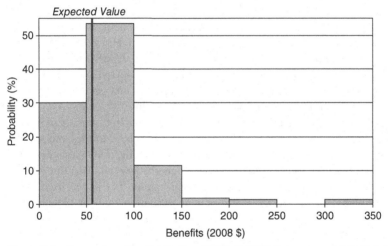

Figure 7.4   Energy benefits distribution (2013, CAISO Ratepayer—LMP Only)

7.1). There is a 70% chance that the annual energy benefits in 2013 exceed $50 million. There is a 5% probability that the project would yield an annual ratepayer benefit between $150 and $350 million, indicating that PVD2 would provide significant insurance value against extreme events.

We now ask: which uncertainty (load, gas price, hydro, mark-up) affects benefits the most? One way to answer this is to compare cases that differ in just one variables. For instance, we can compare different gas cases (BLBM, BBBM, and BHBM, Table 7.1). There are essentially no benefits to CAISO ratepayers (LMP only) if gas prices are low (BLBM), while the highest gas prices (BBBM) yield almost $80 M of benefits in 2013. This latter amount is roughly equivalent to a

$20/MWh price difference between coal and gas-fired power for 1000 MW of imports for half of the year; clearly, imports from coal-burning regions are more valuable if gas prices are higher.

Meanwhile, comparing Case BBBB with BBBH in Table 7.1 shows that moving from a moderate to a high mark-up increases the societal benefits by only 1.7 $M/yr (compared to a $45.3 M/yr base), but changes the California ratepayer benefit by an order of magnitude more (from 98.7 to 124.5 $M/yr, for the LMP + contract path metric). This can be interpreted as follows. The PVD2 project helps mitigate market power in California by bringing in competitive supply, and these benefits are greater if more market power is exercised. The benefits accrue primarily to California ratepayers, in the form of lower bills; from a societal point of view, however, those benefits are largely offset by a loss of producer surplus (profit), so that the effect on net societal benefits (fuel savings) is smaller. (This conclusion is borne out by the result that California ratepayer benefits always exceed societal benefits in Table 7.1, implying that some other parties will be worse off if the line is built.)

A more systematic way to explore the effect of the uncertainties is to perform a linear regression of the benefit estimates in Table 7.1 against the uncertain variables (coded as L = 1, M = 2, and H = 3). For societal benefits (SB) and California ratepayer benefits (CRP) (based on LMP + contract path), we get:

$$SB = 6.7 + 5.7LD + 35.5GP - 32.9HY + 8.0MU, R^2 = 0.89 \tag{7.7}$$
$$(26)(5.9) \quad (5.8) \quad (6.7) \quad (7.3)$$

$$CRB = -184.2 + 63.9LD + 98.3GP - 80.0HY + 60.9MU, R^2 = 0.82 \tag{7.8}$$
$$(118.1)(27.0) \quad (35.5) \quad (29.6) \quad (32.7)$$

where LD = load, GP = gas price, HY = hydro, and MU = mark-up. The numbers in parentheses are the standard errors of the coefficients. All the coefficients have the anticipated signs: benefits increase when load, gas prices, and mark-ups are higher, and decrease when there is more hydropower. At a 5% level of significance (one-tailed test), only GP and HY significantly affect societal benefits, but all uncertainties significantly affect California ratepayer benefits. Note that the four variables have the same order of effect on ratepayer benefits. For instance, going from M to H load increases CRB by 63.9 $M/yr, while going from M to H mark-up increases CRB by 60.9 $M/yr. This result highlights the importance of TEAM Principle 4: the need to consider market-based pricing assessments of transmission benefits.[9]

Not considered in the expected value calculations are the 16 contingency cases in which losses of transmission or generation capacity stress the system. With one exception, each contingency case results in benefits to CAISO ratepayers (LMP + contract path) of over $100 M/year, if the contingency is assumed to last the entire year

---

[9] As another indicator of the importance of market power mitigation benefits, we can compare solutions based on no mark-up (marginal cost bidding) [32] with the solutions in Table 7.1. That comparison shows that the market power case yields 6% higher societal benefits and 92% higher CAISO ratepayer benefits (LMP only) (39% higher if LMP + contract path), assuming the B case for all four uncertainties. However, the percent increase in benefits resulting from considering market power mitigation is appreciably higher under "high stress" conditions (i.e., H loads and gas prices, with dry hydro conditions).

[32], assuming base load, gas price, and hydro conditions. Under other conditions, the benefits can be even higher. This indicates that the insurance value of the PVD2 line would be even greater than indicated by the right hand tail of Fig. 7.4.

7.3.4.3 Benefit Category 2: Operational Benefits

Production cost simulations may not capture all the operational costs that are incurred in managing the electric grid. This is especially true if generation unit commitment costs and ramp rate limits are not explicitly modeled, as in the case of PLEXOS. Thus, costs required to meet an N-1 and relevant N-2 planning contingency criteria will be underestimated. This implies that some operational benefits of the PVD2 upgrade may be overlooked. Including such constraints in large network models may be possible in the future, in which case these benefits would automatically be incorporated in the energy benefits of Tables 7.1–7.3.

For contingencies that do not involve the outage of the PVD2 line, the extra import capacity on the new line reduces the need for internal CAISO on-line generation. Regarding PVD2 line outages, the CAISO operators tell us that they keep a number of units on minimum load to protect against an outage of the present (PVD1) line. In addition to committing units, and the corresponding payment of minimum load cost compensation (MLCC), re-dispatch of units is needed to address real-time congestion which is not resolved in Day–Ahead congestion management. To estimate these operational benefits, we performed a detailed review of historical MLCC and real-time redispatch costs. Accounting for other upgrades that are being implemented, we estimate that the PVD2 upgrade would result in the following reductions: 5.3% of MLCC associated with the Southern California "SCIT" nomogram; 22.5% of the system MLCC, 72% of the nuclear MLCC, and about 12.5% of the re-dispatch cost, resulting in a total annual savings of $20M in 2008 dollars.

7.3.4.4 Benefit Category 3: Capacity Benefit

One approach to analyzing transmission-generation interactions that is consistent with the fifth principle of TEAM (address resource interactions) is to add generation where simulated energy prices indicate it is profitable, and then recalculate the market equilibrium. Such an "endogenous generation investment" analysis was undertaken in the CAISO's application of TEAM to Path 26 [13]. Alternatively, scenarios of changes in generation siting that are broadly consistent with how a transmission investment would change investment incentives could be used, which was done in the Sunrise analysis in Section 7.4, below.

In the PVD2 study, a simpler approach was taken to assess changes in generation investment and the resulting benefits. Because sensitivity analyses showed that energy prices in both California and external markets would not be significantly affected by shifts in generation investment that might occur as a result of installing that line, the energy market benefits would not be altered if generation investment was modeled as endogeous. Therefore, so that study resources could be focused on other issues, the energy market analysis was based on simpler siting assumptions that were the same with and without PVD2. Then a separate analysis estimated the capacity cost savings that would result from shifting an amount of generation investment equivalent to PVD2's firm capacity from southern California to Arizona.

We derived capacity benefits using the assumption that California will continue to have a resource adequacy requirement and that Arizona can be the source of contracted capacity to serve California load. A key assumption for these savings is that the future cost of capacity in Arizona will be less than the cost in California for two reasons: lower capital and fixed operating costs for peakers and, for the early years of the project, a greater resource surplus in Arizona than in California. We expect the demand for capacity, and the resulting price, to be less in Arizona.

We estimate that the differential fixed costs for peakers to be $15/kW/yr in 2008 dollars. If we further assume that firm summer capacity is available for the entire 1200 MW upgrade, the capacity benefit would be $18 M million per year in 2008 dollars. To be conservative, we discount this amount by one-third, and further assume that the benefits will be split equally between the buyers and sellers of capacity. Thus, we estimate a societal benefit of 12 $M/yr and a CAISO ratepayer benefit of 6 $M/yr.

7.3.4.5 Benefit Category 4: Loss Savings PLEXOS used a linearized DC power flow model without losses, so loss savings are omitted in the energy savings of Tables 7.2 and 7.3. (A version of PLEXOS is available that considers losses [20], but was not applied here.) In practice, we expect PVD2 to decrease transmission losses. To estimate loss savings, we used the computed power flows before and after the upgrade, yielding an estimated reduction in losses worth $2 million annually. This estimate implicitly accounts for the interplay between increased losses due to heavier power transfers, and loss reduction due to redistribution of these power flows among existing and new transmission paths.

7.3.4.6 Benefit Category 5: Emissions The PVD2 upgrade allows more efficient Arizona gas-fired generation to displace less-efficient and higher-emission California gas-fired generation. In theory, NO_x allowance prices should depend on energy market conditions. But PLEXOS does not presently simulate the NO_x allowances markets in the WECC, in part due to a lack of emission rate data. Therefore, the results of the model were subjected to post-processing to estimate how much NO_x emissions would decline as a result of the upgrade. Based on the generation shifts, we estimated a NO_x reduction of 390 tons per year, which at typical allowance prices is worth $2.2 million/yr. Half that amount is considered a CAISO ratepayer benefit.

7.3.4.7 Summary of Results In Table 7.3, we summarize our findings and determine an overall benefit-cost ratio for the societal, modified societal, and CAISO ratepayer perspectives. The ratios are positive in every case, but most strongly so for the last perspective (CAISO ratepayer, considering contract path effects). These values depend on the assumed scenarios and their probabilities; as Table 7.1 shows, there is considerable uncertainty concerning these benefits, implying some probability that benefits in any particular year might be less than the cost.

On the other hand, the calculations in Table 7.3 also do not consider the generator and transmission contingency cases, which, as indicated earlier, provide additional insurance value.

7.3.5 Resource Alternatives

Consistent with fifth TEAM principle ("resource alternatives"), we need to consider alternatives to the project in the form of generation (both renewable and fossil-fueled), demand-side management (DSM), and transmission resources.

DSM and renewables are, however, not viewed as alternatives. To the extent that demand-side management (DSM) or renewable resources are technically and economically feasible, these resources should be fully developed. Only when contributions from DSM and renewable resources are maximized should traditional resources be considered. Hence, we focused on thermal generation and transmission alternatives.

In today's market, the most likely generation alternative is a new combined-cycle (CC) generating plant. The question for this analysis is whether the CAISO should promote the PVD2 upgrade, or recommend building new CC's in the CAISO area, or both. An analysis of CC construction costs, based on an assumption that fixed costs would be 10% less in Arizona, shows that when combined with the levelized cost of the PVD2 upgrade, an Arizona facility is 10% more expensive than one in California. At a 90% capacity factor, the Arizona facility is 4% more expensive. By itself, though, this information is incomplete. Other important factors include interconnection costs for fuel and transmission—which will be significantly greater in California—and the limited ability to site resources in CAISO urban areas due to siting opposition. Thus, we believe that local generating options as well as transmission solutions need to be aggressively pursued. Building PVD2 does not preclude the construction of local facilities, as California needs to add 5000 MW or more in the next five years due to load growth and generation retirement.

Turning to transmission, the Southwest Transmission Expansion Plan [33] evaluated 26 potential transmission upgrade plans during 2003. Six alternatives were subjected to further technical and economic analysis. The PVD2 500 kV line was a component of two of those. The analysis concluded that three other alternatives were not viable due to reasons such as lack of project sponsorship, inadequate technical performance, or poor economics. The last of the six alternatives included a variant of the PVD2 line with alternative termination points. We expect that none of these variants to significantly change the scope of the proposed PVD2 project.

During the TEAM review process, some stakeholders suggested an alternative ("EOR9000") that involved upgrading series capacitors on the Perkins-Mead and Navajo-Crystal 500 kV lines between Arizona and Nevada. We ran PLEXOS sensitivity cases with EOR9000 and found that it and PVD2 are complements rather than substitutes. That is, each generally increases the benefits of implementing the other.

7.4 RECENT APPLICATIONS OF TEAM TO RENEWABLES

The most recent applications of the TEAM methodology illustrate its flexibility. It has been used to evaluate proposed transmission additions designed to deliver California renewable energy sources, including the Sunrise and Tehachapi projects [35, 36]. California has ambitious target of producing 20% of its energy from renewable sources by 2010 and 33% by 2020. New transmission infrastructure appears necessary to bring that energy to market.

The scope of these applications was more restricted than the PVD2 study because these were internal California projects, unlike PVD2 which was designed to import power from the Southwest. The most restrictive was the Tehachapi study; in that case, the relatively low cost of the wind resource being accessed meant that the study could be framed as a cost-effectiveness study (how best to access a resource that would be developed in any case), without having to consider generation alternatives. Furthermore, market power effects would not differ among the alternatives, since the same amount of power would be brought to market. On the other hand, in the Sunrise case, the project would allow external resources to substitute for costly new turbine-based generation within the San Diego load pocket. Therefore, Principle 5 (transmission-generation-load management substitution) became more important, and that TEAM analysis was more involved.

7.5 CONCLUSION

Based on our application of TEAM to the Palo Verde-Devers 2 transmission line proposal, we conclude that the methodology and its five guiding principles have substantially enhanced the CAISO's ability to fulfill its responsibility to evaluate and recommend transmission expansion projects.

The results of the case study demonstrate that the methodology produces the comprehensive analytical information that project proponents and review authorities need to make informed decisions in shaping California's transmission infrastructure. The TEAM approach advances this objective by creating a framework to examine a project from multiple viewpoints—from those of the overall western interconnection, to the consumer or transmission line owner. Equally important, the methodology provides a flexible mechanism to identify a range of risks and rewards associated with the project under diverse contingency and market conditions.

The PVD2 application of the TEAM methodology shows that a significant amount of the benefits of a transmission line can arise from market power mitigation by making markets more accessible. It may, in theory, be possible to obtain mitigation benefits by instead regulating generator bidding more stringently without incurring the investment cost of a line. However, we believe that in the long run, addressing the structural issues that create market power (e.g., grid infrastructure) is a superior approach to relying on regulatory intervention. While it is true that in the absence of adequate infrastructure improvements, long-term market power concerns would likely be addressed by more stringent market power mitigation provisions, such provisions may not be very effective (e.g., the California Experience). Moreover, relying upon excessive market power mitigation rules to compensate for infrastructure deficiencies may have other detrimental impacts in terms of discouraging new generation investment or demand response. In light of this, we believe it is useful to examine the market power mitigation benefits of a transmission project under the assumption that market power mitigation rules in the absence of the project would be essentially the same. However, we also recommend that the benefit cost analysis include cost-based bidding scenarios as well so that policy makers can consider the sensitivity of the results to market power assumptions. If a project cannot be justified based on the cost-based bidding scenarios but can be based on the market power scenarios, the policy maker can weigh this information based on their particular policy objectives with regard to fostering market competition and their confidence in the effectiveness of any current or future market power mitigation rules. A policy environment that is more oriented toward regulation could always decide against a transmission project that cannot be justified under cost-based bidding scenarios, whereas a policy environment more oriented towards developing highly competitive wholesale energy markets and minimizing regulatory intervention may decide differently.

An important question is: what is the practical effect of the large modeling effort required by TEAM? For the PVD2 study, this can be gauged by comparing the average benefits, which consider the results of multiple scenarios and the market power analyses, with the benefits under the base scenario without market power. The latter benefit estimate can be viewed as an approximation of what a simpler analytical effort might yield. The expected benefits to CAISO ratepayers (LMP only) from the full analysis ($39 M) is twice the results of the scenario with no market power and base hydro, gas, and load values ($20 M, [32, Table H.1]). Given that most of the benefit-cost ratios for the line were less than 2 (Table 7.3), this shows that the effort expended to consider uncertainty and market power made an important difference in the PVD2 analysis.

Although greater transparency and more careful analysis may increase public understanding and acceptance of transmission proposals, it does not guarantee that beneficial proposals will be approved. Indeed, despite the societal and CAISO benefits of PVD2, the Arizona Corporation Commission declined to approve it in May 2007 because it perceived that Arizona consumers would not benefit from the line. The TEAM methodology's emphasis on the distribution of benefits informed these and other proceedings, and will likely contribute to future consideration of cost-sharing arrangements for the proposed facility. That the line has an overall positive societal net benefit implies that such an arrangement should be possible that benefits both Arizona and California ratepayers.

ACKNOWLEDGMENTS

B.F. Hobbs was partially supported by NSF grant ECS-0621920. We gratefully acknowledge helpful comments by the referees.

REFERENCES

1. National Transmission Grid Study. U.S. Department of Energy, Washington, DC, May 2002.
2. Baldick R, Kahn E. Transmission planning issues in a competitive economic environment. *IEEE Transactions on Power Systems* 1993;8(4):1497–1503.
3. Buygi MO, Balzer G, Shanechi HM, Shahidehpour M. Market-based transmission expansion planning. *IEEE Transactions on Power Systems* 2004;19(4):2060–2067.

4. de la Torre T, Feltes JW, San Roman TG, Merrill HM. Deregulation, privatization, and competition: transmission planning under uncertainty. *IEEE Transactions on Power Systems* 1999;14(2):460–465.

5. Fang R, Hill DJ. A new strategy for transmission expansion in competitive electricity markets. *IEEE Transactions on Power Systems* 2003;18(1):374–380.

6. Hirst E. U.S. transmission capacity: a review of transmission plans. *The Electricity Journal* 2004;17(7):65–79.

7. Hirst E, Kirby B. Expanding transmission capacity: a proposed planning process. *The Electricity Journal* 2002;15(8):54–59.

8. Shahidehpour M. Investing in expansion, the many issues that cloud transmission planning. *IEEE Power & Energy Mag* 2004;2(1):14–18.

9. Thomas RJ, Whitehead JT, Outhred H, Mount TD. Transmission system planning—the old world meets the new. *Proceedings of the IEEE* 2005;93(11):2026–2035.

10. Wolak FA, Barber B, Bushnell J, Hobbs BF. Comments on the London Economics methodology for assessing the benefits of transmission expansions. Opinion of the California ISO Market Surveillance Committee, Oct. 7, 2002, www.caiso.com/docs/09003a6080/1b/4f/09003a60801b4fa9.pdf

11. Consortium of Electric Reliability Technology Solutions. Economic evaluation of transmission interconnection in a restructured market. Prepared for the California Energy Commission, 2004.

12. U.S. Department of Energy. Transmission Bottleneck Project Report. Washington, DC, March 2003.

13. California ISO, Transmission Economic Assessment Methodology (TEAM), Folsom, CA, June 2004. www.caiso.com/docs/2003/03/18/2003031815303519270.html

14. Sheffrin AY. Wires in the air: five key principles for proving economic need. *The Electricity Journal* 2005;18(2):62–77.

15. Awad M, et al. The California ISO Transmission Economic Assessment Methodology (TEAM): Principles and application to path 26. In: Proc. 2006 IEEE PES General Meeting, Montreal, Canada, June 2006.

16. California Public Utilities Commission. Order instituting rulemaking on policies and practices for the Commission's transmission assessment process. R.04-01-026. San Francisco, USA, 2003.

17. California Energy Commission, 2003 Integrated Energy Policy Report. Nov. 12, 2003.

18. Robinson CF. Comments of the California independent system operator on the proposed opinion on methodology for economic assessment of transmission Projects. Investigation 05-06-041. California Public Utilities Commission, July 10, 2006.

19. GE Energy, MAPSTM software—for informed economic decisions. 2007, www.gepower.com/prod_serv/products/utility_software/en/downloads/10320.pdf

20. PLEXOS for Power Systems. Adelaide, Australia, 2007, www.plexossolutions.com, www.plexos.info.

21. Sauma EE, Oren SS. Proactive planning and valuation of transmission investments in restructured electricity markets. *Journal of Regulatory Economics* 2006;30(3):261–290.

22. Hobbs BF, Rijkers and FAM. Strategic generation with conjectured transmission price responses in a mixed transmission pricing system-Part I. *IEEE Transactions on Power Systems* 2004;19(2):707–717.

23. Sheffrin AY, Chen J, Hobbs BF. Watching Watts to prevent abuse of power. *IEEE Power & Energy Magazine* 2004;2(4):58–65.

24. Borenstein S, Bushnell J, Wolak FA. Measuring market inefficiencies in California's deregulated electricity industry. *American Economic Review* 2002;92(5):1376–1405.

25. Joskow PL, Kahn E. A quantitative analysis of pricing behavior in California's wholesale electricity market during summer 2000. *The Energy Journal* 2002;23(4):1–35.

26. Wolfram C. Measuring duopoly power in the British electricity spot market. *American Economic Review* 1999;89:805–826.

27. London Economics International, LLC, A proposed methodology for evaluating the economic benefits in a restructured wholesale electricity market, Prepared for the CAISO, Feb. 28, 2003.

28. DePillis, Jr. MS. Uses of market simulation by market system operators. In: Proc. 2006 IEEE PES Transmission & Distribution Meeting, Dallas, May 24, 2006.

29. PJM Interconnection, LLC, Regional transmission expansion plan, 2006. www.pjm.com/planning/downloads/200660410-rtep-sec-5.pdf

30. Vaiani A, Bresesti P, Vailati R. A method to assess benefits of transmission expansions in the Italian electricity market. In: Proc. 2007. IEEE Gen. Meeting, Tampa, FL, June 24–26, 2007.
31. Schweppe FC, Caramanis MC, Tabors RD, Bohn RE. *Spot pricing of electricity, Appendix A.* Boston: Kluwer;1988.
32. CAISO. Board report, economic evaluation of the Palo Verde-Devers Line No. 2—main report and technical appendices. Departments of Market Analysis & Grid Planning, Feb. 2005. www.caiso.com/docs/2005/01/19/2005011914572217739.html
33. STEP Planning Process, www1.caiso.com/docs/2004/03/08/2004030814004810105.doc
34. Green RJ, Newbery DM. Competition in the British electricity spot market. *Journal of Political Economy* 1992;100(5):929–953.
35. Saracino N. Rebuttal testimony of the California Independent System Operator Corporation. In the Matter of the Application of San Diego Gas & Electric Company for a Certificate of Public Convenience and Necessity for the Sunrise Powerlink Transmission Project, Application 06-08-010, California Public Utilities Commission, June 15, 2007.
36. CAISO. CAISO South regional transmission plan for 2006, Part II: Findings and recommendation on the Tehachapi transmission project. 18 November 2006.
37. Baldick R. Variation of distribution factors with loading. *IEEE Transactions on Power Systems* 2003;18(4):1316–1323.
38. Hobbs BF, Drayton G, Fisher EB, Lise W. Improved transmission representations in oligopolistic market models: quadratic losses, phase shifters, and DC lines. *IEEE Transactions on Power Systems* 2008;23(3):1018–1029.
39. Samuelson PA. Spatial price equilibrium and linear programming. *American Economic Review* 1952;42(3):284–303.

Books in the IEEE Press Series on Power Engineering

Principles of Electric Machines with Power Electronic Applications, Second Edition
M. E. El-Hawary

Pulse Width Modulation for Power Converters: Principles and Practice
D. Grahame Holmes and Thomas Lipo

Analysis of Electric Machinery and Drive Systems, Second Edition
Paul C. Krause, Oleg Wasynczuk, and Scott D. Sudhoff

Risk Assessment for Power Systems: Models, Methods, and Applications
Wenyuan Li

Optimization Principles: Practical Applications to the Operations of Markets of the Electric Power Industry
Narayan S. Rau

Electric Economics: Regulation and Deregulation
Geoffrey Rothwell and Tomas Gomez

Electric Power Systems: Analysis and Control
Fabio Saccomanno

Electrical Insulation for Rotating Machines: Design, Evaluation, Aging, Testing, and Repair
Greg Stone, Edward A. Boulter, Ian Culbert, and Hussein Dhirani

Signal Processing of Power Quality Disturbances
Math H. J. Bollen and Irene Y. H. Gu

Instantaneous Power Theory and Applications to Power Conditioning
Hirofumi Akagi, Edson H. Watanabe, and Mauricio Aredes

Maintaining Mission Critical Systems in a 24/7 Environment
Peter M. Curtis

Elements of Tidal-Electric Engineering
Robert H. Clark

Handbook of Large Turbo-Generator Operation Maintenance, Second Edition
Geoff Klempner and Isidor Kerszenbaum

Introduction to Electrical Power Systems
Mohamed E. El-Hawary

Modeling and Control of Fuel Cells: Disturbed Generation Applications
M. Hashem Nehrir and Caisheng Wang

Power Distribution System Reliability: Practical Methods and Applications
Ali A. Chowdhury and Don O. Koval

Economic Market Design and Planning for Electric Power Systems
James Momoh and Lamine Mili

INDEX

Restructured Electric Power Systems: Analysis of Electricity Markets with Equilibrium Models,
Edited by Xiao-Ping Zhang
Copyright © 2010 Institute of Electrical and Electronics Engineers

Printed in the United States
By Bookmasters